Associative Polymers
in Aqueous Media

ACS SYMPOSIUM SERIES **765**

Associative Polymers in Aqueous Media

J. Edward Glass, EDITOR
North Dakota State University

American Chemical Society, Washington, DC

Library of Congress Cataloging-in-Publication Data

Associative polymers in aqueous media / J. Edward Glass, editor.

    p. cm.—(ACS symposium series, ISSN 0097-6156 ; 765)

Includes bibliographical references and index.

ISBN 0-8412-3659-3

1. Water soluble polymers—Congresses.

I. Glass, J. E. (J. Edward), 1937–   . II. Series.

QD382..W3 A77   2000
547'.70454—dc21   00–30597

The paper used in this publication meets the minimum requirements of American National Standard for Information Sciences—Permanence of Paper for Printed Library Materials, ANSI Z39.48–1984.

Copyright © 2000 American Chemical Society

Distributed by Oxford University Press

All Rights Reserved. Reprographic copying beyond that permitted by Sections 107 or 108 of the U.S. Copyright Act is allowed for internal use only, provided that a per-chapter fee of $20.00 plus $0.50 per page is paid to the Copyright Clearance Center, Inc., 222 Rosewood Drive, Danvers, MA 01923, USA. Republication or reproduction for sale of pages in this book is permitted only under license from ACS. Direct these and other permission requests to ACS Copyright Office, Publications Division, 1155 16th St., N.W., Washington, DC 20036.

The citation of trade names and/or names of manufacturers in this publication is not to be construed as an endorsement or as approval by ACS of the commercial products or services referenced herein; nor should the mere reference herein to any drawing, specification, chemical process, or other data be regarded as a license or as a conveyance of any right or permission to the holder, reader, or any other person or corporation, to manufacture, reproduce, use, or sell any patented invention or copyrighted work that may in any way be related thereto. Registered names, trademarks, etc., used in this publication, even without specific indication thereof, are not to be considered unprotected by law.

PRINTED IN THE UNITED STATES OF AMERICA

Foreword

THE ACS SYMPOSIUM SERIES was first published in 1974 to provide a mechanism for publishing symposia quickly in book form. The purpose of the series is to publish timely, comprehensive books developed from ACS sponsored symposia based on current scientific research. Occasionally, books are developed from symposia sponsored by other organizations when the topic is of keen interest to the chemistry audience.

Before agreeing to publish a book, the proposed table of contents is reviewed for appropriate and comprehensive coverage and for interest to the audience. Some papers may be excluded in order to better focus the book; others may be added to provide comprehensiveness. When appropriate, overview or introductory chapters are added. Drafts of chapters are peer-reviewed prior to final acceptance or rejection, and manuscripts are prepared in camera-ready format.

As a rule, only original research papers and original review papers are included in the volumes. Verbatim reproductions of previously published papers are not accepted.

ACS BOOKS DEPARTMENT

Contents

Preface xi

Block Copolymers and Dendrimers

1. Amphiphilic Polyoxyalkylene Triblock Copolymers: Self-Assembly,
   Phase Behaviors, and New Applications 2
   Tianbo Liu, Chunhung Wu, Yi Xie, Dehai Liang, Shuiqin Zhou,
   Vaughn M. Nace, and Benjamin Chu

2. PEP–PEO Block Copolymers as Model System for the Investigation
   of Micellization in Aqueous Solution 21
   J. Allgaier, A. Poppe, L. Willner, J. Stellbrink, and D. Richter

3. Amphiphilic Block Copolymers as Surfactants in Emulsion
   Polymerization 37
   Matthias Gerst, Horst Schuch, and Dieter Urban

4. The Preparation of Well-Defined Water Soluble–Swellable (Co)Polymers
   by Atom Transfer Radical Polymerization 52
   Krzysztof Matyjaszewski, Scott G. Gaynor, Jian Qiu, Kathryn Beers,
   Simion Coca, Kelly Davis, Andreas Muhlebach, Jianhui Xia, and Xuan Zhang

5. Hybrid Dendritic Capsules: Properties and Binding Capabilities
   of Amphiphilic Copolymers with Linear Dendritic Architecture 72
   Ivan Gitsov

Surfactant-Modified, Water-Soluble Polymers

HMPAM and HEUR Polymers

6. Inter- and Intra-Molecular Aggregation of Associating Polymers in Water 95
   Joseph Selb and Françoise Candau

7. Collapsed and Extended Polysoaps 109
   O. V. Borisov and A. Halperin

8. Solution Structure and Shear Thickening Behavior of Ionomers and Hydrophobically Associating Polymers 127
   Srinivas Nomula, Sharon Ma, and Stuart L. Cooper

9. Determination of Aggregation Numbers in Aqueous Solutions of Hydrophobically Modified Polymers by Fluorescent Probe Techniques 143
   Olga Vorobyova and Mitchell A. Winnik

10. Behavior of Branched-Terminal, Hydrophobe-Modified, Ethoxylated Urethanes 163
    Peter T. Elliott, Linlin Xing, Wylie H. Wetzel, and J. Edward Glass

11. Synthesis and Characterization of Well Defined End-Functionalized Hydrocarbon and Perfluorocarbon Derivatives of Polyethyleneoxide and Poly(N,N-dimethylacrylamide) 179
    Thieo E. Hogen-Esch, Huashi Zhang, and David Xie

Adsorption Studies: POE, HEUR, and HMPAM

12. Dynamics in Adsorbed Polymer Layers 206
    Maria M. Santore, Zengli Fu, and Ervin Mubarekyan

13. Dispersions Containing PEO with C_{16} Hydrophobes: Adsorption and Rheology 221
    Q. T. Pham, J. C. Thibeault, W. Lau, and W. B. Russel

14. Force Study of Adsorbed Layers of Hydrophobically Modified Polyacrylamide 239
    P. T. Starkey, H. T. Davis, M. V. Tirrell, J.-F. Argillier, A. Audibert, and J. Lecourtier

15. Complexations of Beta-Cyclodextrin with Surfactants and Hydrophobically Modified Ethoxylated Urethanes: Analytical Application in Adsorption Measurements 254
    Zeying Ma and J. Edward Glass

16. Association Thickener by Host–Guest Interaction of β-Cyclodextrin Polymers and Guest Polymers 271
    Gerhard Wenz, Meik Weickenmeier, and Jürgen Huff

HMHEC and HMEHEC Polymers

17. Fluorescent Labels: Versatile Tools for Studying the Association of Amphiphilic Polymers in Water 286
    Françoise M. Winnik, Sudarshi T. A. Regismond, and Dan F. Anghel

18. The Adsorption and Surface Dilatational Rheology of Unmodified and Hydrophobically Modified EHEC, Measured by Means of Axisymmetric Drop Shape Analysis 303
    Rolf Myrvold, Finn Knut Hansen, and Björn Lindman

19. How Much Surfactant Binds to an Associating Polymer? The HMHEC/SDS Case Revisited 317
    Lennart Piculell, Susanne Nilsson, Jesper Sjöström, and Krister Thuresson

HASE Thickeners

20. Determination of the Thickening Mechanism of a Hydrophobically Modified Alkali Soluble Emulsion Using Dynamic Viscosity Measurements 338
    C. M. Miller, K. R. Olesen, and G. D. Shay

21. The Network Strength and Junction Density of a Model HASE Polymer in Non-Ionic Surfactant Solutions 351
    W. P. Seng, K. C. Tam, R. D. Jenkins, and D. R. Bassett

22. Rheology of a HASE Associative Polymer and Its Interaction with Non-Ionic Surfactants 369
    R. J. English, R. D. Jenkins, D. R. Bassett, and Saad A. Khan

Indexes

Author Index 383

Subject Index 384

Preface

Associative polymers in aqueous media represent an area of significant industrial importance. This maturing technology is discussed in this text. In the first section, five chapters discuss block copolymers and dendrimers. Chapter 1 is also intended to be an overview of the area of ethylene oxide block copolymers. A unique variation of oxyethylene block polymers with alkane segments is provided in Chapter 2. An application-oriented study of acrylic acid block-copolymers is discussed in Chapter 3. A discussion of living free radical polymerizations as an economical means of obtaining these and other water-soluble block polymers is presented in Chapter 4. The section is completed (Chapter 5) with a discussion on water-soluble dendrimer block polymers, capable of forming micellar aggregates.

The second section of the book, surfactant-modified, water-soluble polymers, updates information provided in earlier volumes: *Water-Soluble Polymers: Beauty with Performance, Polymers in Aqueous Media: Performance through Association,* and *Hydrophilic Polymers: Performance with Environmental Acceptance.* The unique use of surfactants to control sequence distributions in the chain growth polymerization of hydrophobic monomers, in hydrophobically modified (HM) acrylamide polymers (PAM) is discussed in Chapter 6. This is followed by a theoretical discussion on the influence of associations on the conformations of HMPAM and HM-ethoxylated urethanes (HEURs) in flow (Chapter 7). A comparison of an HEUR associative thickener with ionomer technology is considered in Chapter 8. A review of information gained from fluorescence studies of associative thickeners is presented in Chapter 9, followed by a study of terminal-branched hydrophode HEURs (Chapter 10) and terminal mono- and disubstituted HEURs and HMPAM polymers, addressing fluorinated hydrophobes (Chapter 11).

The adsorption behavior of POE, HEUR, and HMPAM polymers is discussed in the second subsection (Chapters 12–15). The unique application of beta-cyclodextrin in adsorption and in viscosity mediation is discussed in Chapters 15 and 16. In the third subsection, studies on hydrophobe modified hydroxyethyl cellulose derivative are discussed (Chapters 17–19). Recent advances in fluorescence, dilational surface viscosities, and the nature of surfactant binding are reported. In the last subsection, the solution behavior of HASE thickeners is discussed in terms of the nature of their solubility and viscosity dependence on neutralization and interaction with surfactants.

Acknowledgments

I am grateful for the efforts of the reviewers of the individual chapters of this text and to the following organizations for their financial support of the speakers at the Boston symposium: Air Products, Hercules, Rohm and Haas, Union Carbide, Rhodia, the American Chemical Society (ACS) Division of Colloid and Surface Chemistry, and the ACS Petroleum Research Institute. The editor feels special thanks for the Aqualon division of Hercules, which has provided financial support for all six symposia organized, since 1984.

A high percentage of participants in the first symposium organized in 1984 were industrial. That has certainly changed during the past 15 years. An attempt is being made to reverse this trend in the last symposium to be organized in Chicago in 2001, on "Water-Borne Coatings." In the spirit of the industrial–academic liaison, the editor expresses gratitude to the following for the support in keeping this interface active in the area of water-soluble polymers:

Jack Estes at Amoco and Leon Miles at ARCO production research
Ernest Just and Arjun Sau at Aqualon-Hercules
Bob Buchacek and Jack Dickinson of the pigments division of DuPont
Ray Fernando of Armstrong World Industries

J. EDWARD GLASS
Polymers and Coatings Department
North Dakota State University
Fargo, ND 58105

Block Copolymers and Dendrimers

Chapter 1

Amphiphilic Polyoxyalkylene Triblock Copolymers: Self-Assembly, Phase Behaviors, and New Applications

Tianbo Liu[1], Chunhung Wu[2], Yi Xie[1], Dehai Liang[1], Shuiqin Zhou[1], Vaughn M. Nace[3], and Benjamin Chu[1,4,5]

[1]Department of Chemistry, State University of New York at Stony Brook, Stony Brook, NY 11794–3400
[2]Chemistry Department, Tamkang University, Tamsui, 25137, Taiwan
[3]Dow Chemical Company, Texas Operations, Freeport, TX 77451
[4]Department of Materials Science and Engineering, State University of New York at Stony Brook, Stony Brook, NY 11794–2275

The study on the micellization behavior of amphiphilic block copolymers in aqueous media has reached the stage where semi-quantitative analysis and prediction of micellar parameters and phase behavior are possible. The deeper understanding has provided an incentive to explore new applications on these self-assembled systems. In this chapter, we summarize the self-assembly and phase behavior of polyoxyalkylene triblock copolymers containing E and P, or E and B blocks (with E, P, B being polyoxyethylene, polyoxypropylene and polyoxybutylene, respectively), in aqueous solution. Laser light scattering was used to characterize micellar parameters in the dilute solution regime and small-angle X-ray scattering was employed to study gel-like ordered packings of the micelles at high polymer concentrations. The hydrophobic blocks form micellar cores and are mainly responsible for the formation of micelles (cmc, aggregation number, and thermodynamic parameters). The hydrophilic block length determines the micellar size and intermicellar interactions. The co-association behavior of a mixture of two different block copolymers in aqueous solution has also been elucidated. The effect of block length on phase separation (clouding temperature) can be discussed on the basis of experimentally determined empirical rules. At high polymer concentrations, unique gel-like micellar packings and their spatial supramolecular rearrangements (e.g., bcc, hexagonal, lamellar) occurred. The phase diagrams could be very rich, especially with the addition of another immiscible solvent. These gel-like materials with ordered hydrophilic-hydrophobic

[5]Corresponding author.

packings gave access to new applications in analytical and biological sciences – the structure served as a medium in capillary gel electrophoresis, e.g., for double-stranded DNA analysis with single base-pair resolution.

Introduction and Retrospection

The self-assembly behavior of block copolymers in selective solvents (mostly in aqueous solution) has become a major topic during the last several decades. Many in-depth reviews and books have appeared, citing over 1,000 references.[1-5] Based on these works, a brief retrospection is given in this section.

The major interest in the field has been concentrated on the characterization of the self-assembly behavior of block copolymers, by using different physical experimental methods, such as scattering techniques, viscosity, size-exclusion chromatography, NMR and electron microscopy.[1] The dominant assembly behavior of block copolymers in selective solvents is micellization, with the less-soluble block(s) forming compact micellar cores and the soluble block(s) forming extended micellar shells.[6] Other association procedures, such as an open-associated network that is mainly found in triblock copolymers in a solvent selectively good for the middle block, have also been reported.[7]

For diblock copolymers and triblock copolymers in a solvent selectively good for the end blocks, star-like micelles will form; and for triblock copolymers in a solvent selectively good for the middle block, flower-like micelles can be expected.[8] They have certain critical micelle concentrations (cmc) and aggregation numbers (n_w) which, for EPE type triblocks, show obvious temperature dependence.[1-3] The cmc is defined as the polymer concentration, above which the formation of micelles occurs. In aqueous solution, larger cmc and smaller n_w are found at lower temperatures, e.g., the hydrophobicity of the less-soluble block became higher with increasing temperature due to the unique hydrogen-bond breaking process. On the contrary, in nonpolar organic solvents, the less-soluble block can be dissolved better with increasing temperature, making the micellization process more difficult at higher temperatures.[5]

Wu et al. studied the micellization of Pluronic L64 (polyoxyethylene-polyoxypropylene-polyoxyethylene, $E_{13}P_{30}E_{13}$) triblock copolymer in a mixture of two immiscible solvents, water and o-xylene, which were good solvents for E and P blocks, respectively.[9-11] Katime and co-workers also systematically studied the self-association of polystyrene/poly(ethylene-propylene) diblock copolymers in different selective organic solvents and solvent mixtures,[12-15] where the micellar parameters changed dramatically with solvents.

Another topic drawing continuous interest is the micellization of block copolymers in the presence of electrolytes, e.g., inorganic salts.[3,5] Generally, the inorganic salts can be divided into two groups based on their effects on the micellization of block copolymers in aqueous solution. One kind of salt, including Na^+, K^+, Ca^{2+} etc., decreases the solubility of copolymers in aqueous solution and therefore makes micellization easier to happen. It is known as the "salting-out effect". Some salts, like SCN^- and urea, have the opposite function known as the "salting-in effect".[16-19]

The temperature-concentration phase diagrams of polyoxyalkylene block copolymers in aqueous solution have been widely reported in the literature and have been reviewed.[3] Figure 1 is a schematic plot which shows the phase behaviors of polyoxyalkylene triblock copolymers in aqueous solution. At low temperatures and polymer concentrations, the copolymer chains exist as unimers. Increasing either temperature or polymer concentration can induce micelle formation. The critical conditions are called critical micelle temperature (cmt) or critical micelle concentration (cmc), respectively. The star-like micelles tend to pack into gel-like systems with ordered structures at higher micellar concentrations. Cubic structures, both body-centered cubic (BCC) and face-centered cubic (FCC) packings, have been reported.[20-23] Gast and co-workers presented the conclusion that it was the weight ratio of the hydrophobic part to the hydrophilic part that determined the micellar packing.[24] Flower-like micelles tend to form open network structures at high polymer concentrations.

The colloidal behavior of block copolymers in selective solvents, especially in aqueous solution, has found extensive applications, for examples, as surfactants and solubilizers in the cosmetic industry and as drug carriers in the pharmaceutical industry.[1,2,5]

The study of commercial polyoxyalkylene triblock copolymers in aqueous solution has been a topic of great interest. The samples were synthesized by BASF, the Dow Chemical Company[25] and the University of Manchester[26-27]. The availability of comparatively large amounts of samples with different block lengths and chain architecture affords a more quantitative and systematic study. Consequently, a more thorough understanding on the self-assembly behavior of such kinds of block copolymers can be used to explore new applications in different fields. In this paper, recent studies on the polyoxyalkylene triblock copolymers (including $E_nP_mE_n$, $P_nE_mP_n$, $E_nB_mE_n$ and $B_nE_mB_n$, with E, P and B being oxyethylene, oxypropylene and oxybutylene, respectively) in aqueous solution, including their self-assembly and phase behavior, as well as a new application to capillary electrophoresis, are presented.

Self-Association

Critical Micelle Concentration (cmc) and Aggregation Number (n_w)

Except for some $P_nE_mP_n$ triblock copolymers,[28] most of the polyoxyalkylene triblock copolymers choose a closed-associated mechanism in aqueous solution by showing the existence of fixed cmc and n_w at certain temperatures.[26,29-33] Laser light scattering (LLS) is a typical technique to determine the cmc by detecting a sudden increase in the scattered intensity with increasing polymer concentration. Booth and co-workers reported that the cmc values of $E_{102}P_{37}$, $E_{52}P_{34}E_{52}$ and $P_{19}E_{113}P_{13}$ were 0.8, 34 and 80 g/dm^3, respectively.[34] For the $P_nE_mP_n$ (or $B_nE_mB_n$) triblock copolymers, the formation of closed-associated micelles required the soluble middle block to form a loop structure and went back into the same micellar core. Such an entropy-loss

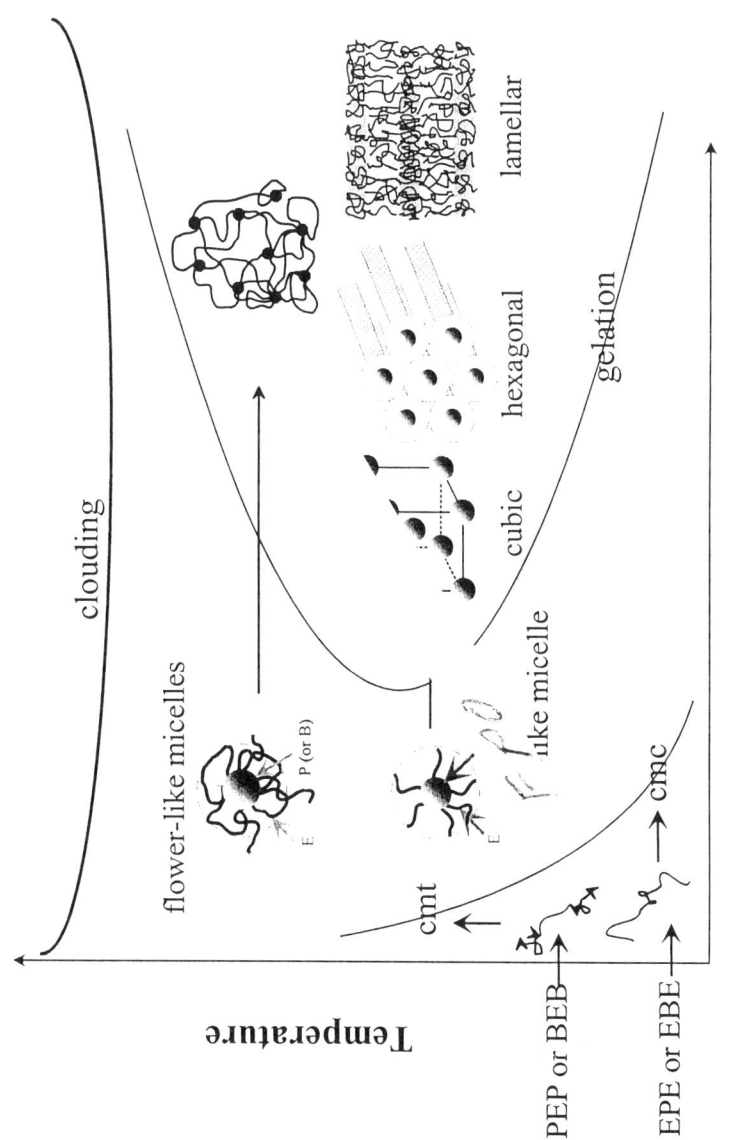

Figure 1. Schematic plot of temperature-polymer concentration behaviors of polyoxyalkylene triblock copolymers in aqueous solution.

process made the copolymers harder to micellize, so that for this kind of flower-like micelle, a higher cmc and a lower n_w could be expected.

In our study on the association behavior of $B_nE_mB_n$ (including $B_4E_{41}B_4$, $B_5E_{39}B_5$, $B_5E_{91}B_5$, $B_6E_{46}B_6$, $B_7E_{40}B_7$, $B_7E_{23}B_7$, $B_{10}E_{271}B_{10}$ and $B_{12}E_{271}B_{12}$) triblock copolymers in aqueous solution, we reported that the hydrophobic end B blocks played a major role in determining the cmc of copolymers.[35] By plotting the logarithmic values of cmc (in units of mol/L) versus the average number of B units in one copolymer chain, a linear relationship found that the cmc value became smaller with increasing B units (N_B), as shown in Figure 2. In Figure 2a, the number of E units in the copolymer chains were randomly distributed, suggesting that the effect of E block was very small. The slope in Figure 2a denotes the transfer energy, which can be defined as the energy used for transferring one hydrophobic unit from solution into the micellar core. This transfer energy increases with increasing temperature, which shows that the B block becomes more hydrophobic at higher temperatures. The length of the end B blocks also determines the n_w of micelles.[36] A similar linear relationship can be found by plotting n_w versus N_B, similar to Figure 2a.

Thermodynamic Parameters

The thermodynamic process of micellization can be expressed by the equation:

$$\Delta H° = d[\ln(cmc)]/d(1/T) \qquad (1)$$

with cmc in units of mol/L and T being the temperature in Kelvin. $\Delta G°$ and $\Delta S°$ can be calculated accordingly. The micellization process in aqueous solution usually appears to be an entropy-driven process, i.e., positive $\Delta S°$ and $\Delta H°$ can be expected.[35] This is owing to the fact that micellization in water is the process of breaking hydrogen bonds between polymer chains and water molecules and then destroying these ordered structures.[37] We reported that for the micellization of $B_nE_mB_n$ triblock copolymers, the ΔH values increased with increasing B block length, suggesting that the micellization process was thermodynamically controlled by the end hydrophobic blocks.

Another approach can further elucidate the effects of the total polymer chain length and the block length ratio on ΔH which was demonstrated by Armstrong et al.[38] and by us[35] in $E_nP_mE_n$ and $B_nE_mB_n$ triblock copolymers, respectively. By plotting the ΔH of micellization per average monomer unit (including N_B and N_E) versus the number of hydrophobic unit, ΔH approached zero when the ratio of N_B (or N_P)/N_E went to zero. These observations proved that it was the hydrophobic part of the copolymer that was responsible for the micellization.

Hydrodynamic Radius (R_h) and Intermicellar Interactions

The R_h of micelles can be measured by dynamic light scattering (DLS) or gradient-pulsed NMR. Figure 2b shows a linear relationship between the R_h of

$B_nE_mB_n$ triblock copolymer micelles in aqueous solution and the product of $N_B^{4/25}N_E^{3/5}$.[35] The latter came from the scaling theory.[39]

The apparent hydrodynamic radius $R_{h,app}$ is related to the apparent diffusion coefficient D_{app} of the particles in solution, which can be measured from the DLS technique, with

$$R_{h,app} = kT/6\pi\eta_0 D_{app} \qquad (2)$$

where k is the Boltzmann constant and η_0 is the solvent viscosity. The polymer concentration dependence of R_h values provides the information on intermicellar interactions: $D_{app} = D_0(1+k_Dc)$ and $R_{h,app} = R_{h,0}/(1+k_Dc)$ where D_0 and $R_{h,0}$ are the diffusion coefficient and R_h at zero polymer concentrations, respectively, and k_D is a parameter denoting intermicellar interactions. For star-like micelles for which the middle block of a triblock copolymer forms the micellar core, the $R_{h,app}$ decreases with the increasing copolymer concentration (positive k_D), suggesting that the interactions among micelles are basically repulsive, which can be successfully approximated by the hard sphere model. On the contrary, for the flower-like micelles, the reversed rule, indicating an attractive interaction (negative k_D), has been found. For the flower-like micelles, certain amounts of hydrophobic end blocks could not go back to the micellar cores to form looping structures, which was obviously entropy unfavorable. They remained dangling in solution and this kind of sticky end blocks had a strong inclination to aggregate. Therefore, the intermicellar interactions became attractive. Following ten Brinke and Hadziioannou,[40] the entropy penalty due to the looping formation is given by:

$$G_{loop}/kT = (3/2)\beta \ln N_\beta \qquad (3)$$

where N_β is the number of repeat units in the middle block. The coefficient β that depends on the length of the two constitutive blocks has a limiting value of 0.3-0.4. From eq.3 it is clear that a longer middle block will lead to a larger entropy loss. This theoretical expression has been qualitatively proven by the experimental fact that for both $B_{10}E_{271}B_{10}$ and $B_{12}E_{260}B_{12}$,[35] the k_D values were very large (-124 and -139 cm^3/g, respectively) when compared with other $B_nE_mB_n$ triblock copolymers with shorter E block lengths (around -2 to -7 cm^3/g), indicating that the entropy penalties for these two copolymers were extremely large.

Micellization of Two Mixed Triblock Copolymers in Aqueous Solution

Recently we reported the LLS study of the micellization process of a more complicated system: a mixture of two amphiphilic triblock copolymers, F127 ($E_{99}P_{69}E_{99}$) and B19-5000 ($E_{45}B_{14}E_{45}$), in aqueous solution.[41] Booth and co-workers had done some work on this subject with the characterization of a mixed micelle formed by $E_{41}B_8$ and $B_{12}E_{76}B_{12}$ block copolymers in aqueous solution.[42] The purpose of our work was to further clarify some basic features and to provide background for new applications. Both block copolymers tended to self-associate into micelles in water. The micellization $E_{99}P_{69}E_{99}$ had a very strong temperature dependence, i.e., it

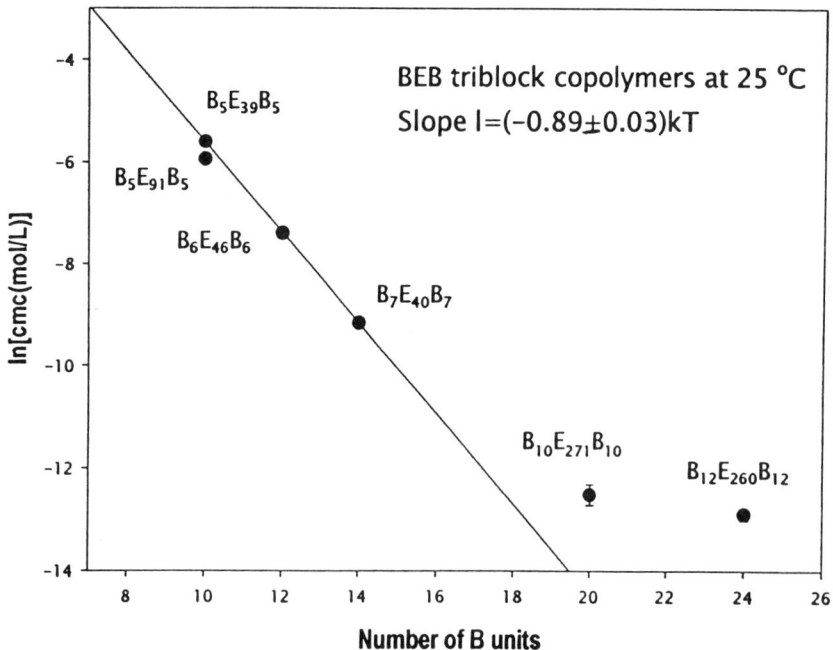

Figure 2a. Plot of logarithmic cmc values of $B_nE_mB_n$ triblock copolymers in aqueous solution versus number of B units in a polymer chain at 25 °C. The value of $B_7E_{40}B_7$ was calculated by extrapolating the data at lower temperatures. (Cited from ref. 35, Figure 2a)

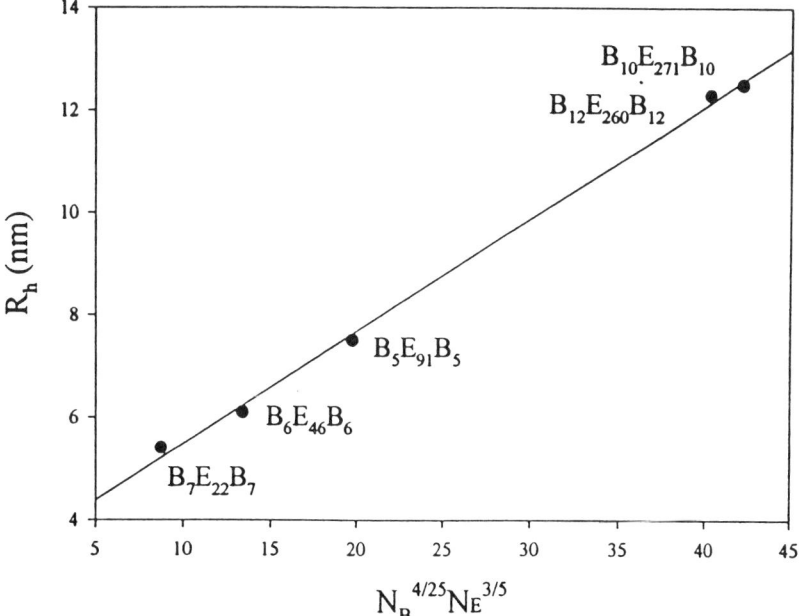

Figure 2b. Plot of hydrodynamic radius (R_h) of $B_nE_mB_n$ triblock copolymers versus $N_B^{4/25}N_E^{3/5}$. (Cited from ref. 35, Figure 9b)

had very high cmc values at low temperatures (53 mg/mL at 20 °C) but very low cmc values at higher temperatures (0.03 mg/mL at 35 °C). On the contrary, the cmc of $E_{45}B_{14}E_{45}$ showed only a weak temperature dependence, changed from 1.11 mg/mL at 20 °C to 0.28 mg/mL at 35 °C. The two block copolymers had the same cmc value at 30 °C (0.40 mg/mL). The R_h of the $E_{99}P_{69}E_{99}$ and $E_{45}B_{14}E_{45}$, measured by DLS, were 11.5 and 6.0 nm, respectively.

It was found that at temperatures where one cmc was much higher than the other, the cmc of the 1:1 wt% mixed triblock copolymer solution was the same as the one with the smaller cmc. Meanwhile, the R_h value of the micelles was also the same as the R_h of the pure micelles with the smaller cmc. This observation suggested that the copolymer with the smaller cmc still self-associated first, and the other copolymer did not get involved at this polymer concentration. However, DLS results suggested that chains from the other copolymer gradually joined into micelles before its cmc, because the apparent R_h of the micelles obviously increased with increasing polymer concentration. In comparison with the pure micelles, R_h decreased slowly with increasing polymer concentration (positive k_D value).

At a temperature of around 30 °C, the two copolymers had similar cmc values. The cmc of the mixed copolymer solution had an apparent cmc even smaller than the smaller cmc of pure copolymer, as shown in Figure 3, suggesting that the copolymer with the higher cmc value also contributed partly to the micellar formation due to the sample polydispersity. We used a parameter β to denote the effect of the copolymer with higher cmc on the cmc of the mixed solution. β could be understood as the ratio of the cmc of mixed polymer solution versus the smaller cmc of pure copolymer. If the cmc of one pure copolymer was much higher than the other one, β had the value of 1, suggesting that the micellization of mixed polymer solution was determined by only one kind of copolymer, as shown in Figure 4 in the regions of less than 27 °C and higher than 33 °C. At 30 °C, the value of β reached its minimum (0.5), indicating that both copolymers had the same cmc and contributed equally to the association in the mixed polymer solution (Figure 4). This temperature was called "co-association point".

Phase Behavior

The phase behavior of polyoxyalkylene triblock copolymers in aqueous solution has been well studied because most of the applications of these copolymers were in aqueous solution. It has been found that the solubility of block copolymers decreased with increasing temperature and, above certain temperatures, phase separation occurred so that the copolymer could no longer be dissolved in water. This temperature is called the "cloud-point temperature" and it shows only a very small concentration dependence in the low polymer concentration regime (Figure 1). At higher polymer concentrations, the micelles tend to pack together to form a gel-like system with ordered structures. The first gel-like structure is usually cubic, formed by the ordered packing of individual micelles. At higher polymer concentrations, the rearrangement of the hydrophilic and hydrophobic regions leads to the formation of

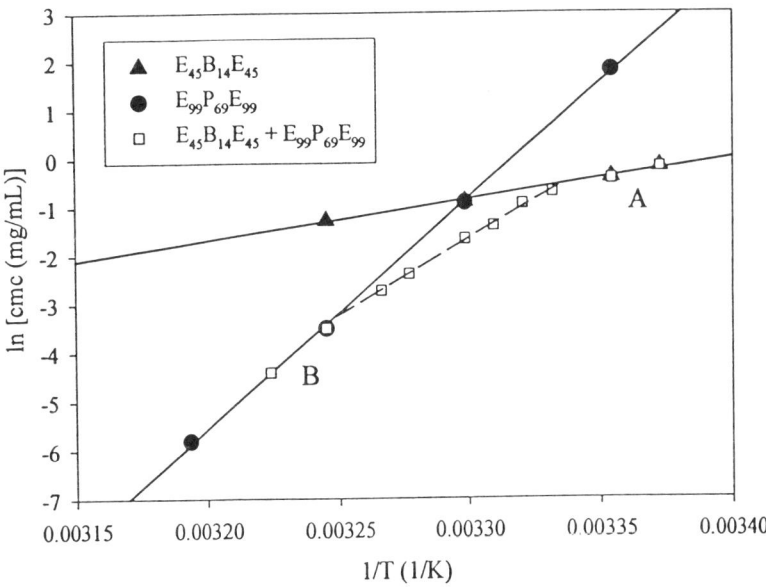

Figure 3. Plots of the cmc values of pure $E_{45}B_{14}E_{45}$ and $E_{99}P_{69}E_{99}$ triblock copolymers versus 1/T(K) to determine the co-association point. The open squares represent cmc values of 1:1 weight ratio mixed $E_{45}B_{14}E_{45}$ and $E_{99}P_{69}E_{99}$ triblock copolymer solution. (Cited from ref. 41, Figure 10)

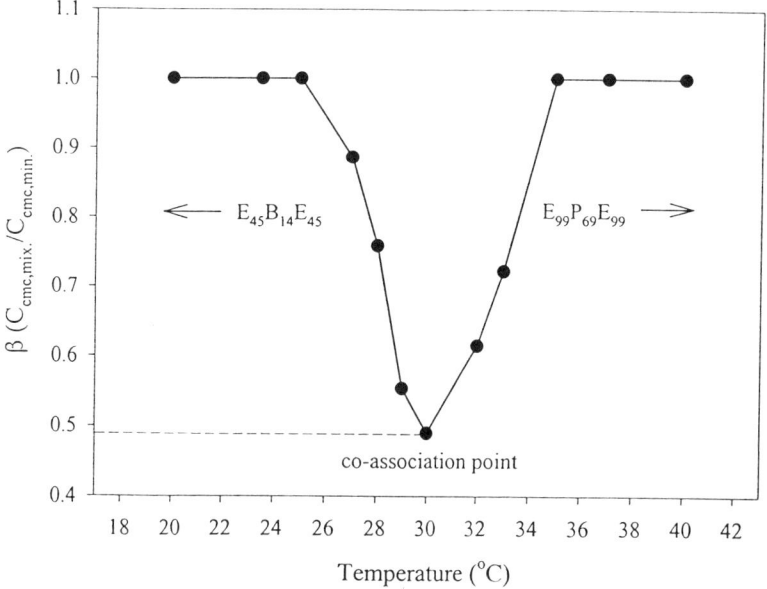

Figure 4. Temperature dependence of of 1:1 weight ratio mixed $E_{45}B_{14}E_{45}$ and $E_{99}P_{69}E_{99}$ triblock copolymer solutions. The co-association point is the point where reaches its minimum value (0.5). (Cited from ref. 41, Figure 12)

hexagonal and lamellar structures. Bicontinuous cubic structures could also be observed sometimes between the hexagonal and lamellar regions.[3]

Cloud-point Temperature (T_{cl})

The cloud-point temperatures of the polyoxyalkylene triblock copolymers in aqueous solution have been widely reported in the literature and commercial reports. However, the existence of some impurities in the samples during synthesis, which often consist of some very hydrophobic parts, seriously affects the determination of T_{cl}. Further sample purification by using a non-polar organic solvent (e.g., hexane) to remove these hydrophobic parts has been proven to be crucial to get the precise T_{cl} data. We have reported that for the commercial sample Pluronic 25R-8 ($P_{15}E_{155}P_{15}$, BASF Corp.), the measured T_{cl} changed from 40 °C to 78 °C after purification, and the sample after purification showed a much narrower polydispersity.[43]

Because of the limitation on the available samples and the purity of the samples, only limited T_{cl} data has been reported with qualitative conclusions, e.g., increasing the weight percentage of the hydrophobic block might decrease T_{cl} and vice versa, until we reported a systematic study on the effects of different block lengths on the T_{cl} of $B_nE_mB_n$ and $P_nE_mP_n$ triblock copolymers in aqueous solution. It was found that the end B (or P) hydrophobic blocks had a very strong negative effect on the T_{cl}. A linear increase of N_B (or N_P) led to an exponential decrease in (T_{cl}-273K), where the constant 273 is the melting temperature of water that had to be subtracted from the total temperature (in Kelvin). On the contrary, the hydrophilic E block seemed to have only a small positive effect. The above observation was made based on the T_{cl} of two series of $B_nE_mB_n$ triblock copolymers: one having around 40 E units and the other having 12 B units on each end block. A general mathematical expression[43]

$$T_{cl,BEB}(K) = 188\exp(-0.14N_B) + 10.6\ln(N_E) + 233 \qquad (4)$$

and also a three-dimension plot (Figure 5) can be set up to calculate the T_{cl} of any $B_nE_mB_n$ triblock copolymers in aqueous solution. The parameter λ (= 0.14) in eq. 4 depended on the nature of the hydrophobic block and it had the value of -0.14 for B unit. The calculated value fit very well to the measured data points, indicating that the expression was reasonable.

Similar expressions could also be made for $P_nE_mP_n$ triblock copolymers by introducing a λ value of -0.032:[43]

$$T_{cl,PEP}(K) = 116\exp(-0.032N_P) + 16.5\ln(N_E) + 233 \qquad (5)$$

where the values of other coefficients were also changed from eq.4. Eq. 5 could also satisfy all the known experimental data.

Booth and co-workers compared the effect of per hydrophobic unit on the thermodynamic parameters (e.g., H) of $E_nB_mE_n$ and $E_nP_mE_n$ triblock copolymers in aqueous solution and indicated that one B unit was comparable to 3 – 4 P units.[44] similar conclusion can also be drawn according to the slope in Figure 2. The slope for

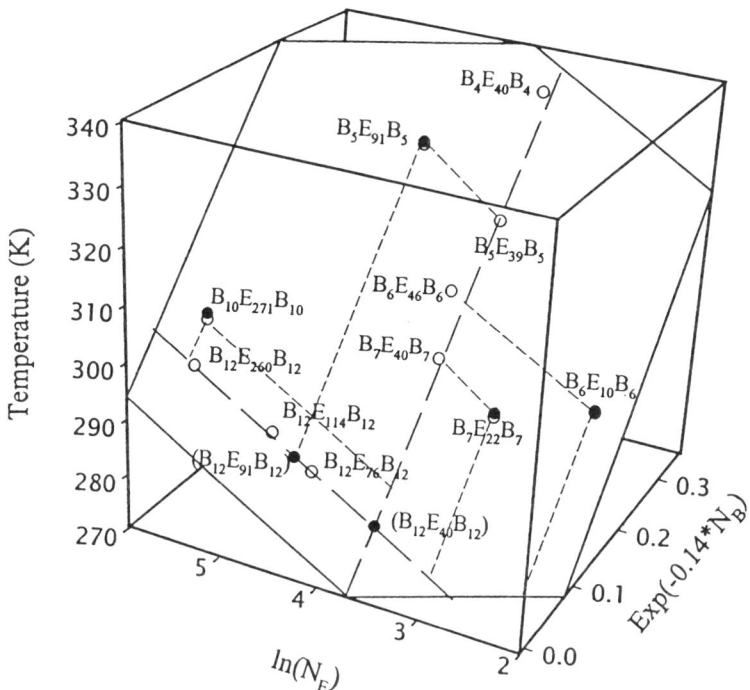

Figure 5. Three-dimensional diagram of T_{cl} of 1 wt % $B_nE_mB_n$ triblock copolymers in aqueous solution. The open circles represent experimental data and the filled circles represent calculated results. (Cited from ref. 43, Figure 5)

$B_nE_mB_n$ copolymer micelles is about 4 times higher than that of $P_nE_mP_n$ copolymer micelles.[35]

Another way was presented by us when we compared the effects of B and P blocks on the cloud-point temperatures of block copolymers. From the study of mixed copolymer micelles,[41] it was clear that the relative hydrophobicity of B units over P units altered with changing temperature. From the coefficients of the exponential parts in eqs. (4) and (5), it was clear to conclude that one B unit acted as about 4.4 P units at phase separated temperature.

Alexandridis et al. studied the micellization and the clouding behaviors of Pluronic L64 ($E_{13}P_{30}E_{13}$) triblock copolymers in aqueous solution with different salts.[45] The presence of LiCl, KCl, NaCl and NaBr decreased both the cmt and the T_{cl} (in the order $Cl^- > Br^-$ and $Na^+ > K^+ > Li^+$) whereas the addition of NaSCN and urea resulted in a cmt and T_{cl} increase (in the order SCN > urea). During this study, they reported the relation between the cmt and T_{cl}: cmt (no salt) - cmt (salt) = T_{cl} (no salt) - T_{cl} (salt), i.e., the effects of inorganic salt on the cmt and the T_{cl} were the same. This empirical relation suggested that there existed an obvious relation between micellization, the "microphase separation", and clouding, the macrophase separation.

Phase Diagram

Most of $B_nE_mB_n$ triblock copolymers have been reported to have no structure in aqueous solution,[46] except $B_7E_{40}B_7$, for which Booth and co-workers reported that a cubic structure was found in its aqueous solution at high polymer concentrations.[47]

Introducing another immscible solvent (e.g., xylene) into copolymer/water systems can make the phase diagrams significantly more complicated. Alexandridis et al. made a series of experiments on the ternary phase diagram of triblock copolymers in a mixture of water and p-xylene at 25 °C.[48-59] The first one was the phase diagram of Pluronic L64 ($E_{13}P_{30}E_{13}$)/water/p-xylene. Normal (oil-in-water) and reversed (water-in-oil) hexagonal, lamellar and bicontinuous cubic structures as well as the their mixed structures were reported.[48] Similar studies on other polyoxyalkylene triblock copolymers/water/p-xylene ternary systems showed similar features. They also reported that for the ternary Pluronic P84 ($E_{19}P_{43}E_{19}$)/water/p-xylene system at 25 °C, 9 different phases could be found in one phase diagram: normal and reverse cubic, normal and reverse hexagonal, normal and reverse bicontinuous cubic and lamellar.[59]

Holmqvist et al. also studied the phase behaviors of triblock copolymers in a mixture of two miscible solvents: water and alkanol.[60] Rich phase diagrams which contained solution, cubic, hexagonal and lamellar regions were found for the $E_nP_mE_n$ and $E_nT_mE_n$ (T = polytetrahydrofuran) triblock copolymers in the mixed solvent. However, their gel-like regions have become much smaller when compared with the phase diagrams of triblock copolymers in a mixture of two immiscible solvents.

The phase diagrams of block copolymers in the presence of inorganic salts has also drawn attention. The inorganic ions which could decrease the cmc of block copolymers in aqueous solution (salting-out effect) could also decrease the gelation concentration of the block copolymers and vice versa.[61] Recently, we reported the formation of polymer-salt complexes in the L64/water/CdCl$_2$ ternary system. Cd^{2+} has the effect of decreasing the solubility of L64 in water.[61] In the L64/water/CdCl$_2$ ternary phase diagram at different temperatures and a constant molar ratio of L64 to CdCl$_2$ (1:2.56), another two-phase region could be detected at very high polymer concentrations in which CdCl$_2$ could not be totally dissolved. New scattering peaks could be found under these conditions which suggested formation of possibly new lamellar structures with an inter-domain distance of about 18 nm.

Separation Medium for DNA Capillary Electrophoresis

Polyoxyalkylene triblock copolymers have already been found to be very useful as surfactants in the cosmetic industry and as drug carriers in the pharmaceutical industry. Here we present a new application: as a separation medium in DNA capillary electrophoresis.

The spherical micelles could be packed together to form a gel-like system, having a sieving ability as a separation medium. One example was given by our group[62-64] and by Hill et al,[65] using Pluronic F127 ($E_{99}P_{69}E_{99}$) in a 1XTBE (Tris-borate-EDTA) buffer solution. The self-association behavior of F127 in 1XTBE was very similar to that in aqueous solution, with the cmc, n_w and R_h being about the same. The micelles formed a gel-like system with a very high viscosity at polymer concentrations greater than 15 wt% at room temperatures. The micellization of F127 had very strong temperature dependence at polymer concentrations ranging from 15 wt% to about 40 wt%; the F127 in 1XTBE buffer solution system would be a solution at low temperatures but a gel-like material at room temperatures. The low-viscosity solution at low temperatures made it easier to be injected into small diameter capillaries. After being injected into the capillary, gel-like materials which were capable for DNA separation, could form by gradually increasing the temperature to about 25 °C. 21.2 w/v% F127/1XTBE has been found to be effective in separating double-stranded DNA fragments. The elctropherogram is shown in Figure 6.

Conclusions

The self-assembly behavior of amphiphilic block copolymers in aqueous solution has shown to be complex but very rich in the structures and phases that such systems could exist. The factors such as chain length, chain ratio and chain architecture can be used to affect the micellization parameters, including the cmc and n_w, as well as the gel-like supramolecular structure. The micellization behavior in more complicated cases, e.g., in the presence of electrolytes and in mixed triblock copolymers, has also been reported.

Current interests have shifted to possible new applications in biological and materials sciences.

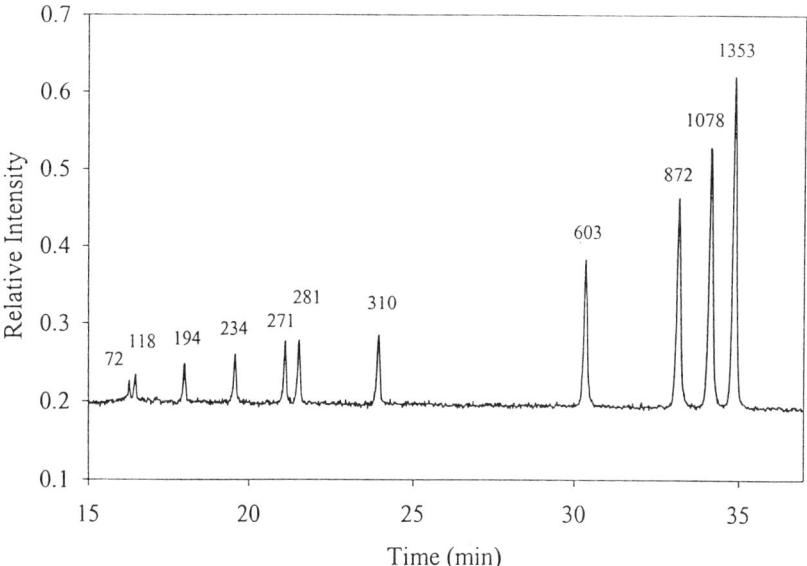

Figure 6. Electropherograms of ϕX174 DNA-Hae III digest in capillary electrophoresis at 25 °C by using 21.2 (wt/v)% F127 in 1X TBE buffer as a separation medium. Peak identifications from right to left in base pairs are: 1353, 1078, 872, 603, 310, 281, 271, 234, 194, 72, 118. (Cited from ref. 62, Figure 16a)

Acknowledgements

BC gratefully acknowledges support of this work by the National Science Foundation (Polymers Program DMR9612386), the U.S. Army Research Office (DAAG559710022), the Department of Energy (DEFG0286ER45237.015 and DEFG0299ER45760), and the National Human Genome Research Institute (2R01HG0138604).

References

1. Tuzar Z. and Kratochvil, P., *Micelles of Block and Graft copolymers in solutions in Surface and Colloid Science*, Vol. 15, E. Matijevic Ed., Plenum Press, New York, **1993**.
2. Kon-no, K., *Properties and Applications of Reversed Micelles in Surface and Colliod Science*, Vol. 15, E. Matijevic Ed., Plenum Press, New York, **1993**.
3. Chu, B. and Zhou, Z., *Physical Chemistry of Polyoxyalkylene Block Copolymer Surfactants in Nonionic Surfactants*, V. M. Nace Ed., Marcel Dekker, Inc., New York, **1996**.
4. Chu, B. *Langmuir* **1995**, 11, 414.
5. Alexandridis, P.; Lindman, B. Ed., Amphiphilic Block Copolymers: Self-Assembly and Applications. **1998**.
6. Zhou, Z.; Chu, B. *J. Colloid Interface Sci.* **1988**, 126, 171.
7. Raspaud, E.; Lairez, D.; Adam, M.; Carton, J.-P. *Macromolecules* **1994**, 27, 2956.
8. Balsara, N. P.; Tirrell, M.; Lodge, T. P. *Macromolecules* **1991**, 24, 1975.
9. Wu, G.; Zhou, Z.; Chu, B. *Macromolecules* **1993**, 26, 2117.
10. Wu, G.; Chu, B. *Macromolecules* **1994**, 27, 1766.
11. Wu, G.; Zhou, Z.; Chu, B. *J. polym. Sci. Part B: Polymer Physics* **1993**, 31, 2035.
12. Quintana, J. R.; Villacampa, M.; Munoz, M.; Andrio, A.; Katime, I. A. *Macromolecules* **1992**, 25, 3125; 3129.
13. Quintana, J. R.; Janez, M. D.; Villacampa, M.; Katime, I. A. *Macromolecules* **1995**, 28, 4139; 4144.
14. Quintana, J. R.; Salazar, R. A.; Villacampa, M.; Katime, I. A. *Makromol. Chem.* **1993**, 194, 2497.
15. Quintana, J. R.; Salazar, R. A.; Katime, I. A. *Makromol. Chem. Phys.* **1995**, 196, 1625.
16. Binana-Limbele, W.; Van Os, N. M.; Rupert, L. A. M.; Zana, R. *J. Colloid Interface Sci.* **1991**, Vol. 144, 458.
17. Bahadur, P.; Pandya, K.; Almgren, M.; Li, P.; Stilbs, P. *Colloid & Polymer Sci.* **1993**, 657.
18. Jorgensen, E. B.; Hvidt, S.; Brown, W.; Schillen, K. *Macromolecules* **1997**, 30, 2355.
19. Alexandridis, P.; Athanassiou, V.; Hatton, T. A. *Langmuir* **1995**, 11, 2442.
20. Wanka, G.; Hoffmann, H.; Ulbricht, W. *Macromolecules* **1994**, 27, 4145.

21. Zhang, K. Ph.D. Thesis, University of Lund, **1994**.
22. Mortensen, K. *Progr. Colliod Polym. Sci.* **1993**, 91, 69.
23. Mortensen, K. *Progr. Colliod Polym. Sci.* **1993**, 93, 72.
24. G.A. McCornell, A.P. Gast, J.S. Huang and S.D. Smith, *Phys. Rev. Lett.*, **1993**, 71, Number 13, 2102.
25. Dow Chemical Co., Freeport, TX, U.S.A. Technical Literature, B-Series.
26. Yang, Z.; Pickard, S.; Deng, N.-J.; Barlow, R. J.; Attwood, D.; Booth, C. *Macromolecules* **1994**, 27, 2371.
27. Yang, Y.-W.; Yang, Z.; Zhou, Z.; Attwood, D.; Booth, C. *Macromolecules* **1996**, 29, 670.
28. Mortensen, K.; Brown, W.; Jorgensen, E. *Macromolecules* **1994**, 27, 5654.
29. Zhou, Z.; Chu, B.; Nace, V. M. *Langmuir*, **1996**, Vol. 12, No. 21, 5016.
30. Zhou, Z.; Yang, Y.-W.; Booth, C.; Chu, B. *Macromolecules* **1996**, 29, 8357.
31. Zhou, Z.; Yang, Y.-W.; Nace, V.M.; Booth, C.; Chu, B. *Macromolecules* **1996**, 29, 3663.
32. Liu, T.; Zhou, Z.; Wu, C.; Chu, B.; Schneider, D.; Nace, V. M. *J. Phys. Chem. B*: **1997**, 101, 8808.
33. Chu, B.; Liu, T.; Wu, C.; Zhou , Z.; Nace, V. M. *Makromolecular Symposia*, **1997**, 118, 221.
34. Altinok, H.; Yu, G.-E.; Nixon, S. K.; Gorry, P. A.; Attwood, D.; Booth, C. *Langmuir* **1997**, 13, 5837.
35. Liu, T.; Zhou, Z.; Wu, C.; Nace, V. M.; Chu, B. *Journal of Physical Chemistry B*, **1998**, 102, 2875.
36. Liu, T.; Zhou, Z.; Wu, C.; Nace, V. M.; Chu, B. *Macromolecules*, **1997**, 30, 7624.
37. Mejiro, K.; Takasawa, Y.; Kawahashi, N.; Tabata, Y.; Ueno, M. *J. Colloid Interface Sci.*, **1981**, 83, 50.
38. Armstrong, J. K.; Parsonage, J.; Chowdhry, B. Z.; Leharne, S.; Mitchell, J. C.; Beezer, A. E.; Laggner, P. *J. Phys. Chem.* **1993**, 97, 3904.
39. Halperin, A. *Macromolecules* **1987**, 20, 2943.
40. ten Brinke, G.; Hadziioannou, G. *Macromolecules* **1987**, 20, 486.
41. Liu, T.; Nace, V. M.; Chu, B. *Langmuir*, **1999**, 15, 3109.
42. Yang, Z.; Yang, Y.-W.; Zhou, Z.; Attwood, D.; Booth, C. *J. Chem. Soc. Faraday Trans.*, **1996**, 92(2), 257.
43. Liu, T.; Nace, V. M.; Chu, B. *J. Phys. Chem. B.* **1997**, 101, 8074.
44. Bedells, A. D.; Arafeh, R. M.; Yang, Z.; Attwood, D.; Heatley, F.; Padget, J. C.; Booth, C. *J. Chem. Soc., Faraday Trans.* **1993**, 89, 1235.
45. Alexandridis, P.; Holzwarth, J. F. *Langmuir* **1997**, 13, 6074.
46. Yang, Y,-W.; Deng, N.-J.; Yu, G.-E.; Zhou, Z.; Attwood, D.; Booth, C. *Langmuir* **1995**, 11, 4703.
47. Yang, Y.-W.; Ali-Adib, Z.; McKeown, N. B.; Ryan, A. J.; Attwood, D.; Booth, C. *Langmuir* **1997**, 13, 1860.
48. Alexandridis, P.; Olsson, U.; Lindman, B. *Macromolecules* **1995**, 28, 7700.
49. Alexandridis, P.; Olsson, U.; Lindman, B. *J. Phys. Chem.*, **1996**, 100, 280.
50. Alexandridis, P.; Zhou, D.; Khan, A. *Langmuir* **1996**, 12, 2690.
51. Alexandridis, P.; Olsson, U.; Lindman, B. *Langmuir* **1996**, 12, 1419.
52. Holmqvist, P.; Alexandridis, P.; Lindman, B. *Macromolecules* **1997**, 30, 6788.

53. Alexandridis, P.; Andersson, K. *J. Phys. Chem. B* **1997**, 101, 8103.
54. Holmqvist, P.; Alexandridis, P.; Lindman, B. *Langmuir* **1997**, 13, 2471.
55. Alexandridis, P.; Olsson, U.; Lindman, B. *Langmuir* **1997**, 13, 23.
56. Holmqvist, P.; Alexandridis, P.; Lindman, B. *J. Phys. Chem. B* **1998**, 102, 1149.
57. Mays, H.; Almgren, M.; Brown, W.; Alexandridis, P. *Langmuir* **1998**, 14, 723.
58. Schmidt, G.; Richtering, W.; Lindner, P.; Alexandridis, P. *Macromolecules* **1998**, 31, 2293.
59. Alexandridis, P.; Olsson, U.; Lindman, B. *Langmuir* **1998**, 14, 2627.
60. Holmqvist, P., Alexandridis, P., Lindman, B. *Langmuir,* **1997**, 13, 2471.
61. Liu, T.; Xie, Y.; Liang, D.; Zhou, S.; Jassal, C.; McNabb, M.; Hall, C.; Chuang, E.; Chu, B. *Langmuir* **1998**, 14, 7539.
62. Wu, C.; Liu, T.; Chu, B.; Schneider, D. K.; Graziano, V. *Macromolecules* **1997**, 30, 4574.
63. Wu, C.; Liu, T.; Chu, B. *J. Non-crystal Solids,* **1998**, 235-237, 605.
64. Wu, C.; Liu, T.; Chu, B. *Electrophoresis* **1998**, 19, 231.
65. R. L. Rill, B. R. Locke, Y. Liu, D. H. Van Winkle, *Proc. Natl. Acad. Sci. USA* **1998**, 95, 1534.
66. Mann, S. *Nature* **1993**, Vol. 365, 499. And the references there in.
67. Zhao, D.; Feng, J.; Huo, Q.; Melosh, N.; Fredrickson, G. H.; Chmelka, B. F.; Stucky, G. D. *Science* **1998**, 279, 548.

Chapter 2

PEP–PEO Block Copolymers as Model System for the Investigation of Micellization in Aqueous Solution

J. Allgaier, A. Poppe, L. Willner, J. Stellbrink, and D. Richter

Institut für Festkörperforschung, Forschungszentrum Jülich GmbH, 52425 Jülich, Germany

In this work we describe the micellization behavior of new poly(ethylene-propylene)-*block*-polyethylene oxide (PEP-PEO) block copolymers in water. PEP is a highly non-polar aliphatic polyhydrocarbon and insoluble in water. The water soluble PEO is polar and strongly incompatible with PEP. In a first part the synthesis and polymer characterization of the materials is described. Then the micellar properties of a series of block copolymers is presented. For all materials the PEP block molecular weight was kept constant and the PEO block molecular weight was varied over a large range. The investigation of the aggregates in water was carried out by small angle neutron scattering and by dynamic light scattering.

Introduction

Amphiphilic block copolymers form micelles in aqueous media due to the fact that the hydrophobic segments are insoluble in water. During the past years this phenomenon has attracted the attention of many research groups. In particular, polystyrene-*block*-poly(ethylene oxide) (PS-PEO) systems have been investigated intensively in water (*1-7*) and in a few cases in nonpolar solvents (*8*). Recently, the synthesis of amphiphillic polymers containing polydienes and polyalkanes (PA) as the hydrophobic part has been reported (*9-11*). For several reasons these new polymers are extremely attractive as model systems for the investigation of the micellization phenomenon in aqueous solution.

In dilute aqueous solution these amphiphillic block copolymers self-assemble into micelles with PA as the insoluble core and PEO as the soluble shell. As the incompatibility of the PA with PEO and water is extremely high the surface between core and shell is expected to be minimal. For this reason, the core/shell interface in the PA-PEO micelles should be very sharp. Additionally, it is assumed that the micelle core is free from the solvent water and also from PEO. Another consequence of the high incompatibility of the PA with water is minimal solubility of the single block copolymer chains in the solvent which is expressed in a very small critical micelle concentration (c.m.c.). If the hydrophobic block is made from the low T_g materials like 1,4-polybutadiene (PB), 1,4-polyisoprene (PI) or poly(ethylene-propylene) (PEP), the core material of the micelles is in the melt state. In this scenario the dissolution of the polymer is assumed to yield aggregates which easily reach thermodynamic equilibrium. In contrast, high T_g materials like PS as the micellar core can lead to structures being not in thermodynamic equilibrium due to the frozen state of the material.

Another important point in this context is that PA-PEO block copolymers show almost the same chemical design as the well known low molecular weight alkane-oxyethylene (C_nE_m) surfactants. Therefore, it is possible to study the micellization behavior starting with low molecular weight molecules up to high molecular weight polymers within the same chemical system.

For the investigation of micellar solutions different techniques as static light scattering (SLS), dynamic light scattering (DLS) and transmission electron microscopy (TEM) have been used. The use of light scattering techniques is limited to low concentrations. In the case of higher concentrations the appearance of multiple scattering prevents the use of this methods. TEM is restricted to materials free of solvent. Here the question remains to which extent the removal of water changes the structure of the micellar aggregates.

To overcome these difficulties we decided to use small angle neutron scattering (SANS) for our investigations. In addition, this technique is even more powerful for our investigations as the synthetic route for the production of the block copolymers allows one to label selectively either the hydrophobic or the hydrophilic part in the block copolymer with deuterium. In this case SANS enables the selective investigation of either the core or the shell structure in the micelles simply by matching the shell or the core with the appropriate H_2O/D_2O mixtures. This technique is called contrast variation. Beyond that, the SANS technique permits the extraction of information from the scattering data about the shape of the micellar aggregates.

For the investigation of micellar properties in water we chose PEP as the material for the hydrophobic block due to its low T_g of -60°C and its chemical stability. A series of PEP-PEO block copolymers was synthesized for a systematic investigation. For all materials in the series, the PEP block molecular weight was kept constant at 5,000 g/mol and was partially deuterated. The molecular weight of the PEO block was varied between 5,000 g/mol and 50,000 g/mol.

Experimental Section

The synthesis and characterization of the PEP-PEO block copolymers (*10*) as well as the characterization of the micelles by SANS and DLS (*12*) have been given elsewhere in detail.

Results and Discussion

Synthesis of PEP-PEO Block Copolymers

The preparation of PEP-PEO block copolymers in the two step process involves in the first step the synthesis of hydroxyl end functionalized PEP-OH (see Scheme 1). To obtain these polymers, deuterated isoprene was polymerized with *sec-* or *tert-*butyllithium. The functionalization was achieved by reacting the living PI with an excess of ethylene oxide (EO). Under these conditions exclusively one unit of EO is added to the PI chain end, leading to deuterated PI-OH (*13, 14*) The hydrogenation of PI-OH using H_2 with a conventional palladium catalyst yielded PEP-OH. Hydrogen and not deuterium was used for the saturation reaction in order to decrease the scattering length density of the material. This point will be discussed in more detail in the SANS chapter. The deuterium/hydrogen (d/h) compositions of the PEP-OH were measured by ^{1}H-NMR. The results (Table I) clearly indicate that not only saturation of the double bond has taken place which would lead to the theoretical composition d_8h_2. In addition, to some considerable extent the replacement of deuterium by hydrogen has occurred. Variations in d/h composition for the individual PEP-OH are most likely due to slightly different reaction conditions, i.e., reaction temperature, reaction time, and polymer/catalyst ratio. The molecular weight data of the PEP-OH blocks are also collected in Table I. As expected, the molecular weight distributions, obtained by SEC, are narrow and the calculated molecular weights are in good agreement with the measured values.

Table I. Molecular Weight Characterization of the PEP-OH Precursors

sample	composition of PEP d	M_n calculate	M_n titration	M_n VPO	M_w/M_n GPC
PEP5OH	$d_{6.37}h_{3.63}$	5,270	5,180	5,130	1.03
PEP5'OH	$d_{6.69}h_{3.31}$	5,290	5,210		1.03

In the second polymerization step the hydroxyl polymers were deprotonated and transferred into the macroinitiators PEP-OK by the reaction with cumyl potassium. For this purpose protic impurities like methanol, the precipitant for the hydroxyl polymers, had to be removed completely. The impurities would also react with cumylpotassium and initiate the homopolymerization of EO. Therefore, the polymers were dissolved in dry benzene, the solvent was removed under reduced pressure, and the polymers stirred under high vacuum for one day at 100°C. This measure increased the efficiency of the purification process. The complete procedure was repeated twice to remove all volatile protic impurities. It turned out that the high vacuum technique, using all glass apparatus, is the ideal method for this procedure. Thus, the contamination of the polymers with air was excluded, since the reaction products of oxygen and cumylpotassium would also homopolymerize EO. The following titration of the polymers, as solution in THF, with the deeply red colored cumylpotassium was conducted likewise under high vacuum conditions to avoid any contact of the reactants with air or moisture. Since the deprotonation of the hydroxyl polymer is a fast reaction, the decoloration of the added organometallic compound occurred immediately. At the end point of the titration the polymer solution turned slightly orange. This color disappeared after one to two hours, indicating a very slight excess of hydroxyl polymer in the system. Supplementary experiments proved that further addition of small traces of cumylpotassium resulted in a permanent coloration of the solution.

The titration can also be considered as end group analysis to determine M_n of the precursor polymers. The values, obtained by this method, are listed in Table I. They agree very well with the calculated M_n and the vapor pressure osmometry (VPO) measurement of PEP5OH. The results also indicate that the terminal hydroxyl groups remain completely untouched during the hydrogenation reaction.

The resulting macroinitiators PEP-OK were used to polymerize EO. After consumption of all monomer, samples for the SEC analysis were taken directly from the reactor without purifying the polymer. The presence of homopolymer in the products was examined in an additional set of experiments with block copolymers, having different relations of the block molecular weights. The SEC trace of PEP5-PEO2 (calculated block molecular weights $M_n(PEP)$ = 4,630, $M_n(PEO)$ = 1,500) is shown in Figure 1a. The absence of PEO homopolymer indicates that the purification process of the precursor is highly efficient. SEC details of the PEP5-PEO15 system (calculated molecular weights $M_n(PEP)$ = 4,630, $M_n(PEO)$ = 14,600) are given in Figure 1b. Residues of the precursor, arising from slight undertitration with cumylpotassium, are not visible. The fast proton exchange between the protonated and the deprotonated chain ends compared to the slow propagation reaction allows a homogeneous growth of all chains (*15*). In a supplementary experiment it was shown, that even the addition of only 50% of the calculated amount of cumyl potassium did not result in residual PEP-OH (*10*). The titration method is not limited to low molecular weights. Even block copolymers with PI and PEP molecular weights of M_n = 23,000 were synthesized without detectable amounts of homopolymer.

Further evidence for the accuracy of the titration method is given in Table II. The predicted molecular weights fit well with the M_n obtained from combined NMR/end group titration measurements. The molecular weight distributions, obtained by SEC, are in the range usually obtainable by living anionic polymerization under high vacuum conditions.

The purification of the products by precipitation was found to be a complex problem. In most common solvents either the hydrophilic or the hydrophobic block is soluble and causes the formation of micelles without precipitation of the polymer. Finally the use of acetone at -20°C allowed the precipitation and elimination of initiator residues in the polymers.

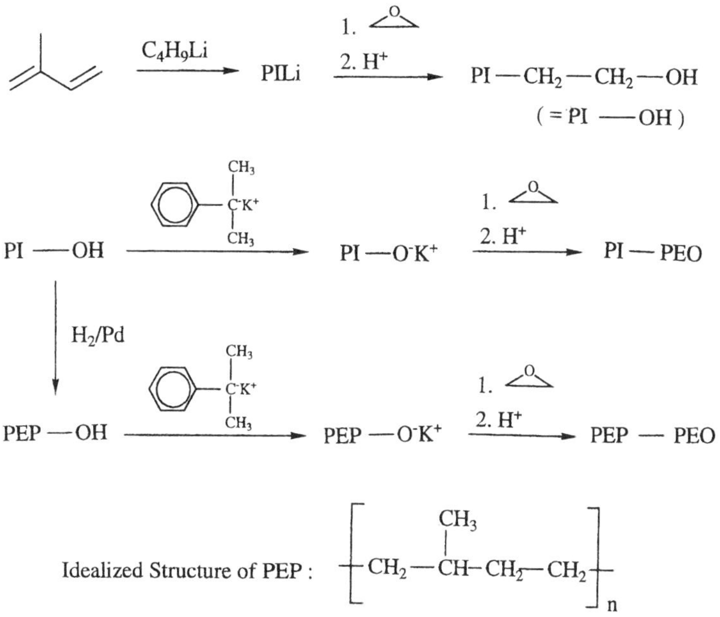

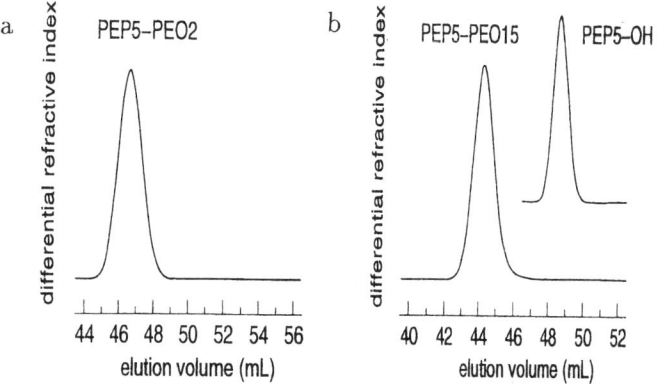

Scheme 1. Synthesis of PEP-PEO block copolymers.

Figure 1. SEC traces of some block copolymers and their corresponding precursors.

Table II. Molecular Weight Characterization of the PEP-PEO Block Copolymers

sample	PEPOH M_n titration	PEO block M_n calculated	M_n NMR	PEP-PEO M_n calculated	M_n NMR	M_w/M_n GPC
PEP5PEO5	5,180	5,780	5,910	11,000	11,100	1.03
PEP5PEO15	5,150	15,600	16,100	20,800	21,300	1.02
PEP5'PEO30	5,210	30,600	33,000	35,900	38,200	1.02
PEP5'PEO50	5,260	53,700	56,800	59,000	62,000	1.02

Micellar Properties of PEP-PEO in Water

Some Introductory Remarks on SANS

The contrast in a SANS experiment is determined by $(\rho_P - \rho_M)^2$, where ρ_P is the scattering length density of the scattering particle and ρ_M is the scattering length density of the matrix. Therefore, $(\rho_P - \rho_M)^2$ is the equivalent to $(dn/dc)^2$ in static light scattering. In general, the scattering length density is given by the expression $\rho = b/v$ whereby b is the scattering length and v is the volume of the scattering particle. The values for ρ are easily accessible as the b values are tabulated for the common isotopes. For a polymer molecule, b is the sum of the scattering length of all individual atoms in the monomer unit. The volume of the monomer unit is calculated from the density of the bulk material and the mass of the monomer unit. For organic molecules, the important fact is, that hydrogen and deuterium have very different scattering lengths and therefore, hydrogenous and deuterated molecules have different scattering length densities. Knowing this, we not only can carry out neutron scattering experiments of deuterium labeled materials in a hydrogenous matrix or reverse. By labeling only a part of the molecule, for example a polymer chain, we are able to examine the scattering behavior of exclusively the labeled sub-structure.

In our micelle study we utilized this technique by labeling the PEP part in the block copolymer with deuterium. If for a neutron scattering experiment of the micellar aggregates a mixture of H_2O and D_2O is used, which has the same scattering length density as hydrogenous PEO (hPEO), there is no scattering contribution of the PEO due to missing contrast. It is important to note that hPEO and H_2O do not have the same scattering length densities and therefore, a mixture of H_2O and D_2O has to be used to match PEO and water exactly (see Scheme 2). The scattering in this scenario, called core contrast, is exclusively produced by the micelle core, consisting of PEP. In the reverse case, the shell contrast, the PEP core is matched with the appropriate H_2O/D_2O mixture. The scattering length density of perdeuterated PEP (dPEP), however, is higher than the value for D_2O. Consequently, dPEP cannot be matched with water, which means that in the scattering experiment in pure D_2O, there will be contributions of not only the shell but also the core. For this reason we decided to use partially deuterated PEP (dhPEP) of the composition $d_{6.5}h_{3.5}$, which can be matched with water (Scheme 2).

By comparing neutron scattering with light scattering another important fact has to be considered. The Q-range accessible in light scattering is between 3×10^{-4} Å^{-1} and 3×10^{-3} Å^{-1} due to the wavelength of visible light. These Q-values can be translated in length scales 1/Q in the order of 300 Å to 3000 Å. To obtain structural information about a particle, the condition $R > 1/Q$ has to be fulfilled. Here R stands for the size of the particle. As the size of the micelles we wish to investigate is in the range of 100 Å to 500 Å, light scattering data of the micellar aggregates will yield only information about the overall size of the particles. Internal information about the shape and structure of the aggregates, however, cannot be obtained by this method. Due to the smaller wavelength of thermal neutrons, the Q-range accessible in a SANS experiment is between 10^{-3} Å^{-1} and 2×10^{-1} Å^{-1}. These values correspond to a length scale of 5 Å to 1000 Å and matches ideally with the requirements for the investigation of the micelles. Therefore, SANS data contain internal information about the structure and shape as well as information about the overall size of the micelles.

SANS Measurements of PEP-PEO Micelles in Water

The procedure for the SANS investigations are discussed in detail for the block copolymer PEP5PEO5 to demonstrate our approach. Along with measurements of the core and shell contrast, the micellar aggregates of PEP5PEO5 were examined in pure D_2O and in the intermediate contrast. For all contrasts the polymer volume fraction ϕ was 1%. Under core or shell contrast uniquely the micelle core or shell contribute to the scattering. Under D_2O and intermediate contrast conditions both parts, the core and the shell, contribute to the overall scattering. Intensity plots of the core and shell contrast are given in Figures 2a and 2b. Here the scattering intensity is expressed in terms of the coherent macroscopic cross section $(d\Sigma/d\Omega)$ (Q). For both contrasts, strong scattering is observed, indicating the formation of large aggregates. The curves exhibit well pronounced secondary maxima, displaying the existence of aggregates with well defined structure. Especially, the sharpness of the minima reveals a narrow size distribution, a sharp core/shell interface and a sharp density cutoff at the periphery of the micelle.

In order to analyze the structure of the micelle, model fitting was performed to the experimental data. As the model, a spherical core/shell structure was used. Thereby, the scattering length density distribution was assumed to be homogeneous over the core and shell, respectively. The sharp density cutoff at the PEP/PEO-water interface and the outer periphery of the micelle was smeared by multiplication with a Gaussian. The scattering amplitude A(Q), which is the Fourier transform of the density profile, for this model is given by

$$A_{M,C}(Q) = \frac{3(\sin(QR_{M,C}) - QR_{M,C}\cos(QR_{M,C}))}{(QR_{M,C})^3} \exp\left(-\frac{\sigma_{M,C}^2 Q^2}{4}\right) \quad (1)$$

where the indices C and M refer either to the core or to the micelle. R_C and R_M are the radii of the core and the micelle, respectively. The smearing of the edges in the core and

28

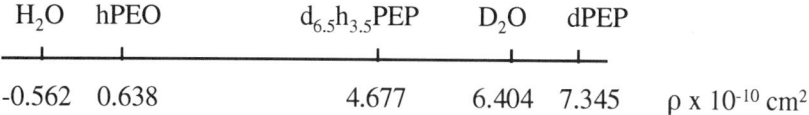

Scheme 2. Scattering length densities ρ for PEP, PEO, and water.

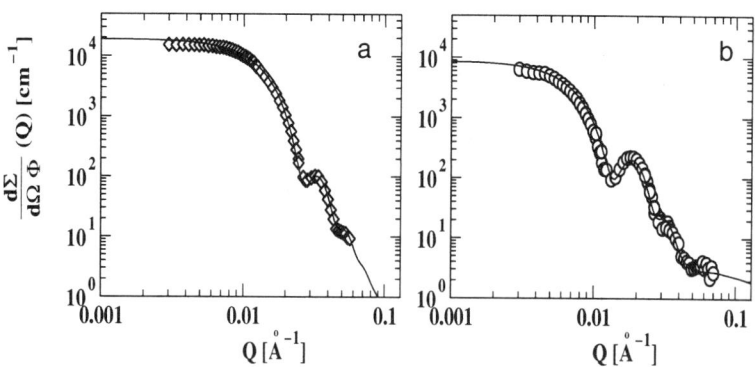

Figure 2. Absolute SANS data under core contrast (2a) and shell contrast (2b) for PEP5PEO5. The solid lines represent the results of the model fits.

shell is given in terms of the parameter $\sigma_{M,C}$. A detailed description of the model is given elsewhere (*12*).

For the calculation, a standard fitting routine was used, which additionally allowed the computation of the different contrasts simultaneously. The following parameters were the independent variables of the model: the core radius R_C, the micelle radius R_M, the volume fraction of water in the shell ϕ_S, and the smearing parameters of the core, σ_C, and the micelle, σ_M.

In the first step of the evaluation, the scattering curve of the core contrast was fitted independently. The scattering length densities of the H_2O/D_2O mixture and of the PEP core were introduced as fixed parameters. The core radius R_C and the smearing of the core-shell interface σ_C were adjustable fitting parameters in the fitting routine. The solid line in Figure 2a presents the resulting fit. We note that on an absolute scale perfect agreement between the data and the theoretical form factor of a sphere is achieved. If we assume that the micelle core consists of PEP only, the micelle aggregation number P can be calculated with the knowledge of R_C using the following equation:

$$P = \frac{(4/3)\pi R_C^3}{v_{PEP} N_{PEP}} \qquad (2)$$

v_{PEP} denotes the volume of a PEP repeat unit and N_{PEP} is the degree of polymerization of the PEP block. The results for P, R_C, and σ_C found using the fitting routine are discussed later.

In a second step all scattering data at different contrasts were fitted simultaneously. Thereby, the parameters obtained in the first step were kept fixed, while the micellar radius, R_M, and the smearing of the outer edge, σ_M, were the variables. As it turned out, the data could not be fitted on an absolute intensity scale. The discussion of this phenomenon is beyond the scope of this article, and is explained elsewhere (*12*).

In order to proceed further, we relaxed the condition of absolute normalization, which was successful in the case of the core contrast, and allowed the cross section to vary in the fit. The good agreement between the fit and the experimental data is documented for the shell contrast in Figure 2b. The scattering data of the D_2O and intermediate contrast, principally, do not contain supplementary information in addition to the core and shell contrast data. However, the accuracy of the results improves by including these additional contrasts into the simultaneous fitting procedure. Details about D_2O and intermediate contrast measurements are not discussed further and are described elsewhere (*12*). The results obtained for R_M and σ_M from the simultaneous fit will be discussed later.

The procedure, explained in case of the block copolymer PEP5PEO5 was now applied to investigate the micelles, formed by the materials PEP5PEO15, PEP5'PEO30, and PEP5'PEO50. Using a step profile smeared by a Gaussian to describe the PEP distribution in the core, for all materials of the PEP5 series good agreement of the core contrast measurement with the model curve was obtained. This fact reveals that the core consists of PEP only with a density equal to the bulk density of PEP. The aggregation numbers P as well as core radii R_C, obtained from the fitting procedure are given in Table III. The values for P and R_C indicate the large dimensions of the micelles. This finding is even more emphasized by the molecular weight of the micelles, M, which easily can be calculated from P and the block copolymer molecular weight. The results for the whole series of block copolymers shows that with increasing PEO chain length the aggregation number decreases constantly. An explanation on the basis of a theoretical model also will be given later. The smearing parameters of the core σ_C, obtained from the fits, were too small to be determined accurately, and had negligible influence on the model calculations. Typical values for σ_C were in the range of 5% to 10% of the core radius. These results indicate that the micelle cores are well defined in shape.

Table III. Characterization of the PEP-PEO Micelles in Water.

sample	P	$M \times 10^6$	$R_C/Å$	$R_M/Å$	$\sigma_M/Å$	x	$R_H/Å$	M_w/M_n
PEP5PEO5	2,430	27.0	176	294	32		339	1.03
PEP5PEO15	1,450	30.8	148	360	75		510	1.03
PEP5'PEO30	1,080	41.3	134	556		1.1	617	1.04
PEP5'PEO50	620	38.4	111	637		1.33	662	1.05

For the description of the micelle shell a step profile smeared by a Gaussian was used successfully in case of PEP5PEO15 in analogy to PEP5PEO5. The smearing parameters σ_M resulting from the fitting procedure are listed in Table III. The calculated PEO density profiles of the shells for the two materials are given in Figures 3a and 3b. In contrast to the core smearing, the smearing of the shell is much stronger and can be determined precisely. Both, the values for σ_M as well as the density profiles reveal that the outer edge of the micelle is more diffuse in case of PEP5PEO15 than for PEP5PEO5.

The scenario is different for the aggregates formed by PEP5'PEO30 and PEP5'PEO50. In these cases, the scattering data could not be fitted using a step profile for the PEO density in the corona. However, the use of a density profile of the r^{-x} type was successful. In this case, the partial scattering amplitude for the shell $A_{Sh}(Q)$ can be written as follows:

$$A_{Sh}(Q) = \frac{4\pi}{C} \int_0^\infty \rho(r) \frac{\sin(Qr)}{Qr} r^2 dr \qquad (3)$$

Here $A_{Sh}(Q)$ is the normalized Fourier transformation of a sphere with the radial density profile $\rho(r)$. The normalization constant is given by:

$$C = \int_0^\infty \rho(r) 4\pi r^2 dr \qquad (4)$$

The radial density profile $\rho(r)$ is given in equation 5. The r^{-x} type function is modified by a Fermi function as cutoff function.

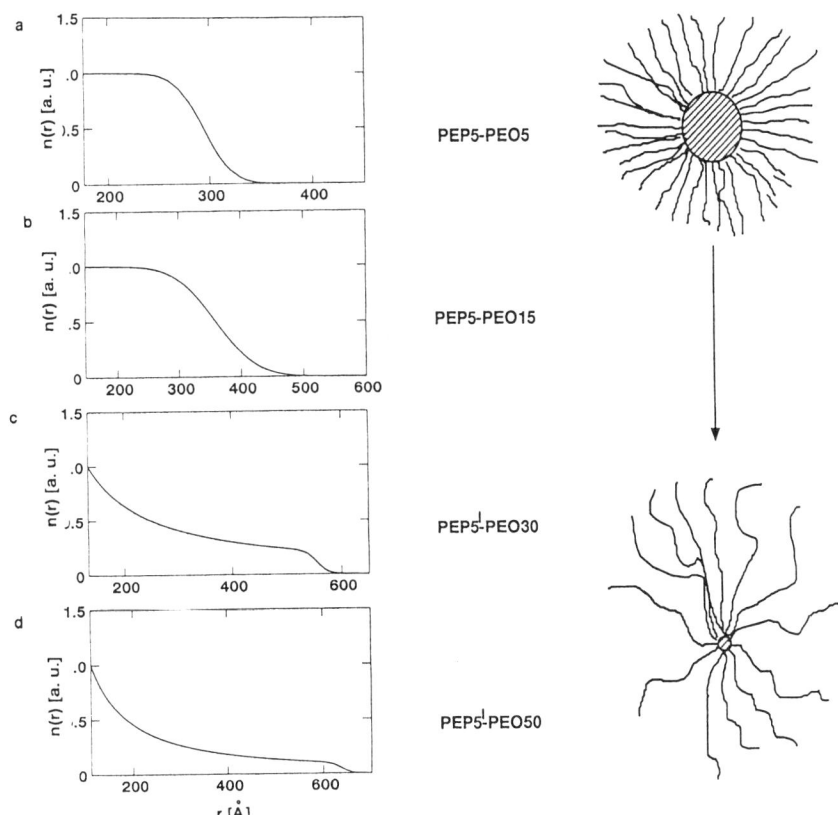

Figure 3. Relative PEO density profiles of the micelle shells for different PEP-PEO block copolymers.

$$\rho(r) = \frac{r^{-x}}{(1+\exp((r-R_M)/\sigma_F))} \tag{5}$$

The exponent x was an adjustable fitting parameter in the fitting routine. σ_F is the smearing parameter of the Fermi function.

With the help of equations 3, 4 and 5 all contrasts of the PEP5'PEO30 and PEP5'PEO50 micelles were fitted simultaneously with R_M and x as the variables. The variation of σ_F had very little influence on the fit and therefore, was kept constant at 10 Å. Now good agreement between the fits and the experimental data was obtained. This is documented in Figure 4, where the shell contrast measurement of PEP5'PEO50 is shown along with the data for the corresponding core contrast. The pattern for the core contrast measurement (Figure 4a) does not exhibit maxima and minima as it is the case for PEP5PEO5. This is due to the low volume fraction of core material in the 1% solution of PEP5'PEO50 and the lower aggregation number. For that reason, the scattering of the micelle cores is low compared to the background scattering and details of the measured curve disappear. The different shape of the shell contrast curve of PEP5'PEO50 (Figure 4b) compared to PEP5PEO5 is due to the different structures of the shells. The less pronounced maxima and minima result from the less defined shape and diffuser outer edge of the micelle shell in case of PEP5'PEO50. The results for R_M and x obtained from the fits for the PEP5'PEO30 and PEP5'PEO50 micelles are given in Table III. The calculated PEO density profiles of the micelle shells are given in Figures 3c and 3d. In contrast to PEP5PEO5 and PEP5PEO15, the distribution of PEO in the corona is much less homogeneous in case of the micelles with the longer PEO chains. The outer edges of the micelles are rather diffuse and it even can be assumed that the cutoff caused by the Fermi function is artificially abrupt. Along with the PEO density profiles, in Figure 3 the different micelle types are visualized for a better understanding.

It is interesting to note, that the value for x, which describes the PEO density profile, is exactly 1.33 in case of PEP5'PEO50. The same value is predicted theoretically for the polymer distribution in star polymers (16). For that reason, the aggregates formed by PEP5'PEO50 can be described as star like micelles. PEP5'PEO30 seems to produce intermediate structures. In this case, the micelles cannot be described fully as star like, on the other hand the PEO in the corona is far away from being distributed homogeneously.

DLS Measurements of the PEP-PEO Micelles in Water

In accordance with the SANS results, the DLS measurements reveal the existence of only one aggregated species of narrow size distribution (17) for each block copolymer. Large scale aggregates due to secondary association as reported for the PS-PEO/water system could not be observed (1). Analysis of the initial decay of the intensity correlation functions (18) reveals a reduced second cumulant, μ_2/Γ^2, characteristic for a narrow size distribution of scatterers: $0.001 < \mu_2/\Gamma^2 < 0.04$, slightly increasing with decreasing scattering angle. It has been shown, that the reduced second cumulant can be related to the conventional polydispersity M_w/M_n by

$$\frac{\mu_2}{\Gamma^2} = v^2(M_w/M_n - 1) \qquad (6)$$

with v the scaling exponent in the relation $D_0 \sim M^v$ (*19*). Using $v = 0.56$ as found for PEO homopolymers (*20*), the polydispersity can be estimated from the high Q-data to $M_w/M_n = 1.03 \pm 0.01$. The sizes of the micelles do not change in the concentration range between $\phi = 0.001\%$ and $\phi = 0.1\%$ due to the linearity of the mutual diffusion coefficient D_m in this concentration range. Moreover, it can be concluded that the critical micelle concentration is well below a polymer volume fraction of $\phi = 0.001\%$. After extrapolation to infinite dilution, the hydrodynamic radii of the micelles R_H were derived from D_0 using the Stokes-Einstein relation. The results for R_H as well as M_w/M_n are listed in Table III. The large size of the PEP-PEO micelles, found in the SANS experiments, is confirmed by the hydrodynamic radii R_H found in the DLS experiments. For the material with the lowest molecular weight, PEP5PEO5, which forms the most compact micelles, the static and dynamic radii are identical within experimental error, as expected for hard spheres. The difference between R_M and R_H is larger in case of the other materials because the outer edge of the micelle is more diffuse.

Thermodynamic reflection of the micellar parameters

The micellar free energy F of a single chain incorporated in a micelle far away from the critical micelle concentration is approximated as a sum of three contributions.

$$F = F_{int} + F_{Sh} + F_C \qquad (7)$$

F_{int} denotes the interfacial contribution to the free energy associated with the core/shell interface and is equivalent to the surface energy of the micelle core material PEP and water.

In this consideration the PEO chains are neglected, because they only cover a small fraction of the core surface. The contributions F_{Sh} and F_C are related to chain stretching in the shell and in the core, respectively. More sophisticated models also consider the loss of entropy when a single polymer chain passes from the solution into the micelle. However, it is important to note, that the only contribution supporting the formation of the micelles is the surface energy, which itself is given by the surface tension γ.

$$F_{int} = \frac{4\pi R_C^2 \gamma}{P_{eq}} \qquad (8)$$

For the system PEP/water a surface tension of $\gamma = 51$ dyn/cm was estimated (*21*). This value is considerably higher than the PS/water surface tension of 35 dyn/cm. If we compare PEP-PEO micelle sizes with the values found for the widely investigated PS-PEO system, we see that PS-PEO micelles of comparable block copolymers are much smaller (*3,22*). This finding is attributed to the higher surface tension of PEP/water. The high surface tension of the PEP/water system also explains the sharp density cutoff at the core surface found in the SANS experiments.

In the course of our study it became apparent that the results for the PEP-PEO system are in astonishingly good accordance with a theory by Nagarajan and Ganesh (*23*). In this model six contributions to the micellar free energy are considered. The reference state for the energy calculations is a single block copolymer chain in infinite dilution. The free energy of the shell and the core consist of two terms. The first term includes entropy contributions of the block copolymer, that are involved when passing from the reference to the micellar state. The second term includes contributions due to chain deformation. The expression for the interfacial free energy, F_{int}, is equal to the one introduced in the former model (equation 8). The basis for the calculations are mean field approximations. For this reason, homogeneous density profiles in core and corona are assumed. To calculate the aggregation number P, the chain length of the PEP and PEO blocks, the Flory-Huggins interaction parameters between the two polymer blocks and water, and the surface tension PEP/water have to be considered. The minimization of the micellar free energy dependent on the aggregation number and the micelle radius then is performed in a numerical procedure as the analytical solution is not possible.

From the calculations for the PEP-PEO/water system the relation $P_{eq} \sim N_{PEO}^{-0.51}$ was obtained, whereby N_{PEO} is the polymerization degree of the PEO polymer block. In Figure 5, the solid line represents the dependence of the calculated aggregation number on the PEO polymerization degree for constant PEP block molecular weight of 5,000 g/mol. The measured aggregation numbers are plotted in the same figure. For the first three polymers of the series, PEP5PEO5, PEP5PEO15, and PEP5'PEO30 the theoretical and the measured results are in very good agreement on an absolute scale. The deviation in case of PEP5'PEO50 can be related to the fact that the Nagarajan-Ganesh theory requires a homogeneous density profile of a dense PEO shell. This condition is certainly not fulfilled for PEP5'PEO50.

Conclusions

In this work we have shown that PEP-PEO block copolymers can be synthesized with narrow molecular weight distribution and with no detectable content of homopolymer. The investigation of the micellar properties was done mainly by SANS using a series of materials with constant PEP molecular weight of 5,000 g/mol and PEO molecular weights varying between 5,000 g/mol and 50,000 g/mol. All materials contained deuterium labeled PEP polymer blocks. It was shown that the micelles are spherical in shape with a core consisting exclusively of PEP and a PEO shell swollen with water. Due to the high incompatibility of PEP with PEO and water the aggregation numbers and micelle dimensions are very large. For the block copolymers with smaller PEO content, the density of PEO in the shell is homogeneous and the density cutoff at the outer edge of the micelle is fairly abrupt. In case of the materials with higher PEO content, the PEO density profile is of hyperbolic shape and the periphery of the micelle is diffuse.

Very good agreement of the experimental results was found with a theory by Nagarajan and Ganesh which is based on mean field approximations. Further work is now in process to extend the study to higher PEO contents in order to study the variation of the micellar shape dependent on the block copolymer composition in more detail. In a second step, PEP-PEO block copolymers with lower molecular weights will be investigated to compare the micellization phenomena of the PEP-PEO block copolymers with low molecular weight C_nE_m amphiphiles of the same chemical structure.

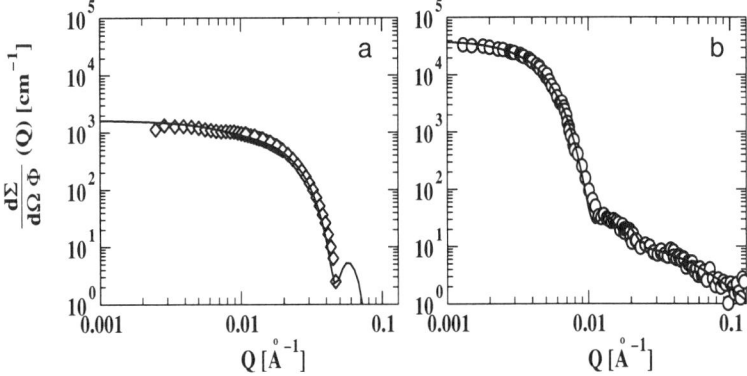

Figure 4. Absolute SANS data under core contrast (4a) and shell contrast (4b) for PEP5'PEO50. The solid lines represent the results of the model fits.

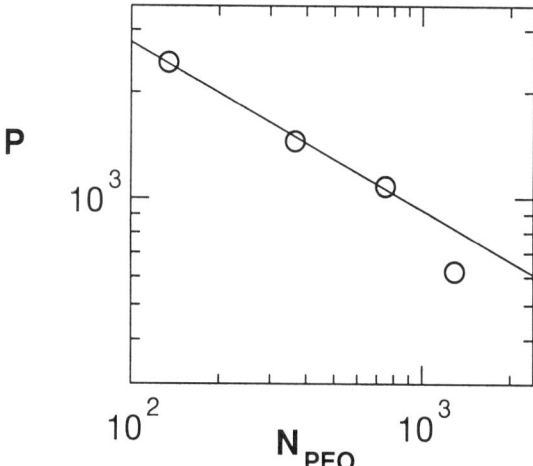

Figure 5. Dependence of the aggregation number P on the PEP polymerization degree N_{PEO}. The solid line represents the calculated dependence by Nagarajan-Ganesh theory.

Acknowledgments. This work was performed in the framework of the project "self-organization of polymers and polymeric glasses", which is financially supported by the Deutsche Forschungsgemeinschaft. M Zähres (Universität Duisburg) is gratefully acknowledged for the NMR measurements, M. Monkenbusch (FZ Jülich) and L. J. Fetters (Exxon Research & Engineering Co., Annandale, NJ) for helpful discussion.

References

1. Xu, R; Winnik, M. A.; Hallett, F. R.; Riess, G.; Croucher, M. D. *Macromolecules*, **1991**, *24*, 87.
2. Wilhelm, M.; Zhao, C.-L.; Wang, Y.; Xu, R.; Winnik, M. A.; Mura, J.-L.; Riess, G.; Croucher, M. D. *Macromolecules*, **1991**, *24*, 1033.
3. Xu, R.; Winnik, M. A.; Chu, B.; Riess, G.; Croucher, M. D. *Macromolecules*, **1992**, *25*, 644.
4. Calderara, F.; Hruska, Z.; Hurtrez, G.; Lerch, J.-P.; Nugay, T.; Riess, G. *Macromolecules*, **1994**, *27*, 1210.
5. Jada, A.; Hurtrez, G.; Siffert, B.; Riess, G. *Macromol. Chem. Phys.*, **1996**, *197*, 3697.
6. Hickl, B.; Ballauff, M.; Jada, A. *Macromolecules*, **1996**, *29*, 4006.
7. Yu, K.; Eisenberg, A. *Macromolecules*, **1996**, *29*, 6359.
8. Gast, A. P.; Vinson, P. K.; Cogan-Farinas, K. A. *Macromolecules* **1993**, *26*, 1774
9. Allgaier, J.; Willner, L.; Richter, D.; German Patent Application, No. P 19634477.8, 1996.
10. Allgaier, J.; Poppe, A.; Willner, L.; Richter, D. *Macromolecules*, **1997**, *30*, 1582.
11. Hillmyer, M. A.; Bates, F. S.; *Macromolecules*, **1996**, *29*, 6994.
12. Poppe, A.; Willner, L.; Allgaier, J.; Stellbrink, J.; Richter, D. *Macromolecules* **1997**, *30*, 7462.
13. Kazanskii, K. S.; Solovyanov, A. A.; Entelis, S. E. *Eur. Polym. J.* **1971**, *7*, 1421.
14. Quirk, R. P.; Ma, J.; *J. Polym. Sci.: Part A: Polym. Chem.* **1988**, *24*, 2031
15. Bayer, U.; Stadler, R. *Macromol. Chem. Phys.* **1994**, *195*, 2709.
16. Daoud, M.; Cotton, J. P. *J. Physique* **1982**, *43*, 531.
17. Brown, W. *Dynamic light scattering*; Oxford University Press, 1993.
18. Kappel, D. E. *J. Chem. Phys.* **1972**, *57*, 4814.
19. Selser, J. C. *Macromolecules* **1979**, *12*, 909.
20. Devanand, K.; Selser, J. C. *Nature*, **1990**, *343*, 739.
21. Landolt; Börnstein. In *Zahlenwerte und Funktionen*; Bartels, J.; Tenbruggencate, P.; Hellwege, K. H.; Schäfer, K.; Schmidt, E., Eds.; Berlin-Göttingen-Heidelberg, 1956.
22. Hurtrez, G.; Dumas, P.; Riess, G. *Polymer Bulletin* **1998**, *40*, 203.
23. Nagarajan, R.; Ganesh, K. *J. Chem. Phys.*, **1989**, *90*, 5843.

Chapter 3

Amphiphilic Block Copolymers as Surfactants in Emulsion Polymerization

Matthias Gerst[1], Horst Schuch[1], and Dieter Urban[2]

[1]BASF AG, Polymer Research Laboratory, Ludwigshafen, Germany
[2]BASF Corporation, Steelcreek Road, Charlotte, NC 28273

The use of amphiphilic block copolymers as surfactants in emulsion polymerization enables one to prepare well stabilized dispersions. In contrast to low molecular weight surfactants new film properties are available, because block copolymers have better compatibility with the polymer particles and lower migration in the resulting film. The amphiphilic block copolymers poly(methylmethacrylate-block-acrylic acid) $P(MMA_m\text{-}b\text{-}AA_n)$, poly(methylmethacrylate-block-methacrylic acid) $P(MMA_m\text{-}b\text{-}MAA_n)$ and poly(isobutylene-block-methacrylic acid) $P(IB_m\text{-}b\text{-}MAA_n)$ were made by living ionic polymerization. In aqueous solution block copolymers of higher molecular weight form spherical "frozen micelles", not exchanging unimers. The micellar aggregation number Z and micellar diameter $2R_h$ in the aqueous medium were obtained by static and dynamic light scattering and turned out to be predictable solely from the two block lengths. Using aqueous solutions of neutralized block copolymers as the only emulsifying agent in emulsion polymerization, stable dispersions were obtained, where the block copolymers act comparable to a seed. The mechanical properties and the water sensitivity of the resulting polymer film change by annealing.

© 2000 American Chemical Society

Introduction

Low molecular weight surfactants are widely used in emulsion polymerization to form and stabilize the polymer particles and to prevent coagulation of the dispersion when applied making e.g. coatings and adhesives. The physical properties of the resulting film, however, often suffer from the presence of surfactants because they favour sensitivity to water or impair adhesion to the substrate. In contrast amphiphilic block copolymers (bc) are expected to prevent such disadvantages due to better compatibility with the polymer particles and lower migration rates. It has already been shown that poly(styrene-block-ethyleneoxide) can be used instead of surfactants to stabilize dispersions [1-3]. Additionally block copolymers can act as blending agents to improve such properties as mechanical strength of the resulting polymer film. Thus the motivation of this work was using block copolymers both to stabilize a dispersion and to create improved film properties.

Synthesis of Block Copolymers

Three types of block copolymers with different molecular weights and block lengths were made by living ionic polymerization [4-8]. All of the block copolymers under investigation have two blocks of different monomers with varying chain length. To achieve the amphiphilic properties a hydrophobic, e.g. a PMMA- or PIB-segment, and a hydrophilic, e.g., a PAA- or PMAA-segment, were choosen. From the combination of these segments the following block copolymers were synthesised:

a) poly(methylmethacrylate-block-acrylic acid), $P(MMA_m\text{-b-}AA_n)$ with m=18-38 and n =26-38;
b) poly(methylmethacrylate-block-methacrylic acid), $P(MMA_m\text{-b-}MAA_n)$ with m=16-67 and n=44-217
c) poly(isobutylene-block-methacrylic acid), $P(IB_m\text{-b-}MAA_n)$ with m=70-134 and n=52-228.

The PMMA-derivatives were produced by living anionic polymerization. In the first step, the initiation of MMA yields a precursor block which could be converted by the addition of the second monomer tBA or tBMA to a block copolymer. The use of tBMA enables one to increase the reaction temperature from -78°C used for tBA to a more convenient level of about -20°C. In order to achieve the amphiphilic properties the tert.-butyl-groups were hydrolysed with hydrochloric acid in dioxane to carboxyl acid groups.

Table 1. Block copolymer data: hydrophobic block length m, hydrophilic block length n, molecular weight Mw in g/mol (GPC), polydispersity Mw/Mn; aggregation data: micelle diameter $2R_h$ in nm, aggregation number Z.

block copolymer	m	n	M_w g/mol	M_w/M_n	Z	$2R_h$ nm
P(MMA_m-b-AA_n)	18	26	3 600	1.42		
	28	30	5 000	1.32		
	38	38	6 600	1.14		
P(MMA_m-b-MAA_n)	16	44	6 800	1.26	18	13
	25	29	6 800	1.37	43	21
	32	69	12 000	1.32	39	17
	58	57	14 200	1.33	250	34
	64	112	17 400	1.09	94	36
	67	217	27 000	1.06	34	34
P(IB_m-b-MAA_n)	70	52	8 900	1.06	188	25
	70	70	10 600	1.07	192	30
	134	145	20 800	1.04	133	57
	134	228	28 000	1.03	144	58

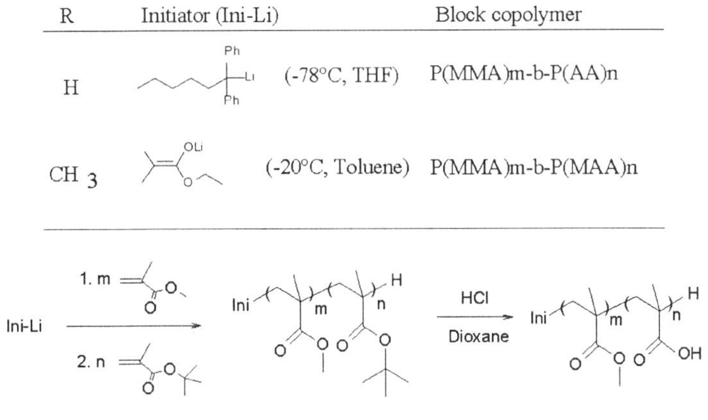

R	Initiator (Ini-Li)		Block copolymer
H	(Ph-CH-Li-Ph structure)	(-78°C, THF)	P(MMA)m-b-P(AA)n
CH_3	(OLi ethyl enolate structure)	(-20°C, Toluene)	P(MMA)m-b-P(MAA)n

Figure 1: Synthesis of PMMA-PAA and PMMA-PMAA block copolymers [9].

In the case of the PIB-b-PMAS block copolymers a different approach was necessary. The initiation of the living polymerization of iso-butylene needed a cationic initiator. The crucial reaction step in the synthesis was the switch over from the cationic to anionic chain end as shown in Fig. 2. After polymerising the tBMA the block copolymer was obtained, as already described, by hydrolysis.

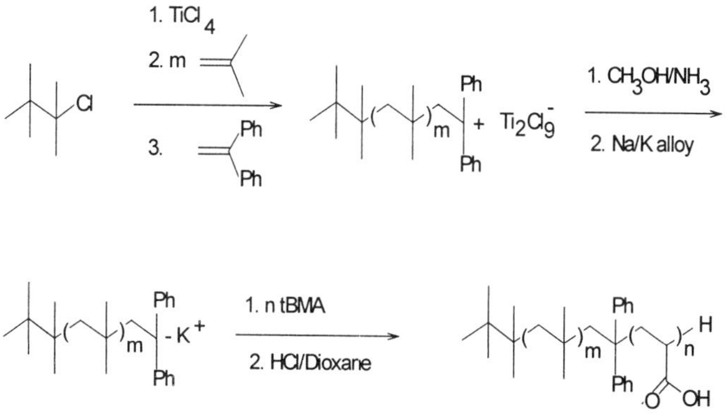

Figure 2: Synthesis of PMMA-PAA and PMMA-PMAA block copolymers [10].

Aggregation of Block Copolymers

The aqueous bc-solutions were prepared solving 5g bc in 50g methanol. After diluting with 50 ml water the bc-solution was neutralized with an equivalent amount of ammonia or sodium hydroxide. In case of the P(IB$_{134}$-b-MAA$_{145/228}$), 500g THF and 500g water were used instead of 50g methanol and 50g water. The organic solvent was removed by distillation and the pH was adjusted to 12 with 0.2 N sodium hydroxide solution. The bc-solutions were concentrated to obtain a solids content of about 3% (w/w). They were used as a stock solution for the following experiments.

The surface tension of aqueous solutions containing P(MMA$_m$-b-AA$_n$) neutralized with ammonia depends on the chain length of the hydrophobic PMMA block [Fig. 3]. Below a chain length of about 20 MMA molecules the block copolymer is surface active and lowers the surface tension of water. At higher chain lengths the surface tension is about that of pure water, indicating that such block copolymers do not have the tendency to be present at the water/air interface.

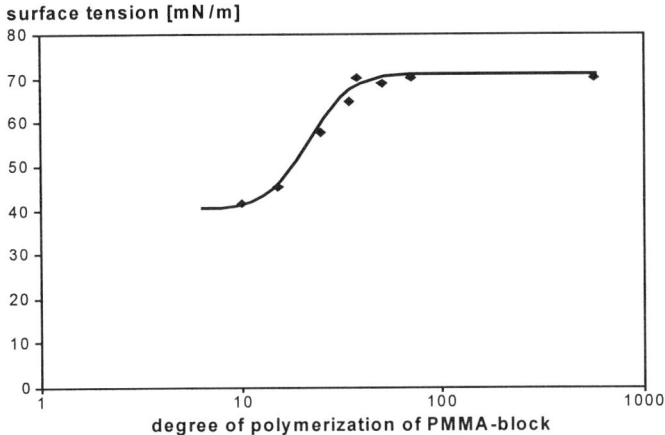

Figure 3: Surface tension of aqueous P(MMA$_m$-b-AA$_n$) solution (0.5% [w/w]) as a function of the hydrophobic block length m.

Thus, when m>20, determination of the critical micelle concentration by a ring tensiometer is not possible. But observing the titration of an aqueous BODIPY-NLS® [8] solution with a block copolymer solution by confocal fluorescence correlation spectroscopy (FCS) [11] shows a strong decrease in mobility of the fluorescent dye. The fluorescent agent is a C_{16}-fatty acid with a fluorescent label. The solubility in water is 1 nM/liter. If only water is present, this concentration leads to 1 dissolved molecule within the measurement volume (the focus volume of the laser beam). After dilution of the micellar solution (containing the fluorescent agent) a 24 hour delay is used before the measurement of the fluorescence correlation time, in order to allow sufficient time for establishing the thermodynamic equilibrium regarding the number of micelles. FCS yields the critical concentration for micellization (CMC), where very small CMC are accessible. The following figure (Figure 4) shows a typical FCS determination of the CMC. When the block copolymer concentration is larger than the CMC, the number of dye molecules within the laser focus volume increases due to the solubilization of dye molecules. For still higher concentrations, the dye fluorescence starts to be quenched. Salt content did not influence the results.

According to these experiments the critical aggregation concentrations of P(MMA$_m$-b-AA$_n$), P(MMA$_m$-b-MAA$_n$) and P(IB$_m$-b-MAA$_n$) are between 0.2 and 200 mg/l in water [Table 2].

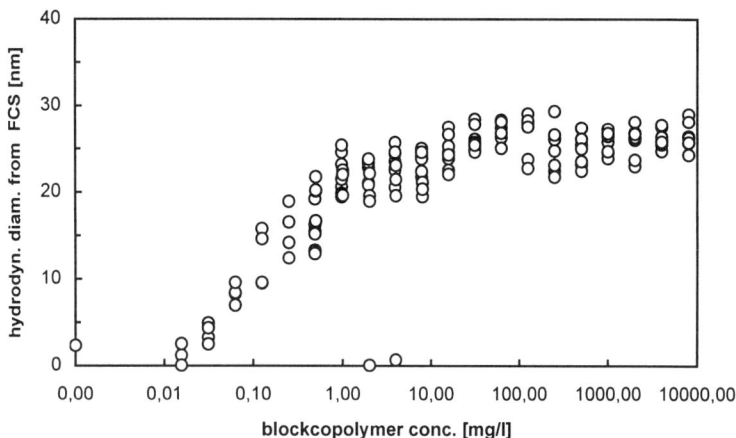

Figure 4a): Example of FCS-measurment of $P(IB_{70}-b-MAA_{70})$.
Hydrodynamic diameter of the particles carrying a dye molecule as a function of the BC concentration. The strong increase shows the onset of micellar aggregation.

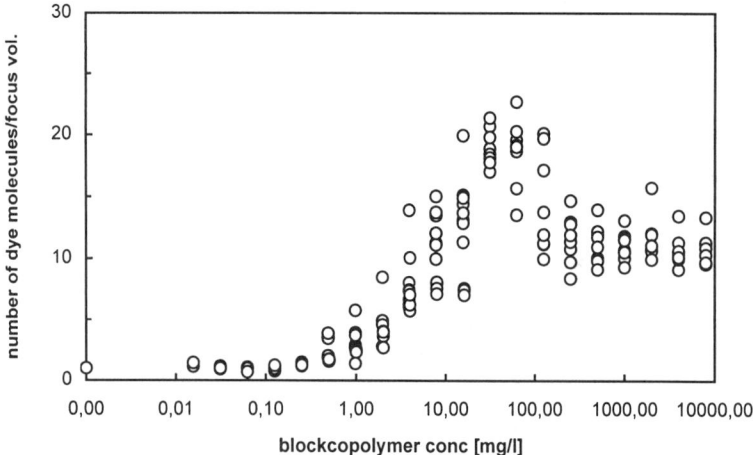

Figure 4b): Example of FCS-measurment of $P(IB_{70}-b-MAA_{70})$.
Number of dye molecules in the laser focus as a function of the block copolymer concentration.

Table 2: Values for critical micellisation concentration (CMC) for P(IB$_m$-b-MAA$_n$)

m	n	[mg/l] [a]	[mg/l] [b]
70	52	0,05	0,3
70	70	0,05	0,2
134	145	0,07	0,3
134	288	0,1	0,3

P(MMA$_m$-b-AA$_n$)

m	n	[mg/l] [a]	[mg/l] [b]
18	26	4-8	-
28	30	-	-
38	38	20-40	-

P(MMA$_m$-b-MAA$_n$)

m	n	[mg/l] [a]	[mg/l] [b]
16	44	50-100	100-200
25	29	0,1-0,5	3-7
32	69	7	70
58	57	0,05-0,1	1
64	112	0,1-0,2	10
67	217	1	10

[a] change of diameter: micelles detectable for concentrations of more than given [mg/l], [b] change of number: micelles detectable for concentrations of more than given [mg/l].

To measure the rate of exchange of block copolymers between aggregates by time resolved fluorescence spectroscopy according to the principle of non radiative energy transfer, [12] separately prepared aqueous solutions of naphthalene (donor) terminated P(MMA$_{20}$-b-AA$_{40}$) and pyrene (acceptor) terminated P(MMA$_{20}$-b-AA$_{44}$) were mixed. Surprisingly neither at room temperature nor at 150°C, which is above the glass transition temperature of the block copolymer, was any exchange observed (which would have led to a shortening of the donor fluorescence decay). This indicates that these aggregates are very stable "frozen micelles".[13]

The aggregation number Z and the micellar diameter $2R_h$ were determined by static and dynamic light scattering. The basics of the light scattering (LS) method are described elsewhere [14]. Here a light scattering goniometer of ALV / Langen, Germany, with a correlator ALV 3000 was used, attached to a Nd:YAG-Laser, 532 nm, 400mW (Adlas). A high laser power is required for a good signal to noise ratio to determine the autocorrelation function $g_2(\tau)$.

The solutions were clarified by a 0,45 μm filter (Millipore). Measurements were performed at scattering angle 30° and 90°, with one concentration of 1 g/l. This reduction of the usual Zimm analysis is reasonable due to the small interaction between the micelles (the second virial coefficient A_2 is small); the concentration of 1 g/l is well above the CMC; and the angular dependence of the scattered light is weak. Thus the extrapolation to scattering angle 0° is straightforward. The salt content was 0.02 n Na_2SO_4 and pH = 10. The exact values were found to be of no critical influence on the results. The experiments with the dynamic light scattering yield the average hydrodynamic diameter $2R_h$ of the colloidal particles. The accuracy is up to ± 1-2 nm, much better than the accuracy for R_g (± 5-10 nm). Also the autocorrelation function $g_2(\tau)$ is obtained, which, by use of the CONTIN-algorithm, gives the particle size distribution (PSD) referring to $2R_h$. The maximum resolution of the PSD is given by the following: In a bimodal distribution the components are recognized if the diameters differ by a factor of three or more. The small diameter is assigned to single micelles, the bigger one to aggregates of micelles which were also detected with the analytical ultracentrifuge (AUC).

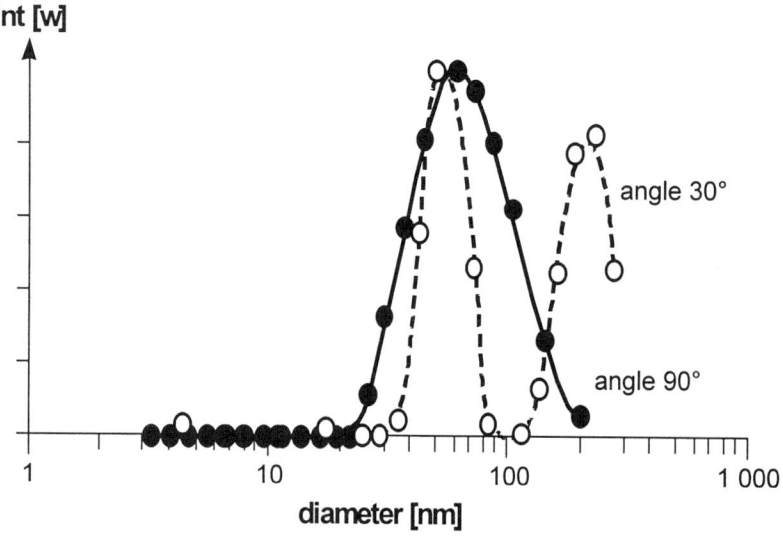

Figure 5: Evaluation of the micelle diameter $2 R_h$ from dynamic light scattering data of $P(PIB_{134}$-b-$MAA_{145})$ in aqueous solution at 25°C using the CONTIN algorithm; - -- scattering angle 30°, —scattering angle 90°.

The experiments of the static light measurements yields the average absolute molar mass M_w of the colloidal particle, here of the micelles, and the radius of gyration R_g (average value). The accuracy in M is 10-20%; which is important for the evaluation of the aggregation number $Z = M_w/M$ (with M: molar mass of single polymer molecule), which are given in figure 1. R_g ($\pm$ 5 - 10 nm) represents an average value, which much is less exact for the size characterization than $2R_h$ for the samples studied. The quantities $2R_h$ and M_w from dynamic and static LS can be used to calculate an average (relative) density of the colloidal particle [15].

Assuming the micelles are spherical, a simple geometric core-shell model can be applied (Figure 6).

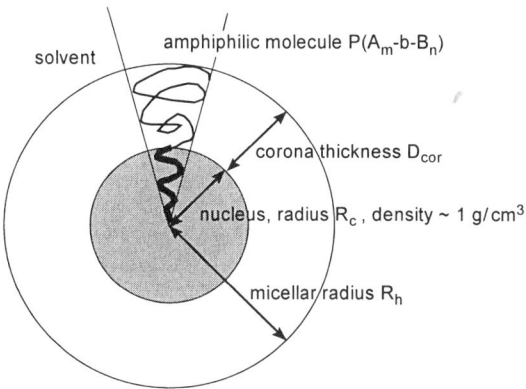

Figure 6: Model of a spherical micelle formed by the block copolymer $P(A_m\text{-}b\text{-}B_n)$

The core is formed by the hydrophobic chains. The corona consists of tethered solvophilic chains. According to S. Förster et al. [16] the aggregation number Z can be calculated from the block lengths m and n by $Z=Z_0\, m^2\, n^{-0.8}$ and the corona thickness by $D_{cor}=D_0\, Z^{0.2}\, n^{0.6}$. The corona thickness D_{cor} was calculated from the measured micellar diameter $2R_h$ and the diameter $2R_C$ of the hydrophobic nucleus: $D_{cor}=R_h\text{-}R_C$. The core density is assumed to be 1 g/cm^3 due to the strong hydrophobic character. The fitting parameter $Z_0=0.9$ is related to the volume of a solvophobic monomer unit and a geometric packaging parameter, $D_0=0.24$ is related to the length of a solvophilic monomeric unit. The data for the reduced aggregation number shows, that the main trend is excellently described by such a simple geometric model, meaning that the steric arrangement dominates the micellar architecture.

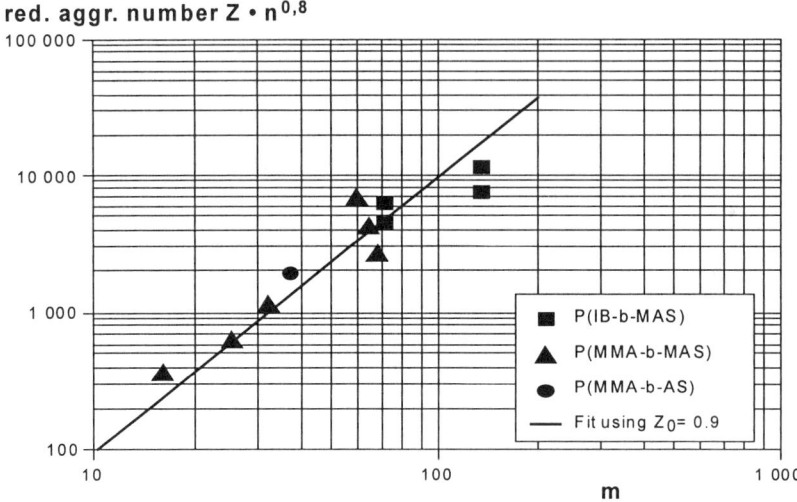

Figure 7: Reduced aggregation number $Z \cdot n0.8$ of block copolymer micelles in aqueous solution as a function of the hydrophobic block length m; the line line is calculated from $Z = Z_0 \cdot m^2 \cdot n^{-0.8}$ using $Z_0 = 0.9$.

Emulsion Polymerization

Aqueous solutions of the neutralized block copolymers were used as the only emulsifying agent in emulsion polymerization of a mixture of methylmethacrylate (49-50%) /n-butylacrylate (49-50%)/ acrylic acid (0-2%), which was performed at 90°C in batch. The solids content of the dispersion after polymerization was 20-30%(w/w), the pH between 5 and 7.5 depending on the amount block copolymer used. The particle size distribution was measured with AUC [17]. The latex diameters ranged from ≈ 220 nm to 36 nm with increasing amount of blockcopolymer.

Three cases have to be differentiated (figure 8): with block copolymers $P(MMA_m\text{-b-}AA_n)$ forming "frozen micelles" a slope of 1 is obtained, indicating that the frozen micelles act similar like a seed where the number of polymer particles is equal to the number of micelles and $n_{BLOCK\ COPOLYMER}$ is constant. [18] A slope higher

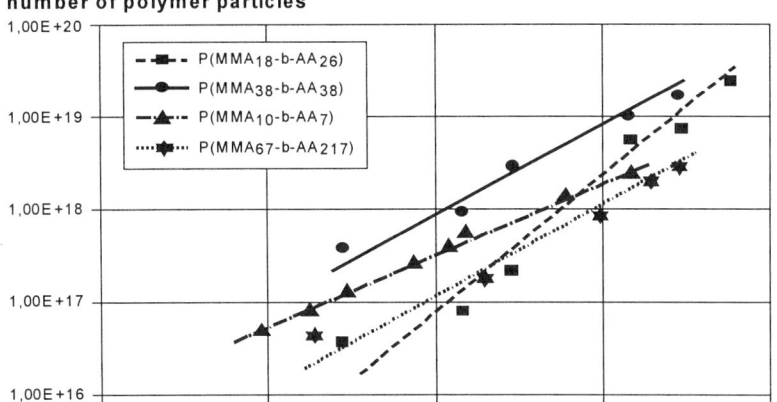

Figure 8: number of polymer particles per kg dispersion as a function of the block copolymer concentration in g/kg dispersion

than 1, already described for $P(S_{10}\text{-b-}EO_{68})$ [3], is obtained with low m.w. $P(MMA_m\text{-b-}AA_n)$. Only $P(MMA_{10}\text{-b-}MAA_7)$ [19], gives a slope lower than 1, which is comparable to surfactants [20].

Film Properties

Model dispersions with a monomer composition similar to coating formulations (BA[50%]-MMA[50%]-copolymer) were prepared with the block copolymers PMMA-b-PAA and PMMA-b-PMAA. The resulting films are equipped with switchable properties by annealing at, e.g., 150°C for 30 min. The restructure of the film morphology and accordingly the mechanical dynamic response in a DMA experiment is clearly visible in the figures 9 and 10. The images from the transmission electron micrograph indicate that the initially almost homogeneously distributed block copolymer $PMMA_{38}\text{-b-}PAA_{38}$ (10% in polymer film) arranged in clusters after annealing. This behaviour results in a significant decrease of the storage modulus E′ in the temperature interval from 30 to 100°C.

Dynamic mechanical analysis of the BA/MMA film made with 10 % $P(MMA_{38}\text{-b-}AA_{38})$ indicates that the polymer morphology is changed by annealing at 150°C. The significant decrease of the storage modulus (figure 10) is due to a rearrangement of the block copolymer phase which is almost homogeneously distributed within the film cast at room temperature. It was confirmed by electron microscopy that, after annealing, the block copolymer is arranged in bigger domains. This behaviour results in a significant softening of the polymer film between 30 and 100°C.

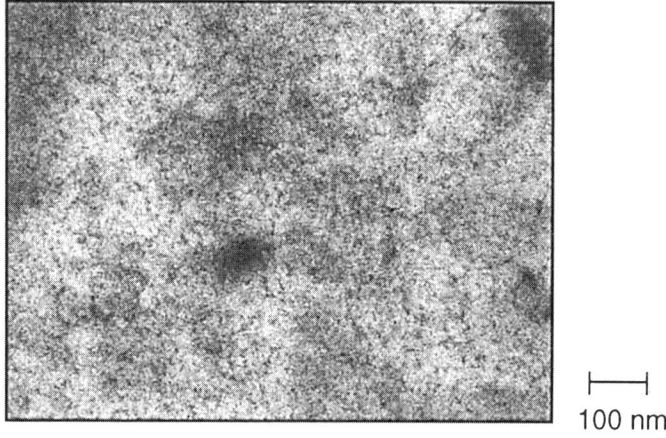

Figure 9a:Transmission electron micograph of a BA[50%]-MMA[50%]-copolymer film with 10% (w/w) blockcopolymer PMMA$_{38}$-b-PAA$_{38}$ in polymer film after annealing at 150°C for 30 min. Films from DMA-analysis, stained with uranyl acetate (TEM-measurements: Dr. Heckmann, BASF AG).

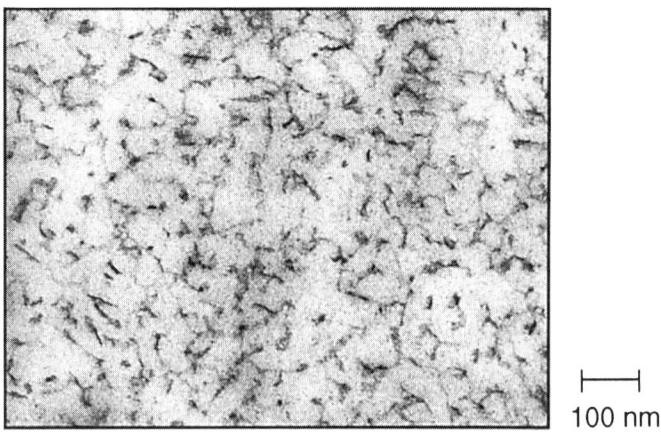

Figure 9b Transmission electron micograph of a BA[50%]-MMA[50%]-copolymer film with 10% (w/w) blockcopolymer PMMA$_{38}$-b-PAA$_{38}$ in polymer film before annealing at 150°C for 30 min. Films from DMA-analysis, stained with uranyl acetate (TEM-measurements: Dr. Heckmann, BASF AG).

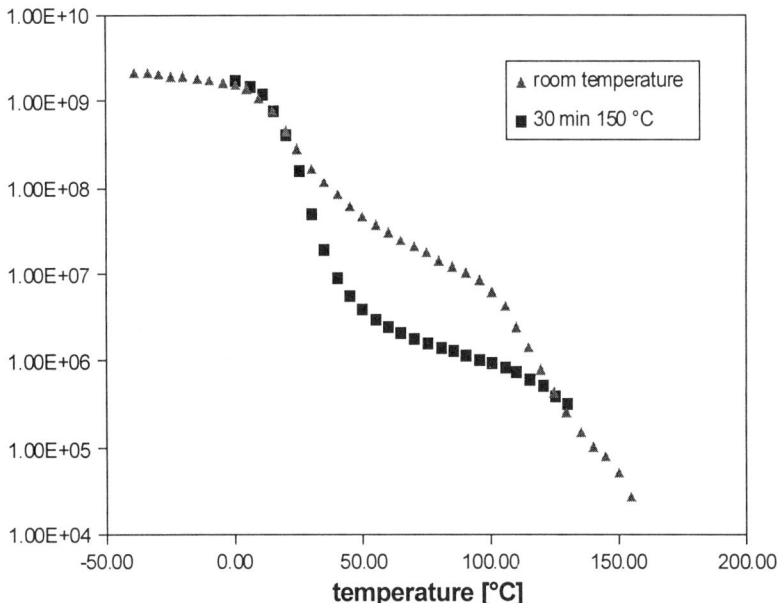

Figure 10: Dynamic mechanical analysis of a polymer film of 50% BA[50%]-MMA[50%]-copolymer containing 10% (w/w) P(MMA$_{38}$-b-AA$_{38}$) before and after annealing at 150°C for 30 min.

An other example of switchable properties is the water uptake of the polymer film, which is significantly decreased after annealing.

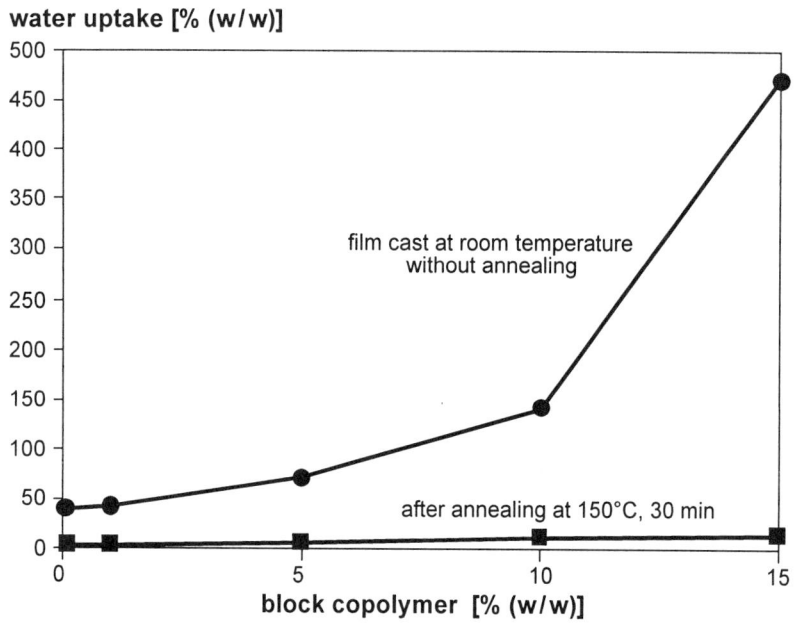

Figure 11: Water uptake of a BA[49%]-MMA[49%]-AA[2%]-copolymer film made with P(MMA$_{67}$-b-MAA$_{217}$) before and after annealing at 150°C during 30 min.

Acknowledgment

We thank T.Rager, W.Meyer, G.Wegner (Max-Planck-Institute for Polymer Research, Mainz), J.Feldthusen, W.Stauf, A.H.E.Müller, (University Mainz) and S.Jüngling for the synthesis of block copolymers, P.Rossmanith and W. Mächtle for measurements with the analytical ultracentrifuge, J.Klingler and W.Schrof for the fluorescence spectroscopy work, W.Heckmann and T.Frechen for electron micrographs, A.Zosel for dynamic mechanical analysis and K. Mathauer for a part of the emulsion polymerization work. Financial support by BMBF (No. 03N30043) is greatfully acknowledged.

References

1. L. Leemans, R. Fayt, Ph. Teyssié, N.C. de Jaeger, *Marcromolecules* 1991, *24*, 5922-5925
2. G.L. Jialanella, E.M. Firer, I. Piirma, *J. Polym. Sci. A: Polym. Chem.* 1992, *30*, 1925-1933
3. M. Berger, W. Richtering, R. Mülhaupt, *Polymer Bulletin*, 1994, *33*, 521-528
4. C.L.Winzor, Z. Mrazek, M.A. Winnik, *Eur. Polym. J.,* 1994, *30*, 121-128
5. T. Rager, *Ph.D. Thesis*, Mainz, 1997, Shaker Verlag Aachen
6. W. Stauf, Diplomarbeit, University of Mainz, 1997
7. J. Feldthusen, B. Iván, A.H.E. Müller, *Macromol.*,1997, *30*, 6989
8. J. Feldthusen, B. Iván, A.H.E. Müller, *Macromol.*,1997, *31*, 578
9. Synthetic work done by T. Rager, W. Meyer, G. Wegner, Max-Planck-Institute for Polymer Research Mainz, Germany; W. Stauf, A. Müller, University Mainz, Germany,S. Jüngling, M. Gerst, Polymers Laboratory, BASF AG, Ludwigshafen, Germany
10. J. Feldthusen, B. Iván, A.H.E. Müller, Makromolecules 30, 6989, (1997), J. Feldhusen, B. Iván, A.H.E. Müller,Macromolecules 31, 578, (1997), J. Feldhusen, B. Iván, A.H.E. Müller, BASF AG Ludwigshafen, (1996), DE 196 10 350 A1 (O.Z. 46668), J. Feldhusen, B. Iván, A.H.E. Müller, BASF AG Ludwigshafen, (1996), DE 196 48 028 A1 (O.Z. 47487)
11. T. Förster, *Z. Naturf.*, 1949, *4a*, 321
12. Z. Tuzar, S.E. Webber, C. Ramireddy, P. Munk, *Polymer Preprints*, 1991, *32 (1)*, 525-526
    P. Munk, C. Ramireddy, M. Tian, S.E. Webber, K. Procházka, Z. Tuzar, Makromol. *Chem., Macromol. Symp.,* 1992, *58*, 195-199
    M. Tian, A. Qin, C. Ramireddy, S.E. Webber, P. Munk, *Langmuir*, 1993, *9*, 1741-1748
13. S.W. Provencher, *Makromol. Chem.*, 1979, *180*, 210
14. M. Schmidt, Simultaneous static and dynamic light scattering, in W. Brown, ed. Dynamic light scattering, Oxford (GB) 1993
15. H. Schuch, J. Klingler, P. Rossmanith, T. Frechen, M. Gerst, A. Müller, J. Feldthusen, results to be published
16. S. Förster, M.Zisenis, E. Wenz, M. Antonietti, *J. Chem. Phys.*, 1996, *104*, 9956-9970
17. W. Mächtle, in *"Analytical Ultracentrifugation in Biochemistry and Polymer Science"*, S.E. Harding, A.J. Rowe and J.C. Horton (eds.), 1992, Royal Soc. of Chem., Cambridge, UK
18. T. Liu, H. Schuch, M. Gerst, B. Chu, *Macromolecules*, 1999, *32*, 6031-6042.
19. Research product of TH Goldschmidt AG, Germany
20. A.S. Dunn in P.A. Lovell, M.S. El-Aasser (ed.), *Emulsion Polymerization and Emulsion Polymers*, J. Wiley & Sons, 1997

Chapter 4

The Preparation of Well-Defined Water Soluble–Swellable (Co)Polymers by Atom Transfer Radical Polymerization

Krzysztof Matyjaszewski, Scott G. Gaynor, Jian Qiu, Kathryn Beers, Simion Coca, Kelly Davis, Andreas Muhlebach, Jianhui Xia, and Xuan Zhang

Center for Macromolecular Engineering, Department of Chemistry, Carnegie Mellon University, 4400 Fifth Avenue, Pittsburgh, PA 15213

The development of atom transfer radical polymerization (ATRP), a controlled/"living" radical polymerization system, has allowed for the preparation of a variety of well defined polymers ($DP_n = \Delta[M]/[I]_o$; $M_w/M_n < 1.5$) based on styrenes, (meth)acrylics, acrylonitrile, and other monomers. This paper is to review the extension of ATRP in preparing well-defined water soluble homopolymers and amphiphilic, water swellable copolymers. The monomers that have been polymerized using ATRP include: 2-hydroxyethyl acrylate, 2-hydroxyethyl methacrylate, 2-(dimethylamino)ethyl methacrylate and 4-vinylpyridine. Well-defined poly(t-butyl acrylate) was also prepared and was transformed to poly(acrylic acid) by hydrolysis. Various amphiphilic block copolymers were prepared with these hydrophilic monomers and other, hydrophobic monomers.

Introduction

The recent explosion of interest in controlled/"living" radical polymerizations is the result of polymer chemists trying to develop new polymeric materials from existing monomers.[1] Controlled/"living" polymerizations offer the possibility of synthesizing polymers with precise control of the end groups, composition, functionality, and architecture of a polymer. Such variability, coupled with the ability of radical polymerization to polymerize a wide variety of monomers, is expected to offer the polymer chemist a virtually limitless array of polymeric materials for nearly every application. Towards this end, we have developed atom transfer radical polymerization (ATRP) as a method for preparing well-defined polymers.*(2-8)*

ATRP is, at its essence, the reversible activation and deactivation of a polymer chain by reaction of a dormant chain end (R-X) with a transition metal catalyst (M_t^n) complexed with a ligand, Scheme 1. The dormant chain end is capped with a group, generally a halogen, that can be removed homolytically by reaction with the transition metal catalyst to form a radical and a transition metal halide ($X-M_t^{n+1}$) (with rate constant k_a). This radical can then initiate the polymerization of the monomer and propagate. As the radical chain propagates, the radical can do one of three things: react with monomer and continue to propagate (with rate constant k_p), react with another radical to terminate by coupling or disproportionation (with rate constant k_t), or react with the transition metal halide to cap the growing chain and reduce the transition metal catalyst to its lower oxidation state (with rate constant, k_d).

Scheme 1

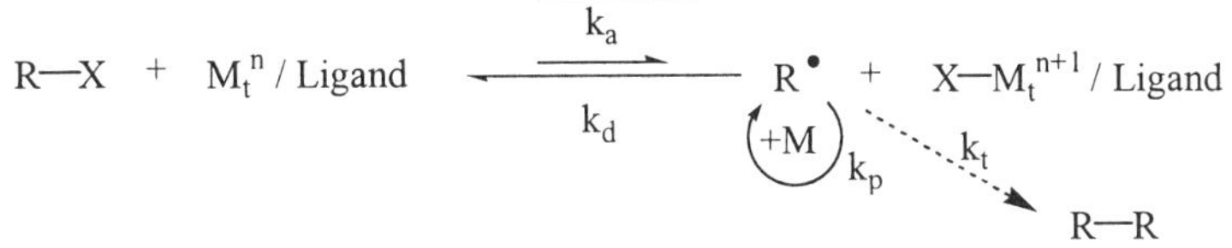

$$R-X + M_t^n / Ligand \underset{k_d}{\overset{k_a}{\rightleftharpoons}} R^\bullet + X-M_t^{n+1} / Ligand$$

$$(+M) \quad k_p \quad k_t \quad \rightarrow R-R$$

In order to obtain a controlled/"living" radical polymerization, termination of the polymer chain by bimolecular radical reactions must be suppressed. To do this, the concentration of the radicals in solution must be kept very low. This is accomplished by attenuation of the equilibrium between the active and dormant states of the growing polymer chain. If the equilibrium is shifted too far towards the active species, the system behaves as a simple redox initiated polymerization and no control is obtained. Contrarily, if the equilibrium is shifted too far towards the dormant species, no polymerization is observed. In this system, some termination (<10% of initiator) during the early stages of the polymerization may occur; the concentration of the deactivator (transition metal halide, $X-M_t^{n+1}$) increases, shifting the equilibrium towards the dormant chains. This phenomenon is known as the persistent radical effect.*(9)*

A variety of initiators have been used with nearly all having the radical adjacent to a group which is capable of stabilizing the radical either by delocalization (phenyl, ketone, cyano groups), or by induction (multiple halogens such as trichloro-). The transition metals can be one of a variety such as copper,*(2-7) (10-14)* iron,*(8,15,16)* ruthenium,*(17,18)* nickel,*(19,20)* rhodium,*(21)* or palladium.*(22)* These transition metals are complexed by a ligand, usually a pyridine derivative,*(2,4,14)* but linear amines*(6)* and phosphines*(8,20)* have also been successfully employed.

This review will cover the use of ATRP to prepare well-defined water soluble polymers. Also, the preparation of amphiphilic copolymers will be discussed, with these being prepared by formation of block copolymers of hydrophobic and hydrophilic monomers.

Water Soluble Homopolymers

The preparation of water soluble polymers by ATRP is sometimes simple and straight forward, other times difficult. The reason for this is that the polar nature of the medium used to conduct the polymerization may shift the equilibrium in unwanted directions, resulting in higher (or lower) radical concentrations. Since high radical concentrations are counter productive, i.e., greater proportion of terminated polymer chains, it may be required to adjust the reaction conditions, temperature, solvent, catalyst concentration, etc., so that a controlled radical polymerization can be observed. The following will outline the conditions used to obtain well-defined polymers and any potential problems that one should be aware of.

2-Hydroxyethyl Acrylate (HEA)

The polymerization of HEA is relatively straight forward.*(23)* This monomer has been polymerized in either bulk or aqueous solution. However, one should take care to remove any residual diacrylate or acrylic acid that may be present in commercially available monomer. Our laboratory has found that it is possible to remove the diacrylate by washing with hexanes; removal of the acrylic acid, as well as any inhibitor, can be achieved by passing the monomer through a column of basic alumina.

The simplest manner of conducting the polymerization is in bulk. When methyl 2-bromopropionate (MBrP) was used as the initiator, with Cu(I)Br complexed with two equivalents of 2,2'-bipyridine (bpy) as the catalyst, the polymerization was observed to obey first order kinetic with respect to monomer up to ninety percent conversion (Figure 1); the reaction was conducted at 90 °C. As can be seen in Figure 2, the molecular weights of the growing polymer increased with conversion, although they were higher than predicted by $DP_n = \Delta[M]/[I]_o$. The difference in the observed and predicted molecular weights can be attributed to the inaccuracy of using linear polyMMA standards for molecular weight determination by SEC in DMF. Use of ^{1}H NMR and MALDI-TOFMS to determine the M_n of lower molecular weight samples revealed that the molecular weights of the polymers were in agreement with theoretical values. Also, it will be noted that the polydispersity (M_w/M_n) of the growing polymer decreased with conversion to a final value of $M_w/M_n \sim 1.2$.

The polymerization can also be conducted in water. Using a 50/50 mixture of water and monomer, by volume, the polymerization was conducted at 90 °C, with a ratio of monomer to initiator of 100:1; the catalyst was in a 1:1 ratio with initiator. After twelve hours, the polymerization had reached 87% conversion and the molecular weight, as determined by SEC in DMF, was $M_n = 14,700$; $M_w/M_n = 1.3$. In all of the polymerizations, the hydroxyl groups appeared to be unaffected during the polymerization (as observed by NMR).

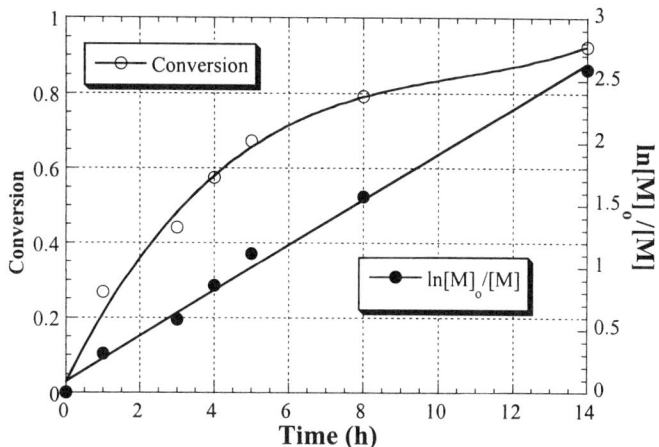

Figure 1. Zero and first order kinetic plots of the bulk polymerization of 2-hydroxyethyl acrylate at 90 °C. $[HEA]_o:[MBrP]_o:[Cu(I)Br/2bpy]_o = 50:1:1$

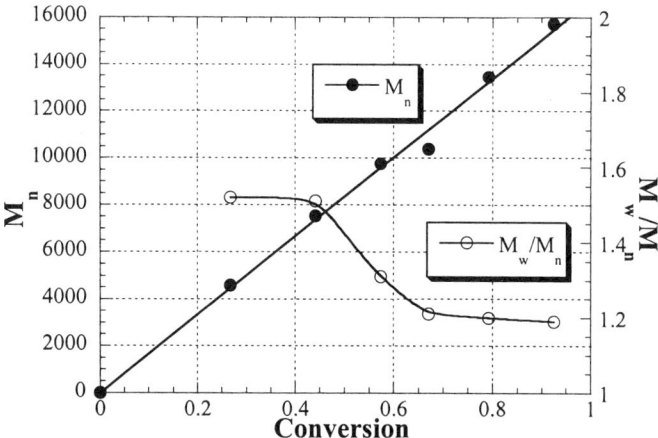

Figure 2. Molecular weight evolution with conversion for the bulk polymerization of 2-hydroxyethyl acrylate at 90 °C. $[HEA]_o:[MBrP]_o:[Cu(I)Br/2bpy]_o = 50:1:1$

2-Hydroxyethyl Methacrylate

2-Hydroxyethyl methacrylate (HEMA) was found to be more difficult to polymerize than HEA, althought their structures are quite similar. When conducted in bulk, the polymerization was very fast, proceeding at room temperature and generating significant amounts of heat. When the polymer was analyzed, it had much higher molecular weights than predicted and the molecular weight distributions were very broad, $M_w/M_n > 1.8$. To reduce the rate of the polymerization, the reaction was performed in DMF solution. Although molecular weights increased linearly with conversion, the polydispersities also increased with conversion and were somewhat higher than desired.

The loss of control of the polymerization could be attributed to one, or a combination of, two factors: the tertiary structure of the methacrylate-halogen end group, and the high polarity of the solvent/monomer. It is known that methacrylates produce higher concentrations of radicals,(3) i.e., MMA has a larger equilibrium constant between active and dormant species. This situation could be aggravated by the use of a highly polar medium and further shift the equilibrium towards the formation of radicals. Should the concentration of the radicals become too great, irreversible termination will be observed, accompanied by loss of control of the polymerization.

A controlled polymerization of HEMA was realized when a mixed solvent system was used: a 70/30 mixture of methyl ethyl ketone and *n*-propanol in a 1:1 ratio with monomer. Additionally, the polymerization was conducted with less catalyst (2:1 initiator:catalyst, initiator = ethyl 2-bromoisobutyrate, EBr*i*B) and was conducted at lower temperatures (50 °C) than for HEA. Although there was a slight curvature in the first order kinetic plot (Figure 3), suggesting some termination, the molecular weights increased linearly with conversion and the polydispersities remained low, $M_w/M_n < 1.4$ (Figure 4).

Trimethylsilyl Protected HEA/HEMA

Treatment of HEA or HEMA with trimethylsilyl chloride, in the presence of base, resulted in the formation of the silyl protected HEA or HEMA, Scheme 2.*(24)* These monomers were prepared because they are not as polar as their respective hydroxyl derivatives, and their polymers are soluble in organic solvents. These monomers were found to polymerize under similar conditions as those employed for other acrylates, or methacrylates, i.e., methyl acrylate or methyl methacrylate. As a representative example, the molecular weight behavior versus conversion of TMS-HEMA is shown in Figure 5. The molecular weights, as determined by SEC in THF using a calibration curve based on linear polyMMA standards, increased linearly with conversion and corresponded very well with predicted molecular weights up to $M_n = 100,000$, while the polydispersities were lower than in a conventional radical polymerization, $M_w/M_n < 1.4$.

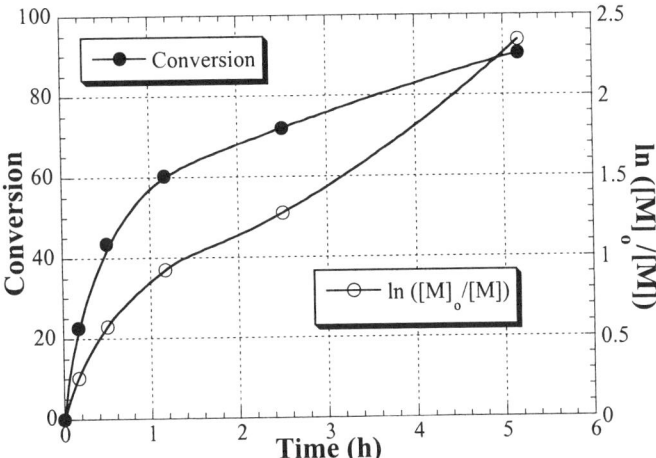

Figure 3. Zero and first order kinetic plots of the polymerization of 2-hydroxyethyl methacrylate at 50 °C.
[MEK/n-Propanol]:[HEMA]$_o$:[EBriB]$_o$:[Cu(I)Br/2bpy]$_o$ = 100:100:1:0.5

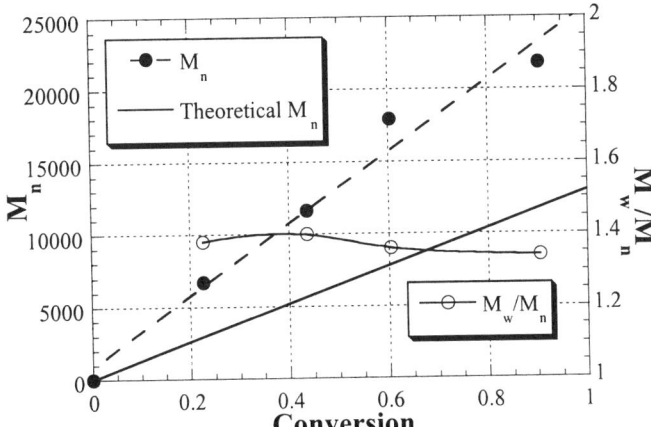

Figure 4. Molecular weight evolution with conversion for the polymerization of 2-hydroxyethyl methacrylate at 50 °C.
[MEK/n-Propanol]:[HEMA]$_o$:[EBriB]$_o$:[Cu(I)Br/2bpy]$_o$ = 100:100:1:0.5

Scheme 2
Protection of the Monomer

R-C(=O)-O-CH₂CH₂-OH →[TMS-Cl, CH₂Cl₂, NEt₃, 0 °C]→ R-C(=O)-O-CH₂CH₂-O-Si(CH₃)₃

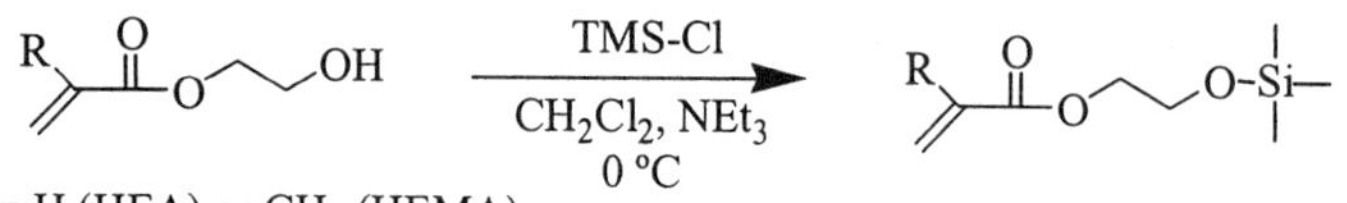

R = H (HEA) or CH₃ (HEMA)

Deprotection of the Monomer

[-CH₂-CR(C(=O)-O-CH₂CH₂-O-Si-)-]$_n$ →[THF, H₂O, KF, TBAF or Dioxane, H₂O, HCl, Δ]→ [-CH₂-CR(C(=O)-O-CH₂CH₂-OH)-]$_n$

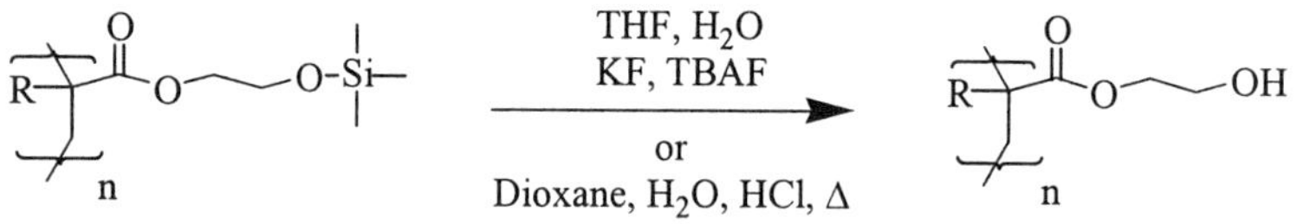

Deprotection of the resulting polymer, either the acrylate or methacrylate, was straightforward. One method of deprotection was the use of "wet" THF, in the presence of potassium fluoride using a catalytic amount of tetrabutylammonium fluoride. The polymer was also hydrolyzed using hydrochloric acid, by refluxing in a dioxane and water mixture. The use of the protected monomers and these facile deprotection methods are useful for cases of compatibility where the unprotected monomers can not be used, or where the use of conventional organic solvents, i.e., not water, are desired (vide infra).

2-(Dimethylamino)ethyl Methacrylate (DMAEMA)

DMAEMA was readily polymerized by ATRP, but to obtain well-defined polymer required that the reaction conditions be optimized.*(25)* For example, when the polymerization was conducted, at 90 °C, in the non-polar solvent toluene, the polymerization was uncontrolled, Table I. However, when the polymerization was conducted in more polar solvents, such as anisole, butyl acetate, ethylene carbonate, or dichlorobenzene, polymers with molecular weights corresponding to predicted theoretical values and with low polydispersities were obtained.

Table I. Effect of Solvent on the Polymerization of DMAEMA[a]

Solvent	Time (h)	Conv.	$M_{n, th}$	$M_{n, SEC}$	M_w/M_n
Toluene	4.25	0.33	6 500	73 600	2.81
Anisole	1.25	0.79	15 800	17 900	1.45
Dichlorobenzene	1.25	0.78	15 500	15 800	1.43
Butyl Acetate	1.50	0.86	17 300	25 300	1.57
Ethylene Carbonate	1.00	0.63	12 500	11 500	1.51

[a] $[CuBr]_o:[HMTETA]_o:[EBriB]_o$ = 23.3 mM, $[DMAEMA]_o$ = 2.96 M

Further optimization of the polymerization was obtained by variation of the ligand used to coordinate with the transition metal. The well known copper(I)bromide/2bpy catalyst afforded well-defined polymer, but even better, results were obtained with less expensive, commercially available, linear amine ligands (N,N,N',N'-tetramethylethylenediamine, TMEDA; N,N,N',N'',N''-pentamethyldiethylenetriamine, PMDETA; 1,1,4,7,10,10-hexamethyltriethylenetetramine HMTETA). The polymer with the lowest polydispersity was obtained using the linear tetramine, HMTETA, Table II.

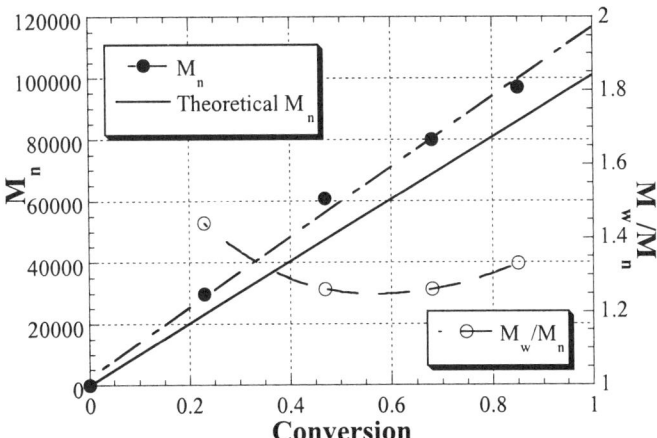

Figure 5. Molecular weight evolution with conversion for the polymerization of TMS-HEMA.

Table II. Effect of Ligand on the Polymerization of DMAEMA[a]

Ligand	Time (h)	Conv.	$M_{n, th}$	$M_{n, SEC}$	M_w/M_n
bpy	1.25	0.85	16 900	20 500	1.55
TMEDA	4.25	0.76	15 100	18 600	1.53
PMDETA	1.00	0.68	13 500	17 600	1.61
HMTETA	1.25	0.79	15 800	17 900	1.45

[a][CuBr]$_o$:[EBriB]$_o$ = 23.3 mM, [DMAEMA]$_o$ = 2.96 M; [Ligand]/[CuBr] = 2 for bpy, TMEDA; [Ligand]/[CuBr] = 1 for PMDETA, HMTETA

Additionally, it was found that by performing the polymerizations at temperatures lower than previously reported for MMA, i.e., 90 °C, polymers with narrower molecular weight distributions could be obtained. Indeed, the best results were obtained with the polymerization being conducted at room temperature, albeit, the polymerization was slower, Table III.

Table III. Effect of Temperature on the Polymerization of DMAEMA[a]

Temperature (°C)	Time (h)	Conv.	$M_{n,\,th}$	$M_{n,\,SEC}$	M_w/M_n
90	1.25	0.78	15 500	15 800	1.43
70	1.25	0.64	12 700	13 100	1.37
50	1.80	0.69	13 800	14 100	1.37
23	4.67	0.67	13 400	18 900	1.25

[a] $[CuBr]_o:[HMTETA]_o:[EBriB]_o = 23.3$ mM, $[DMAEMA]_o = 2.96$ M

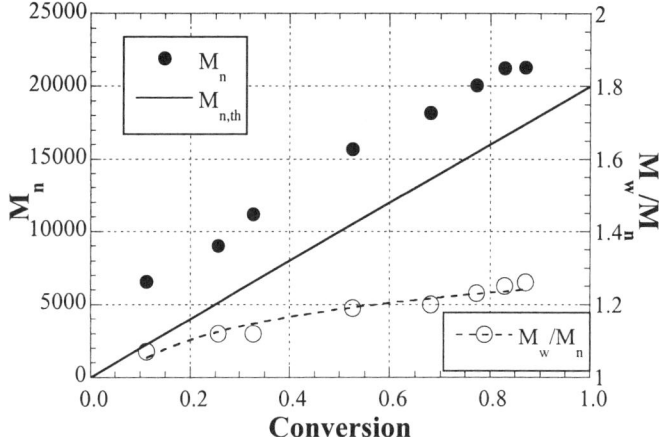

Figure 6. Molecular weight evolution with conversion versus conversion for the polymerization of DMAEMA initiated by 2-bromopropionitrile (BPN) at 50 °C. $[DMAEMA]_o = 2.96$ M, $[CuBr] = [BPN]_o = [HMTETA] = 0.023$ M.

The above reactions were performed using ethyl 2-bromoisobutyrate (EBriB) as the initiator. It was latter discovered that lower polydispersity materials could be obtained by using 2-bromopropionitrile (BPN). This was attributed to BPN having a higher rate of activation than EBriB, resulting in faster and more uniform initiation of the growing polymer chains. The molecular weight behavior versus conversion using this initiator is shown in Figure 6; some of the resulting SEC traces are shown in Figure 7. The low molecular weight tail is likely due to some unavoidable side reactions, i.e., irreversible termination, elimination of halogen end group, etc.

t-Butyl Acrylate

Monomers that contain carboxylic acid functionalities have not been successfully polymerized using ATRP. To overcome this problem, *t*-butyl acrylate was

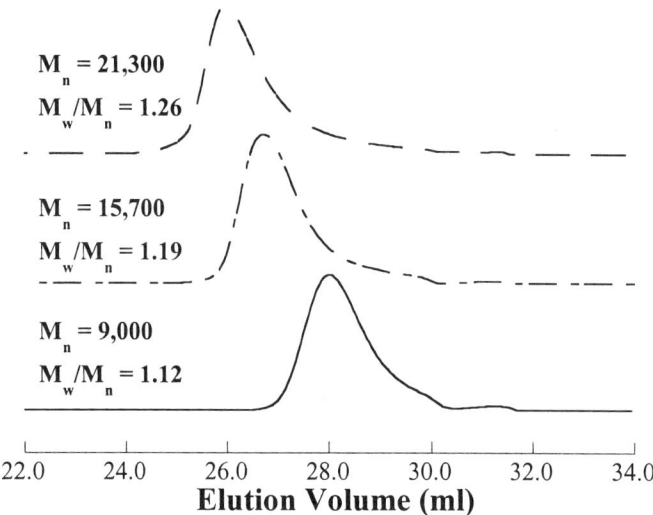

Figure 7. SEC traces obtained in the ATRP of DMAEMA.

polymerized by ATRP; the carboxylic acid is protected by the *t*-butyl group.*(26)* This monomer was easily polymerized. The kinetic and molecular weight plots are shown in Figures 8 and 9, respectively. Deprotection of the *t*-butyl group was achieved by treatment with HCl in dioxane at 60 °C for six hours. The poly(acrylic acid) was easily obtained with the evolution of isobutene, a gaseous byproduct.

4-Vinylpyridine

The last monomer to be discussed here is 4-vinylpyridine (4VP).*(27)* 4-Vinylpyridine has been difficult to polymerize using ATRP; the rates are very slow. This was presumably due to competing complexation of the transition metal by the bpy and the monomer or poly(4-vinylpyridine). Recently we reported the use of a strongly coordinating, tetradentate ligand for use in ATRP.*(28)* This ligand is shown in Figure 10 and is denoted as Me_6-TREN. With this ligand, the polymerization of 4VP was found to proceed at an acceptable rate.

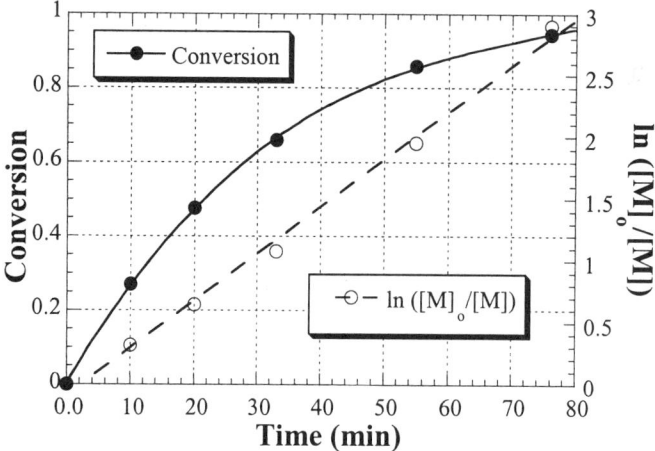

Figure 8. Kinetics of the ATRP of *t*-butyl acrylate in dimethoxybenzene (20% solvent) at 60 °C. $[tBA]_o:[I]_o:[Cu(I)Br]_o:[Cu(II)Br_2]_o:[PMDETA]_o = 50:1:0.5:0.025:0.525$

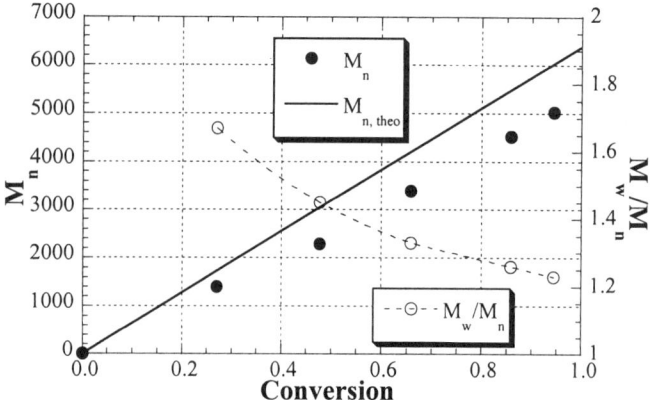

Figure 9. Molecular weight byproduct of the ATRP of *t*-butyl acrylate in dimethoxybenzene (20% solvent) at 60 °C. $[tBA]_o:[I]_o:[Cu(I)Br]_o:[Cu(II)Br_2]_o:[PMDETA]_o = 50:1:0.5:0.025:0.525$

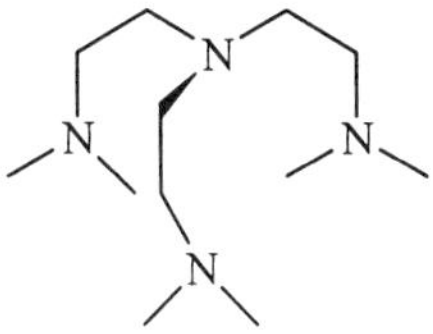

Me₆-TREN

Figure 10. Tris(2-(dimethylamino)ethyl)amine, Me₆-TREN.

Although the polymerization was found to proceed at an acceptable rate using Me₆-TREN, it was soon discovered that the choice of the halogen on the initiator played a significant role in the success of the polymerization. As can be seen in Figure 11, the polymerization appears to stop when 1-phenylethyl bromide is used as the initiator, but the polymerization proceeds to completion when the 1-phenylethyl chloride is employed. The effect of the halogen is also apparent when the molecular weight behavior versus conversion is examined, Figure 12. When bromine was used, the molecular weights were higher than for chlorine, but more significantly, the polydispersities of the polymer increased dramatically as the polymerization proceeded; with 1-phenylethyl chloride, the molecular weight distribution remained below 1.2 throughout the reaction.

The reason for the more efficient polymerization using 1-phenylethyl chloride than with 1-phenylethyl bromide can be explained by the relative rates of quaternization of the halogen end group. If the halogen end group reacts with the nitrogen in the pyridine ring of the monomer, or the pendent pyridine rings in the chain, the quaternary ammonium salt can be formed. The formation of such a complex is detrimental to the ATRP process as the halogen can no longer be homolytically cleaved from the polymer chain end to reinitiate polymerization; this side reaction results in apparent termination of the polymerization. Additionally, elimination of HX from the chain end can occur. The slower rate of quaternization, or elimination, when the halogen is chlorine, than compared to bromine, is significant enough such that the rate is much lower than the overall rate of polymerization allowing for the preparation of well-defined poly(4-vinylpyridine). The small contribution of "termination" is evidenced in the SEC traces of the polymer prepared by using 1-phenylethyl chloride, Figure 13; there is very little tailing in the low molecular weight end of the molecular weight distributions.

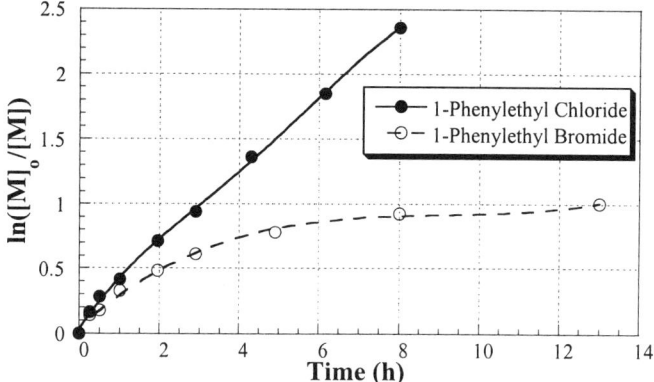

Figure 11. Comparison of the first order kinetic plots for the polymerization of 4VP using different initiators.
[4-VP]$_o$/[1-PEX]$_o$ = 95; [1-PEX]$_o$/[CuBr]$_o$/[TREN-Me$_6$]$_o$ = 1/1/1
40 °C; 50% (v/v) in 2-propanol

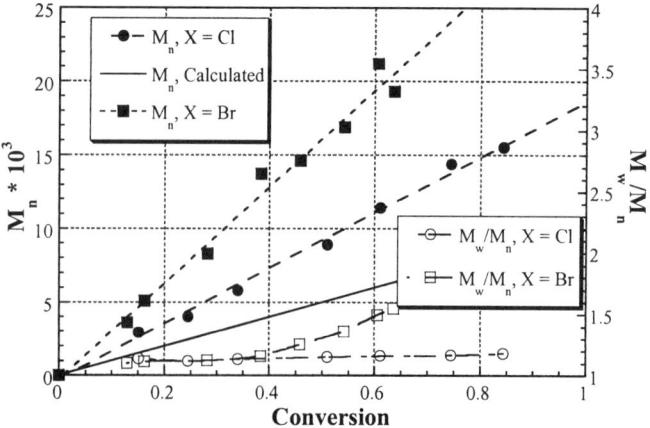

Figure 12. Evolution of the molecular weight with conversion for polymerization of 4VP using different initiators, 1-phenylethyl-X.
[4-VP]$_o$/[1-PEX]$_o$ = 95; [1-PEX]$_o$/[CuBr]$_o$/[TREN-Me$_6$]$_o$ = 1/1/1
40 °C; 50% (v/v) in 2-propanol

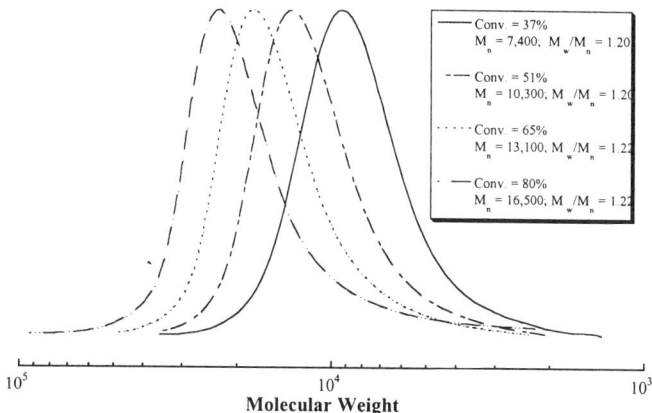

Figure 13. SEC (DMF) traces from the polymerization of 4VP using 1-phenylethyl chloride.

Amphiphilic Block Copolymers

ATRP can be used to prepare a wide variety of block copolymers. Block copolymers have novel physical/mechanical properties when the polymer segments are chosen so that they have different characteristics, i.e., hard/soft, polar/non-polar, hydrophilic/hydrophobic. By combining the work described above with the previous ATRP of styrenes, acrylates, and methacrylates, it is possible to prepare a wide variety of amphiphilic copolymers.

HEA/Butyl Acrylate Block Copolymers

Initially, the trimethylsilyl protected derivative of HEA (TMS-HEA) was used to prepare the block copolymers.*(24)* This was done to simplify the polymerization and characterization procedures as the block copolymer of HEA and BA was expected to be soluble in only very polar organic solvents such as DMF. Using a TMS-HEA macroinitiator ($M_{n,\,NMR}$ = 9,600; $M_{n,\,SEC}$ = 9,400; $M_w/M_{n,\,SEC}$ = 1.2), butyl acrylate (BA) was polymerized to obtain block copolymers with different proportions of BA and TMS-HEA, Table IV. The trimethylsilyl groups were subsequently removed by treatment with HCl in wet THF.

Table IV. Synthesis of Poly(TMS-HEA-*b*-BA) Using a TMS-HEA Macroinitiator[a]

Sample	TMS-HEA:BA (^1H NMR)	$M_{n, Calc}$	$M_{n, SEC}$	M_w/M_n
Macroinitiator	1:0	9 400	9 400	1.20
A	77:23	11 600	12 500	1.21
B	50:50	15 800	16 400	1.20
C	22:78	28 700	31 900	1.21

[a] $[Macroinitiator]_o:[Cu(I)Br]_o:[PMDETA]_o = 1:0.5:0.5$, 80 °C

The preparation of block copolymers of HEA and BA without the use of protecting groups on the HEA was accomplished by first polymerizing BA followed by addition of HEA. The final polymer was comprised of 82 mol% butyl acrylate and had a molecular weight, as determined by SEC using DMF as the eluent, of M_n = 8,630; M_w/M_n = 1.34 ($M_{n, calc}$ = 9,310). Without the use of a protecting group to allow for the polyHEA to be soluble in organic solvents, i.e., butyl acrylate, the organic segment must be prepared first. This is because the poly(butyl acrylate) is soluble in HEA, but polyHEA is not soluble in butyl acrylate. This problem can be overcome if a suitable solvent is chosen, but strongly polar organic solvents, such as DMF, generally pose additional problems for ATRP, i.e., coordination with the catalyst thus changing its reactivity.

Copolymers of DMAEMA

Since it was found that the polymerization of DMAEMA is best conducted in a dichlorobenzene solution, the question of selecting conditions where both segments (hydrophobic/hydrophilic) is not a concern; polyDMAEMA, polystyrene, poly(acrylates) and poly(methacrylates) are soluble in dichlorobenzene. For example, poly(methyl acrylate) was prepared (M_n = 24,100; M_w/M_n = 1.15) as a macroinitiator for the polymerization of DMAEMA. After the polymerization was conducted in dichlorobenzene using the copper(I) bromide/HMTETA catalyst in a 1:1 molar ratio with the poly(methyl acrylate) at 90 °C, the block copolymer was isolated with M_n = 39,200 and M_w/M_n = 1.15.*(29)* Similarly, a polyMMA macroinitiator was prepared by ATRP (M_n = 17,500; M_w/M_n = 1.07) and used for the synthesis of a MMA/DMAEMA block copolymer (M_n = 32,100; M_w/M_n = 1.14). SEC traces of the macroinitiators and resulting block copolymers from these two polymerizations are shown in Figures 14 and 15, respectively.

Additionally, triblock copolymers can be prepared by ATRP. These can be prepared by simply using the diblock copolymers prepared as above as macroinitiators for a third segment. However, although this method is good for ABC type copolymers, the use of this technique to prepare symmetrical ABA block copolymers is difficult as the two, outer A blocks may not be of similar molecular weights. To overcome this problem, a difunctional initiator can be employed.

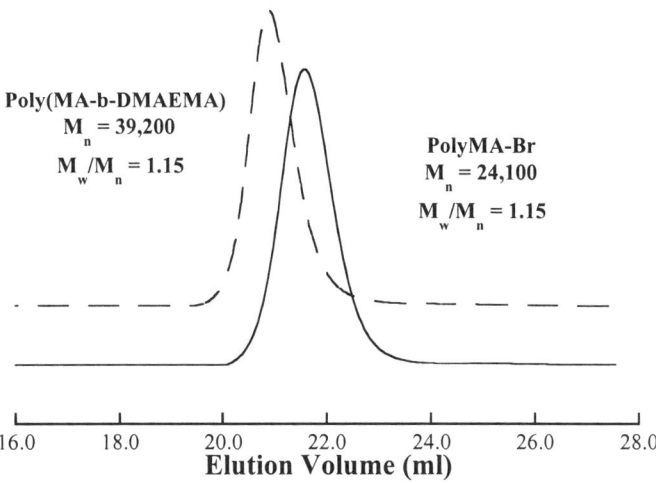

Figure 14. SEC traces of poly(methyl acrylate) macroinitiator and poly(methyl acrylate-*b*-DMAEMA).

By use of a difunctional initiator, the polymer chain grows in two directions, thus allowing for the central B block to be prepared first. This macroinitiator is then used to polymerize the outer A blocks. Such a strategy assures that the resulting polymer is symmetrical, as well as saving a step in the polymerization process. The synthesis of a triblock copolymer was demonstrated using bis(2-bromopropionyloxy)ethane to prepare a polyMMA macroinitiator and then using it to initiate the polymerization of DMAEMA, Scheme 3. The SEC traces of the polyMMA macroinitiator (M_n = 33,000; M_w/M_n = 1.17) and the poly(DMAEMA-*b*-MMA-*b*-DMAEMA) triblock copolymer (M_n = 69,300; M_w/M_n = 1.24) are shown in Figure 16.

Scheme 3

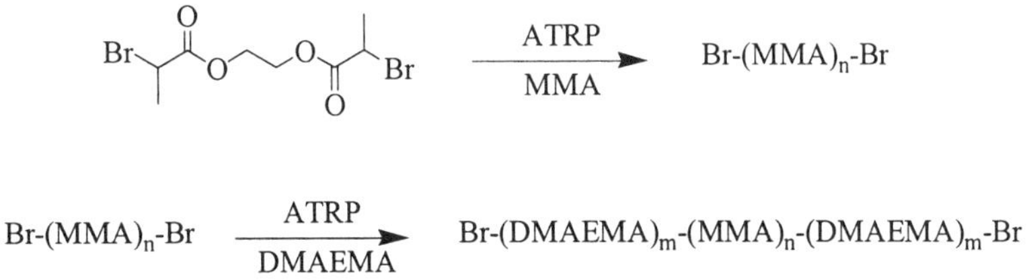

Br-CH(CH₃)-C(O)-O-CH₂CH₂-O-C(O)-CH(CH₃)-Br →(ATRP, MMA)→ Br-(MMA)$_n$-Br

Br-(MMA)$_n$-Br →(ATRP, DMAEMA)→ Br-(DMAEMA)$_m$-(MMA)$_n$-(DMAEMA)$_m$-Br

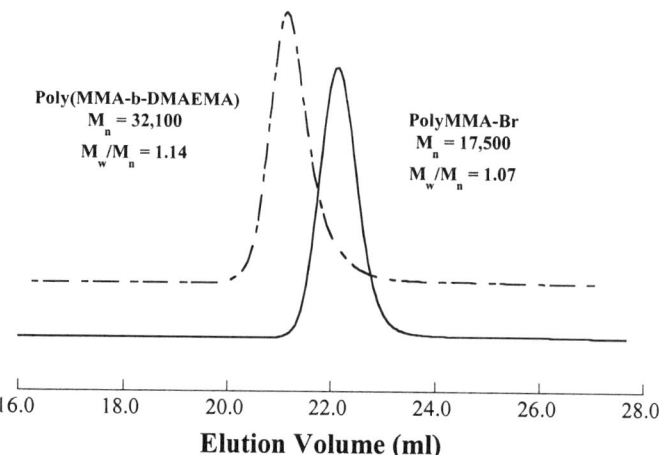

Figure 15. SEC traces of poly(MMA) macroinitiator and poly(MMA-*b*-DMAEMA).

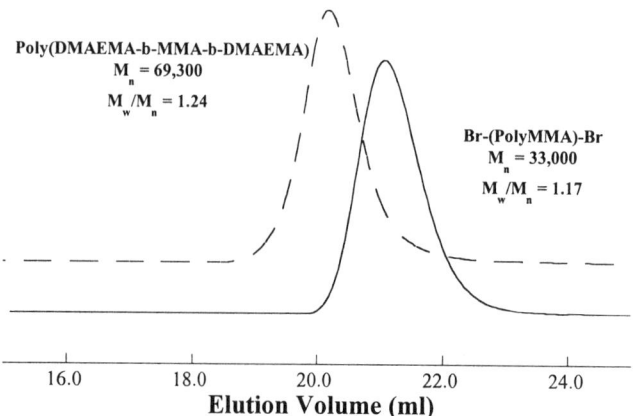

Figure 16. SEC traces of the difunctional poly(MMA) macroinitiator and resulting triblock copolymer, poly(DMAEMA-*b*-MMA-*b*-DMAEMA).

Graft Copolymers of N-Vinylpyrolidinone and Polystyrene

Finally, amphiphilic copolymers of novel architecture have been prepared by synthesizing polystyrene macromonomers (MM) and copolymerizing these with N-vinylpyrolidinone (NVP), Scheme 4.*(30)* The polystyrene macromonomers were prepared by ATRP using vinyl chloroacetate as the initiator. Since the copolymerization reactivity ratios dictate that the radical copolymerization of styrene and vinyl acetate would result in the homopolymerization of styrene; the copolymerization of the vinyl acetate with the styryl radical (the initiator fragment at the polystyrene chain end) would occur only very slowly. Thus, linear polystyrene polymers, of various molecular weights (M_w/M_n < 1.2), with vinyl acetate end groups were prepared.

Scheme 4

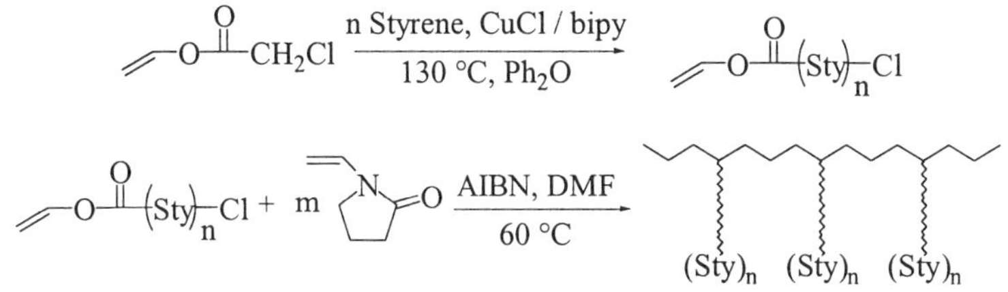

The macromonomers were dissolved in a mixture of DMF and NVP and polymerized using AIBN at 60 °C. Table V lists the results of the polymerization of the macromonomer with NVP: the MM wt% is the final value of the copolymer determined by ^{1}H NMR after purification, the value in parenthesis is the amount initially added to the reaction mixture; the number of grafts/chain was calculated from the weight % of macromonomer and M_n, as determined by SEC in DMF. To see how these materials behaved in aqueous environments, the resulting polymers were dried under vacuum, weighed, and then placed in a sealed container with water. After one week, the polymer samples were removed and re-weighed. As can be seen from Table V, the hydrogel materials did not dissolve buy absorbed large amounts of water.

Table V. Results of Copolymerization of Styrene Macromonomers with N-Vinylpyrolidinone

MM M_n	MM wt%	Copolymer M_n	M_w/M_n	Grafts /Chain	Equil. H_2O Content (%)[a]
5,800	34 (40)	316,000	5.9	18.6	85
	13 (20)	219,000	2.5	4.9	92
	7.7 (10)	185,000	1.8	2.5	97
11,900	40 (50)	65,700	1.6	2.2	82
	30 (30)	83,300	1.8	2.1	93
	24 (20)	114,200	1.8	2.3	92

Conclusion

ATRP is a versatile polymerization method that allows for the preparation of well-defined, water soluble polymers. The monomers that were successfully polymerized include 2-hydroxyethyl acrylate (HEA), 2-hydroxyethyl methacrylate (HEMA), 2-(dimethylamino)ethyl methacrylate (DMAEMA) and 4-vinylpyridine (4VP). Poly(acrylic acid) was prepared by polymerization of *t*-butyl acrylate and removing the *t*-butyl groups by treatment with acid. Additionally, ATRP was used to prepare novel, amphiphilic copolymers by synthesizing block copolymers of one of the water soluble polymers, listed above, with either styrenes, acrylates, or methacrylates.

References

(1) *Controlled Radical Polymerization*; Matyjaszewski, K., Ed.; ACS Symposium Series 865: Washington, D.C., 1998; Vol. 685.
(2) Wang, J.-S.; Matyjaszewski, K. *J. Amer. Chem. Soc.* **1995**, *117*, 5614.
(3) Matyjaszewski, K.; Wang, J.-S. *Macromolecules* **1995**, *28*, 7901.
(4) Matyjaszewski, K.; Patten, T.; Xia, J.; Abernathy, T. *Science* **1996**, *272*, 866.
(5) Matyjaszewski, K.; Patten, T. E.; Xia, J. *J. Amer. Chem. Soc* **1997**, *119*, 674.
(6) Matyjaszewski, K.; Xia, J. H. *Macromolecules* **1997**, *30*, 7697.
(7) Matyjaszewski, K.; Coca, S.; Gaynor, S. G.; Wei, M.; Woodworth, B. E. *Macromolecules* **1997**, *30*, 7348.
(8) Matyjaszewski, K.; Wei, M.; Xia, J.; McDermott, N. E. *Macromolecules* **1997**, *30*, 8161.
(9) Fischer, H. *Macromolecules* **1997**, *30*, 5666.
(10) Percec, V.; Barboiu, B. *Macromolecules* **1995**, *28*, 7970.
(11) Percec, V.; Kim, H.-J.; Barboiu, B. *Macromolecules* **1997**, *30*, 8526.
(12) Haddleton, D. M.; Jasieczek, C. B.; Hannon, M. J.; Shooter, A. J. *Macromolecules* **1997**, *30*, 2190.
(13) Haddleton, D. M.; Crossman, M. C.; Hunt, K. H.; Topping, C.; Waterson, C.; Suddaby, K. G. *Macromolecules* **1997**, *30*, 3992.
(14) Haddleton, D. M.; Kukulj, D.; Duncalf, D. J.; Heming, A. M.; Shooter, A. J. *Macromolecules* **1998**, *31*, 5201.
(15) Ando, T.; Kamigaito, M.; Sawamoto, M. *Macromolecules* **1997**, *30*, 4507.
(16) Moineau, G.; Dubois, P.; Jerome, T.; Senninger, T.; Teyssie, P. *Macromolecules* **1998**, *31*, 545.
(17) Kato, M.; Kamigaito, M.; Sawamoto, M.; Higashimura, T. *Macromolecules* **1995**, *28*, 1721.
(18) Ando, T.; Kato, M.; Kamigaito, M.; Sawamoto, M. *Macromolecules* **1996**, *29*, 1070.
(19) Granel, C.; Dubois, P.; Jerome, R.; Teyssie, P. *Macromolecules* **1996**, *29*, 8576.

(20) Uegaki, H.; Kotani, Y.; Kamigaito, M.; Sawamoto, M. *Macromolecules* **1997**, *30*, 2249.
(21) Moineau, G.; Granel, C.; Dubois, P.; Jerome, R.; Teyssie, P. *Macromolecules* **1998**, *31*, 542.
(22) Lecomte, P.; Drapier, J.; Dubois, P.; Teyssie, P.; Jerome, R. *Macromolecules* **1997**, *30*, 7631.
(23) Coca, S.; Jasieczek, C.; Beers, K. L.; Matyjaszewski, K. *J. Polym. Sci., Polym. Chem. Ed.* **1998**, *36*, 1417.
(24) Mühlebach, A.; Gaynor, S. G.; Matyjaszewski, K. *Macromolecules* **1998**, *31*, 6046.
(25) Zhang, X.; Xia, J.; Matyjaszewski, K. *Macromolecules* **1998**, *31*, 5167.
(26) Coca, S.; Matyjaszewski, K. *Polym. Prepr. (Am. Chem. Soc., Div. Polym. Chem.)* **1997**, *38(1)*, 691.
(27) Xia, J.; Zhang, X.; Matyjaszewski, K. *Macromolecules* **1999**, *in press*,
(28) Xia, J.; Gaynor, S. G.; Matyjaszewski, K. *Macromolecules* **1998**, *31*, 5958.
(29) Zhang, X.; Matyjaszewski, K. *Macromolecules* **1999**, *32*, 1763
(30) Matyjaszewski, K.; Beers, K. L.; Kern, A.; Gaynor, S. G. *J. Polym. Sci., Polym. Chem.* **1998**, *36*, 823.

Chapter 5

Hybrid Dendritic Capsules: Properties and Binding Capabilities of Amphiphilic Copolymers with Linear Dendritic Architecture

Ivan Gitsov[1,2]

[1]Faculty of Chemistry, College of Environmental Science and Forestry, State University of New York, Syracuse, NY 13210
[2]Department of Chemistry and Biochemistry and Cornell Center for Materials Research, Cornell University, Ithaca, NY 14853

This chapter describes the synthesis and the characterization of amphiphilic hybrid ABA block copolymers that self-assemble in aqueous media. The materials under investigation are constructed of poly(ethylene glycol) (PEG), as the water-soluble B block and poly(benzyl ether) monodendrons as the hydrophobic A fragments. The copolymers could be formed by different synthetic schemes that are briefly discussed. The process of self-assembly in aqueous media is investigated by size-exclusion chromatography (SEC. The data obtained indicate that the critical micelle concentrations (cmc) for these systems are between 1×10^{-7} and 5×10^{-6} mol/L and depend on the dendron/PEG ratio. The character of the micellar core is investigated by UV-Vis and fluorescence spectroscopy using different polyaromatic hydrocarbons (PAHs). The PAHs encapsulation results can also serve as a preliminary evaluation of the binding capabilities of the linear-dendritic amphiphiles and their potential application in the biomedical and environmental fields.

Polymers that are capable of association and self-assembly in aqueous media are an important class of materials with significant and diverse application potentials. They can be found in numerous biological processes including molecular recognition, encapsulation and delivery, and also play an important role in many advanced technologies. Self-assembling molecules can be used for the complexation and transport of different substrates into and through hostile environments, they can change the mechanism and stereoselectivity of chemical reactions and enhance their rates (*1*). Amphiphilic block- and graft copolymers are intensively investigated for this purpose because of their distinctive micellar behavior in selective solvents, interesting surface activity and complexation capabilities. The utilization of these materials as emulsifiers in different coatings, as drug carriers with membrane recognition and specific cell targeting, phase-transfer agents and polymer electrolytes is progressing rapidly (*2*). In addition, polymeric amphiphiles form aggregates that tend to be more stable than the micelles assembled of low molecular weight surfactants. Despite their numerous advantages as micelle forming materials, these polymers have also certain limitations. Some of the most difficult problems arise from the uncontrollable flexibility and the relatively broad molecular weight distribution of the building linear blocks resulting in ill defined, compact and entangled cores. This negatively affects their complexation and release capacity and hampers the quantitative evaluation of the structure-properties relationship for these systems.

Recently several groups reported their studies on unimolecular dendritic micelles and offered an ingenious and promising solution to this problem (*3*). Certain unique features distinguish these dendritic systems from the conventional macromolecular amphiphiles - the perfectly branched structure that emanates from a central core, the highly organized interior and the ever-increasing number of functional groups at the periphery (*4*), Scheme 1. A good example for the potential of the dendritic micelles is the entrapment of different "guest" molecules by appropriate modification of the exterior of the "host" dendrimer (*5*). This approach is rather spectacular and encouraging, but it also has limitations. Interestingly, the application of dendritic amphiphiles as phase-transfer agents, drug carriers and nanoscale reactors could be hampered by a feature considered one of their advantages. The highly organized interior of the dendritic unimolecular micelles has a semi-rigid molecular construction with prearranged geometry and void size, Scheme 1; this greatly reduces the number of substrates capable of penetration. In addition, chemical modification of the periphery of the dendrimer is imperative for the entrapment and subsequent release.

It would be interesting to design and test a self-assembling amphiphilic copolymer that combines the advantageous properties of both linear and dendritic amphiphiles. The first attempt in this direction was reported by Newkome et al. in 1986 with the synthesis of dumbbell-shaped dendritic alcohols ("arborols") containing hydrophilic dendritic groups attached to a short hydrophobic alkyl chain (*6*). These compounds exhibit interesting supramolecular behavior and form extended ribbon structures. In our studies we chose poly(ethylene glycol), PEG, as the hydrophilic block because of its proven biocompatibility and solubility in a broad variety of media. A series of hybrid amphiphilic ABA copolymers were produced with PEG as the B block and poly-3,5-dihydroxybenzilic dendrimers as the A blocks (*7*). The structure of the copolymer containing two second-generation monodendrons and PEG is presented in Scheme 2. It was shown that these hybrid macromolecules aggregate in unimolecular or multimolecular micelles depending on the size of the dendritic segments and the length of the PEG block (*7,8*). According to their behavior in aqueous media these systems belong to the group of amphiphilic copolymers that self-assemble through their end-blocks. This class of material is important both from

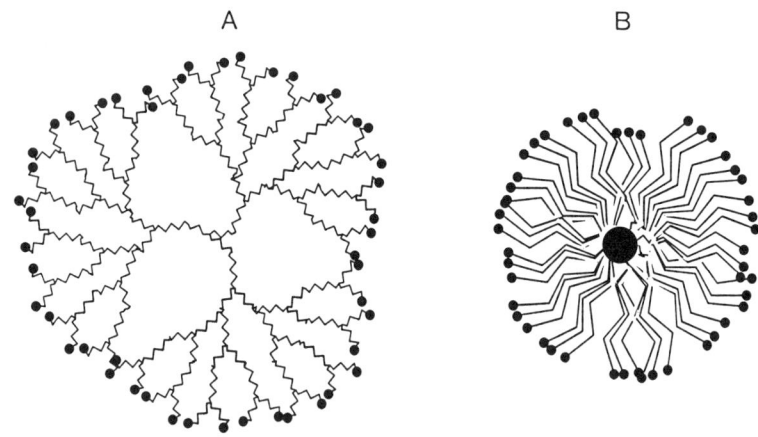

● : Solvent selective group

Scheme 1.
Unimolecular Dendritic Micelle (A) *vs* Conventional Micelle (B)

theoretical and practical points of view because of the entropy constraints, the variety of possible macromolecular assemblies (stars, rosettes, physical networks and others) and the resulting final characteristics of the systems (9).

The first part of the chapter reviews the synthesis of the hybrid linear-dendritic amphiphiles composed of PEG and poly-3,5-dihydroxybenzyl ether monodendrons, and the solution and solid-state properties that are related to their association behavior. The second portion concentrates on the encapsulation studies and information on the character of the dendritic core that can be derived from them.

Experimental Details

Materials. The hybrid linear-dendritic block copolymers, [G-2]-PEG5000-[G-2] were synthesized from poly(ethylene glycol), PEG, with nominal molecular weight 5,000 (Shearwater Polymers, Inc.) and two second-generation poly(benzyl ether) monodendrons (Scheme 2.A) with a benzyl bromide moiety at their 'focal' points (10). The experimental details are given elsewhere (7). The polyaromatic hydrocarbons (PAHs) - phenantrene, pyrene and perylene (all from Aldrich), and fullerene C_{60} (Southern Chemical Group) were used without further purification. The experiments were performed in deionized (DI) water (18.3 MΩ).

Instrumentation. The solubilization experiments were performed in an Aquasonic model 75D sonicator (VWR Scientific), the solutions were clarified by centrifugation using a Sorvell TC6 swingarm centrifuge (DuPont) and the spectroscopic measurements were made in a DU 640B UV-Vis spectrometer (Beckman) and SPEX Fluorolog-3 spectrophotometer (Instruments CA). Size-exclusion chromatography (SEC) analyses were performed in methanol-water (1:1, v/v, with 0.1 M NaN$_3$) at a nominal flow rate 1 ml/min and 30°C. The SEC system consisted of a Waters 510 pump, a U6K universal injector (Waters), a differential viscometric detector R 110 (Viscotek) and a differential refractive index (dRI) detector Refractomonitor IV (Milton-Roy) connected in parallel. The separations were achieved across a set of two Shodex PROTEIN KW 802.5 and 804 columns.

Dissolution of PAHs in water. An excess of the PAH, finely ground was placed in a 25 ml volumetric flask and DI water was added to the full volume of the flask. The mixture was sonicated in the water bath of the sonicator for 6 h at 30°C and was left at the same temperature to equilibrate for 48 h. The clear solution above the PAHs particles was analyzed by UV-Vis spectroscopy in several 24-hour intervals.

Hybrid Copolymer solutions. Solutions of [G-2]-PEG5000-[G-2] in DI water with concentrations 1.1×10^{-7}, 1.1×10^{-6}, 1.1×10^{-5}, 1.1×10^{-4}, and 1.1×10^{-3} mol/L were prepared. The copolymer dissolved gradually and was left to equilibrate for 48 h at room temperature. After that period the resulting clear colorless solutions were analyzed by UV-Vis and fluorescence spectroscopy.

Encapsulation studies. To each solution of the hybrid was added 50 mg of a finely ground PAH. The heterogeneous mixtures were sonicated for 6 h at 30°C and left to equilibrate for 24 h. After recording the UV-Vis and fluorescence spectra the solutions were sonicated for one hour and left to equilibrate for additional 48 h. Spectroscopic analyses were performed again. The mixtures were left at 30°C for 9 days and final spectral analyses were made with each mixture.

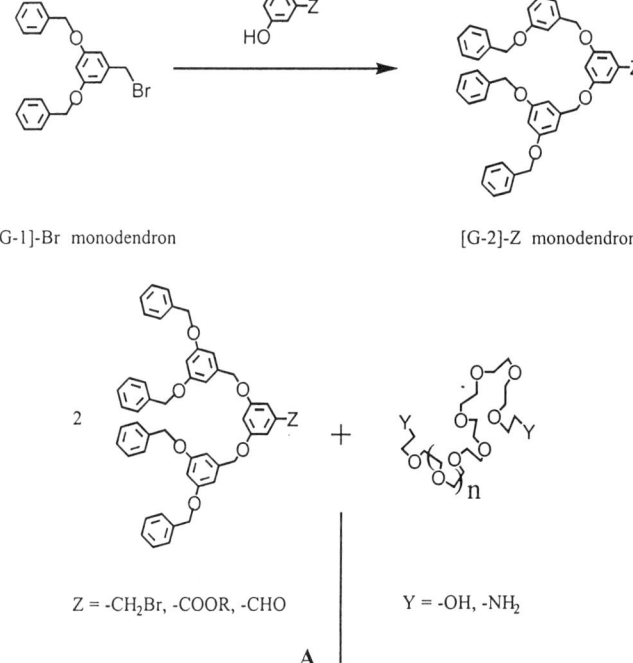

Scheme 2. A
Synthesis of Second-Generation Monodendrons, [G-2]-Z, and Subsequent Formation of ABA Copolymers by Reaction of Preformed Blocks

Spectroscopic procedures. All UV-Vis analyses were made at 30°C from 190 to 600 nm with a scanning speed of 600 nm/min. The changes in the concentrations of the PAHs were calculated using the following molar extinction coefficients: pyrene - ε = 32,300 L×mol^{-1}×cm^{-1} at λ = 319 nm (*11*); phenantrene - ε = 14,400 L×mol^{-1}×cm^{-1} at λ = 293 nm (*11*), perylene - ε = 27,200 L×mol^{-1}×cm^{-1} at λ = 418 nm (*11*) and C$_{60}$ - ε = 175,000 L×mol^{-1}×cm^{-1} at λ = 254 nm (*12*). The steady-state fluorescence spectra were recorded on SPEX Fluorolog-3 spectrophotometer (Instruments CA) from 300 nm to 500 nm with both excitation and emission slits set at a 4 nm bandpass. The excitation wavelengths were: 335.5 nm (pyrene); 282 nm (phenantrene) and 408 nm (perylene).

Results and Discussion

This report will discuss only a selected member of this copolymer series, namely [G-2]-PEG5000-[G-2]. It contains second-generation poly(benzyl ether) monodendrons with well developed branching structure that are easy to make in large quantities and PEG with molecular weight 5,000 that is commercially available from several vendors.

Synthesis of linear-dendritic block copolymers containing PEG. Three distinct methods can be used for the preparation of the linear-dendritic hybrids. The first one is based on the coupling of preformed blocks - PEG and two Fréchet-type monodendrons using different chemical reactions (Scheme 2A). The most effective reaction is the Williamson ether synthesis using a dendritic wedge with a benzyl bromide moiety at the 'focal' point and telechelic PEG containing two hydroxyl end-groups. Within 3 hours the reaction leads practically to full conversion even at only slight excess of the dendritic bromide (*13*). The process can be easily monitored by size-exclusion chromatography with dual (dRI/UV) detection. A transesterification reaction between the dendritic ester and PEG can also be used to form a block copolymer that differs from the previous one only by the nature of the link between the two fragments - an ester *vs* ether group. The reaction is performed under dynamic vacuum in the melt and in the presence of catalytic amounts of tin or cobalt salts (*13*). Alternatively, a reductive amination (*14*) between a dendritic aldehyde and PEG with two peripheral amino groups (Jeffamine® polymers have also been tested) can be employed to produce the targeted hybrid copolymer (*15*). However, the last two methods afford the desired compounds in much lower yields (50 - 80 %). The construction of linear dendritic hybrids based on the reaction of activated PEG and protected monodendrons was also reported (*16*).

The second approach (Scheme 2B) is based on the assumption that dendritic alcoholates, previously reported to initiate the polymerization of ε-caprolactone (*17*), can also trigger the ring-opening polymerization of ethylene oxide. Then the "living" polyether chain can be terminated by a dendritic bromide affording the same ABA block copolymer (*18*). Potential advantages of this method are the ability to grow the hydrophilic block to a specific chain length that is not commercially available and attach dissimilar dendritic fragments at the PEG ends.

The third strategy, Scheme 2C, uses telechelic PEG as the starting block and protected dihydroxybenzyl bromides can be added to grow the dendritic fragments in a typical protection/deprotection sequence (*19*). This approach was initially employed

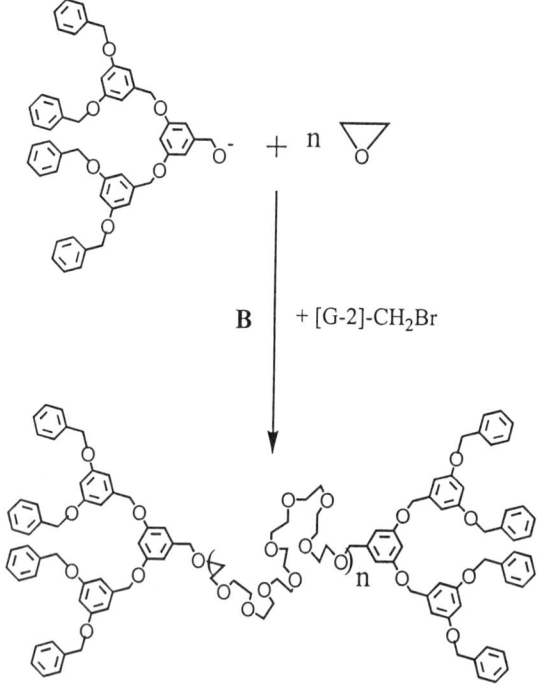

Scheme 2. B
Synthesis of ABA Copolymers Using Monodendrons as Initiators and End-capping Agents

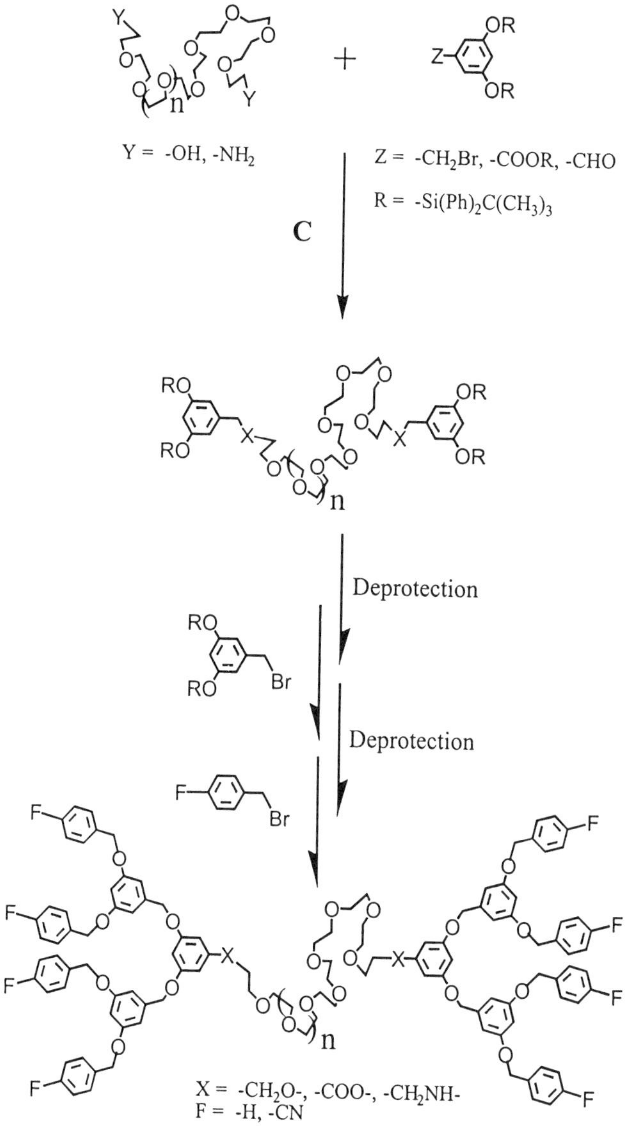
Y = -OH, -NH$_2$
Z = -CH$_2$Br, -COOR, -CHO
R = -Si(Ph)$_2$C(CH$_3$)$_3$
C
Deprotection
Deprotection
X = -CH$_2$O-, -COO-, -CH$_2$NH-
F = -H, -CN

Scheme 2. C
Synthesis of ABA Copolymers using PEG as a Template for Divergent Dendrimer Growth

by Chapman and coworkers for the formation of AB copolymers containing PEG and dendritic poly(α,ε-L-lysine) (20). Methoxy-terminated PEG was used as a template for the growth of the hydrophobic dendritic fragment. One of the advantages of this method is the possibility of forming hybrid amphiphilic copolymers with reactive functional groups at the periphery of the dendritic blocks that can be used later for the attachment of biologically active compounds or fluorescent tags.

It should be noted that so far, very few of the properties of the published linear-dendritic amphiphiles have been tested and very little is known about the mechanism of micelle formation for these systems, the number and position of the segments forming the core and the corona. The character of the microenvironment in the core of the assemblies is completely unknown. All this information is of paramount importance for the future practical application of dendritic amphiphiles as encapsulating agents, platforms for sustained drug release and vessels for chemical reactions on a nanometer scale.

Solution behavior of [G-2]-PEG5,000-[G-2]. It is known that the solubility of the amphiphilic linear-dendritic copolymers in aqueous media depends on the PEG/dendrimer ratio (8). [G-2]-PEG5000-[G-2] dissolves easily in methanol and water forming clear slightly opalescent solutions. SEC in methanol/water reveals that the composition of these solutions is concentration dependent. It should be mentioned that an extensive mass balance study was performed to verify that the composition picture provided by SEC is not obscured by adsorption and retention of fractions on the column packing. In all cases the injected materials could be recovered with an average yield of 99.8 %. At concentrations 1.5×10^{-4} mol/L and higher two distinct fractions are clearly visible in the SEC traces, Figure 1A. The first one that appears at 15.2 mL has an apparent molecular weight of 53,000 (related to PEG standards) indicating the presence of a multimolecular micelle. The second entity is a broad peak between 18 and 22 mL most probably consisting of ill-defined micelles with aggregation numbers between 2 and 4. Decreasing the concentrations of the copolymer from 5×10^{-5} mol/L down to the detection limit (1.5×10^{-6} mol/L) does not change the eluograms drastically, Figure 1B. The high molecular weight peak is still clearly visible along with increasing amounts of unimolecular micelles (21.5 mL). The limited sensitivity of the refractive index detector prevents the detection of the critical micelle concentration (cmc) of [G-2]-PEG5000-[G-2] by SEC. Static surface tension measurements performed in water at 24°C reveal that the onset of the self-assembly for this copolymer (cmc) appears at 1.3×10^{-6} mol/L (21). Heating the solutions at 60°C over extended periods of time does not change their composition, Figure 2. It could be assumed that at these conditions the micellar organization of the copolymer molecules is thermodynamically stable with the majority of the dendritic blocks concentrated in the core.

Solid state properties of [G-2]-PEG5,000-[G-2]. The peculiar organization of the hybrid amphiphile in aqueous medium is further evidenced by crystallization studies. Often the morphology of films cast from solution is a direct result of the specific organization of the polymer in the liquid medium. It can be fine-tuned by the pre-arrangement of the amphiphilic copolymer in selective solvents and by the nature of the substrate. This is an issue of growing importance because of the increased utilization of polymeric surfactants for surface modification (21). Crystals grown on

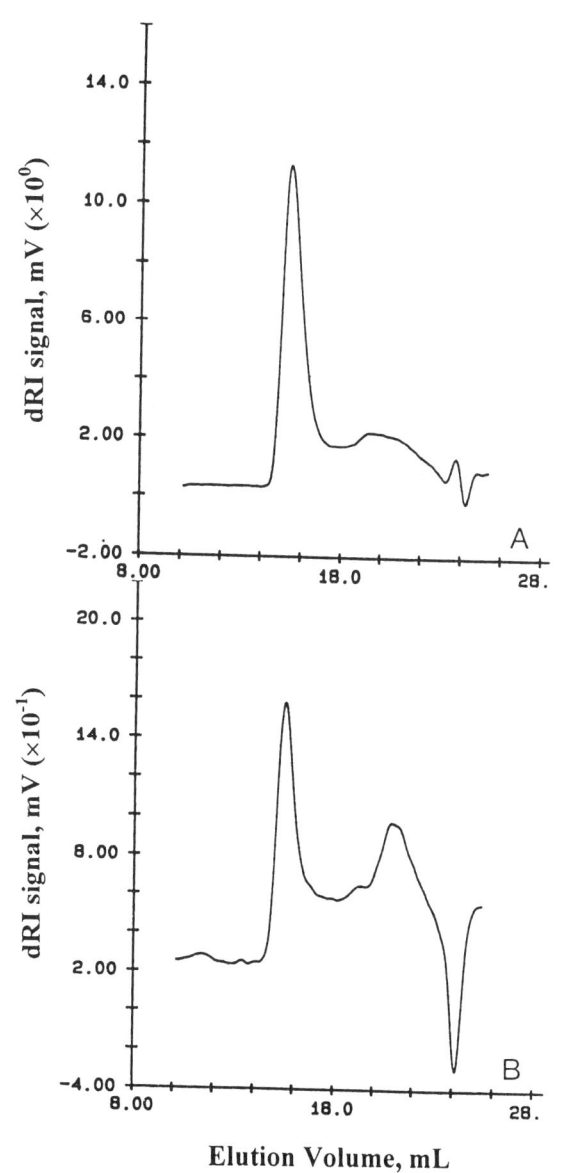

Figure 1. Aqueous size-exclusion chromatography of [G-2]-PEG5000-[G-2] at different concentrations. A - 1.5×10^{-4} mol/L; B - 5×10^{-5} mol/L.

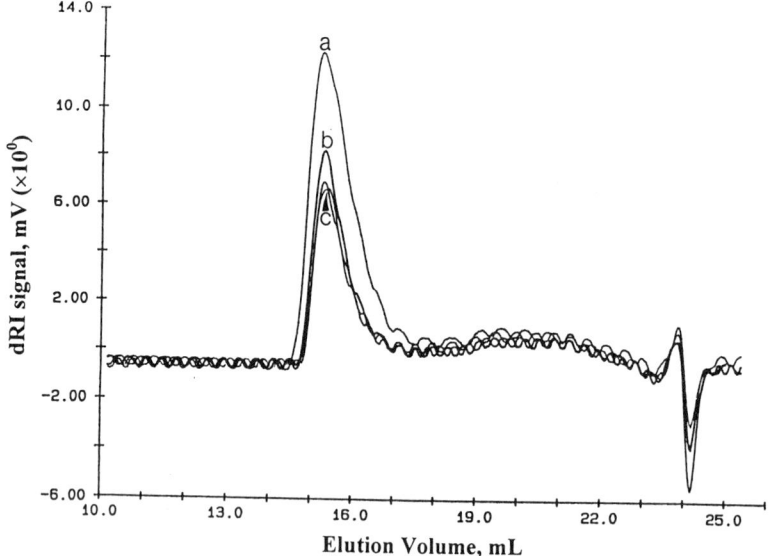

Figure 2. Aqueous size-exclusion chromatography of [G-2]-PEG5000-[G-2] solutions in water after heating at 60°C. a - initial solution; b - after 3 h; c - after 24 h and 7 days.

glass substrates from the melt (Figure 3) or from solutions in solvents good for both blocks (8) have a spherulite form and show a typical Maltese-cross pattern when analyzed by optical microscopy under crossed polarizers. Aqueous solutions with low concentrations (1×10^{-7} - 1×10^{-5} mol/L) yield thin films with similar surface morphology. When the experiment is repeated at higher concentrations in a chamber with controlled humidity the results are surprisingly different. The linear-dendritic hybrid crystallizes in unique star-like structures with sizes between 1 and 7 mm, and platelet morphology (Figure 4). It could be speculated that these formations are superaggregates constructed by a large number of micellar entities. It is evident that the self-assembly of [G-2]-PEG5000-[G-2] in water significantly affects the nucleation of the matrix mesophase (PEG) and the characteristics of its crystallization. This statement, however, needs further verification. The structure of the semi-diluted solutions and the crystals therefrom are currently being investigated by a combination of microscopy and scattering methods.

Encapsulation of polyaromatic hydrocarbons (PAHs) by [G-2]-PEG5000-[G-2] in water. Fluorescence and UV-Vis measurements are probably the most widely used methods to monitor the micellization process, evaluate the character of the core of the assemblies and study the binding of organic molecules to micellar systems in aqueous media. Pyrene and derivatives are the standard probes used in the studies (22-24). The vibronic fluorescence spectra of these compounds and more specifically the intensity ratio of the fourth and first peak in them (I_4/I_1) are very sensitive indicators for the local polarity of the environment surrounding the probe (25). Figure 5 shows the dependence of the pyrene fluorescence on the concentration of [G-2]-PEG5000-[G-2]. It is seen that at concentrations 1.1×10^{-4} mol/L and higher, the normalized spectra undergo a clear shift to longer wavelengths with a simultaneous change in the I_4/I_1 values. At low concentrations the probe is exposed to a relatively polar environment (I_4/I_1 = 0.79 - 0.83). It should be noted, that these values are still significantly higher than the reported values for pure water (0.71) and compounds similar to PEG and the linear-dendritic amphiphile: ethylene glycol (0.62), Brij 35 (0.69) and Triton X-100 (0.72) (25). It is seen, that at higher concentrations the I_4/I_1 ratio increases markedly up to 1.03, a value characteristic for entirely hydrophobic surroundings (25). Surprisingly, no eximer formation ($\lambda_{ex} \approx 480$ nm) is observed in the concentration range studied (Figure 5). The results obtained indicate that pyrene is encapsulated predominantly in the core of the micelles as it is shown schematically in Figure 6. Previous studies by Fréchet and coworkers have proven that poly(benzyl ether) dendrimers show little tendency to interpenetrate (26) and therefore they can form a *non-entangled* highly ordered entity in the center of the assembly. In this way the micellar core is capable of accommodating a *large number of molecules* that can be isolated (compartmentalized) in the internal voids of the dendrimers as well as in the space between the individual dendritic "wedges". This positioning of the pyrene molecules obviously prevents to a large extent their association and logically hampers the eximer formation. The I_4/I_1 value at high concentration clearly excludes the possibility that in the process of self-assembly water molecules might be partially entrapped in the voids between the monodendrons.

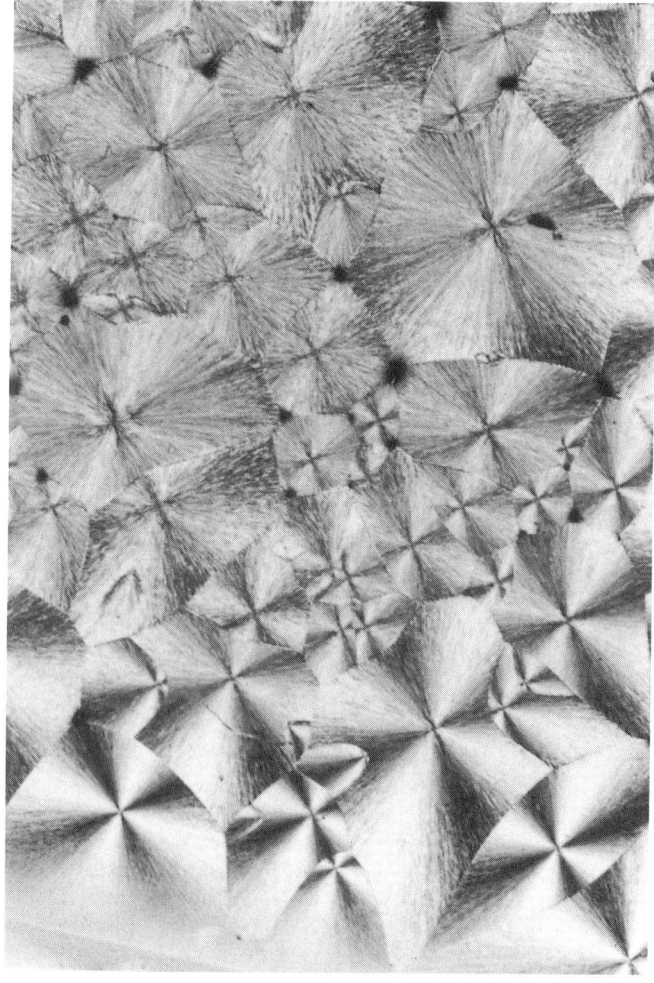

Figure 3. Polarized optical micrograph of [G-2]-PEG5000-[G-2] crystallized from the melt. Magnification 25×.

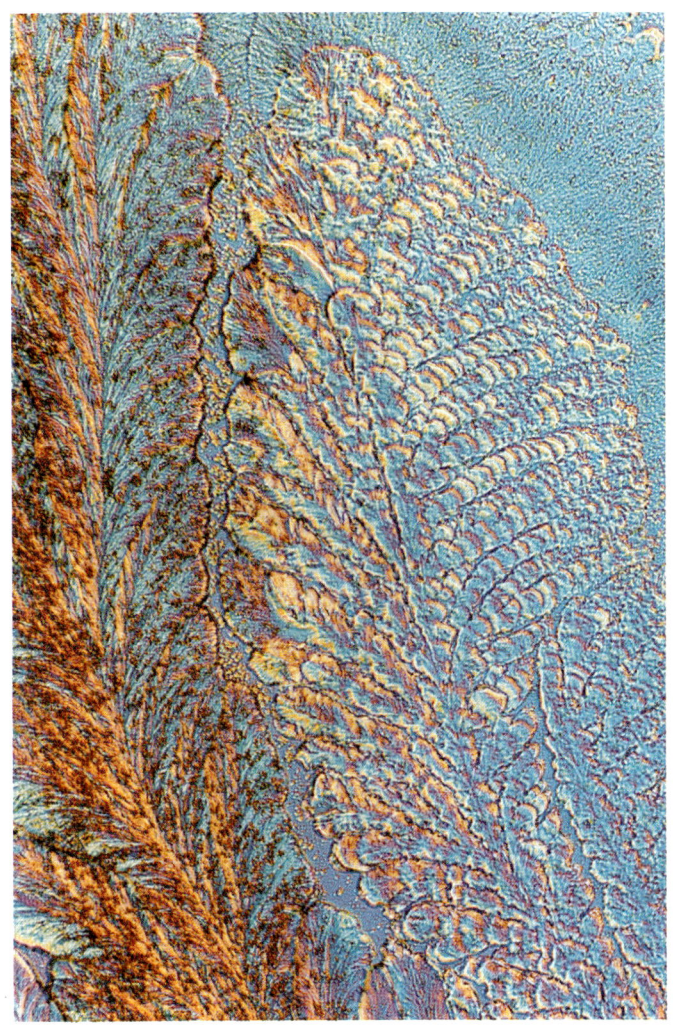

Plate 1. Polarized optical micrograph of [G-2]-PEG5000-[G-2]/C_{60} mixture crystallized from aqueous solution, copolymer concentration 1×10^{-3} mol/L. Magnification 25×.

Plate 2. Polarized optical micrograph of [G-2]-PEG5000-[G-2]/pyrene mixture crystallized from aqueous solution, copolymer concentration 1×10^{-3} mol/L. Magnification 25×.

85

Figure 4. Optical microscopy of [G-2]-PEG5000-[G-2] crystallized from aqueous solutions. Concentration 1×10^{-3} mol/L. A - Dark-field micrograph, magnification 15×; B - polarized micrograph of the same crystal, magnification 25×.

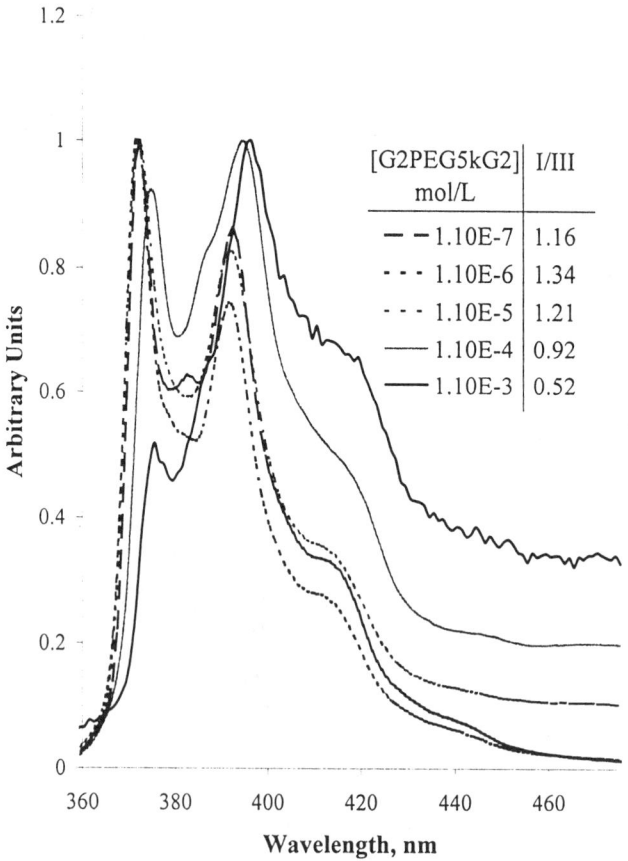

Figure 5. Fluorescence spectroscopy of mixtures of [G-2]-PEG5000-[G-2] and pyrene in water at different concentrations of the hybrid copolymer.

Figure 6. Schematic illustration of pyrene encapsulation by [G-2]-PEG5000-[G-2] in water.

The fluorescence studies can be extended further with binding experiments in order to evaluate the potential of the amphiphilic hybrids as encapsulating agents for removal of dangerous trace contaminants in aqueous waste (27). UV-Vis spectroscopy is used to investigate the entrapment of pyrene and other polycyclic aromatic compounds - phenantrene and perylene, Figure 7. Naturally, the smaller size of phenantrene and its geometry allow the accommodation of more molecules in the core of the micelles, while perylene is obviously excluded from the interior voids of the monodendrons and the quantities that can be bound are significantly lower. Quantitative information for the amounts of PAHs bound in similar systems is rather scarce. However, the data available indicate that the linear-dendritic amphiphiles compare favorably with their dendritic or linear analogues. The aqueous solubility of pyrene achieved with [G-2]-PEG5000-[G-2] is 5×10^{-5} mol/L (0.5 mol per mol polymer) against 9.5×10^{-5} mol/L obtained with surface functionalized fourth-generation poly(benzyl ether) dendrimer (28) (0.45 mol per mol dendrimer) and 1 mol per mol of comb-grafted polystyrene-polyethylene oxide with molecular weight 120,000 (29).

The encapsulation capability of [G-2]-PEG5,000-[G-2] is demonstrated further by the solubilization of large carbon clusters (fullerenes) in water. Incorporation of C_{60} and C_{70} into micelles and vesicles has recently attracted considerable attention (30) because of their potential use as artificial redox mediators in biological systems. The initial experiments show that amphiphilic hybrid linear dendritic copolymers are able to encapsulate significant amounts of C_{60} in water (Plate 1) in sharp contrast to dendritic unimolecular micelles which are not capable of C_{60} binding.

Micellar solutions containing high loads of pyrene and C_{60} are still capable of forming large crystalline structures (Plates 1 and 2). The typical color of the encapsulated compounds is visibly concentrated in the platelets of the crystals, indicating again their incorporation within the frame of the [G-2]-PEG5000-[G-2] assemblies. Their morphology, however, is significantly less organized in comparison with the pure crystals grown of the same copolymer under identical conditions (Figure 4).

Conclusions

The results of this study indicate that amphiphilic linear-dendritic copolymers could successfully complement the common low molecular weight surfactants and amphiphilic block copolymers. In contrast to the known micelle forming linear copolymers the novel hybrid macromolecules have perfectly branched globular segments that form the core of the micelle. This new approach for multimolecular micelles offers the distinct advantage of introducing a highly organized and restrictive entity in the center of the associating system. The potential benefits from this novel design could include increased selectivity to different organic substrates, improved solubility in aqueous and organic media of polycyclic aromatic compounds including fullerenes, a broad array of complexation capabilities, enhanced surface activity and catalytic potential. There is little doubt that such micelles containing dense and highly organized globular fragments will enable the development of novel unique devices and technologies that will find applications in biotechnology, environmental protection and the biomedical field.

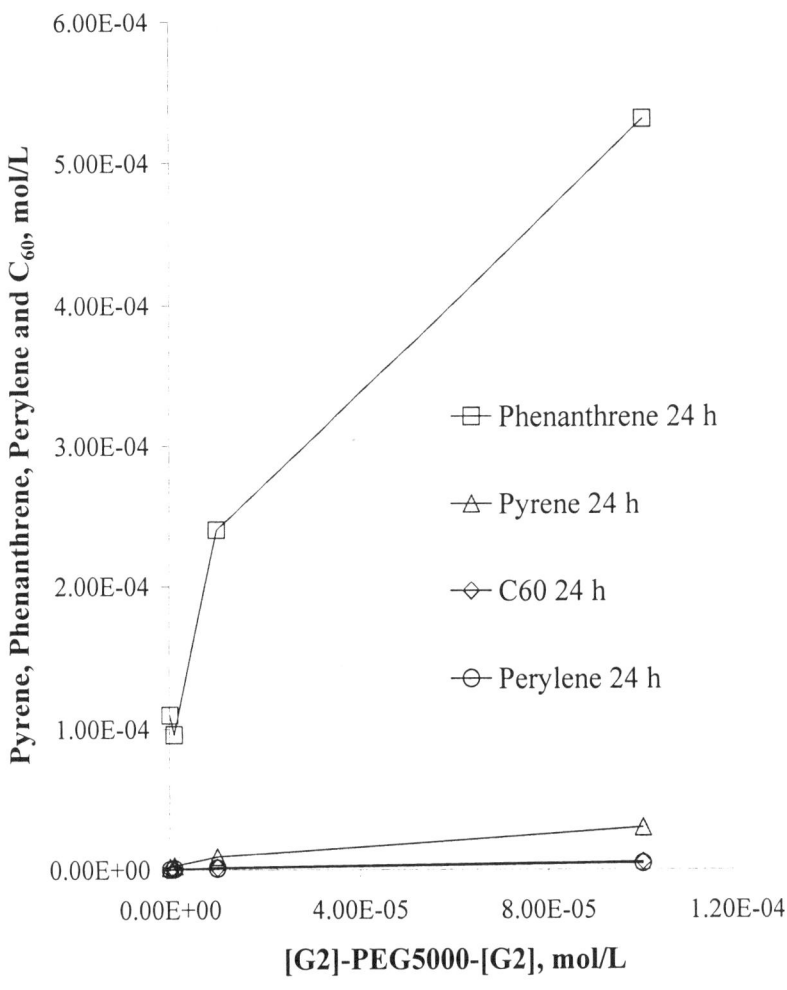

Figure 7. Binding of PAHs and C_{60} by [G-2]-PEG5000-[G-2] in water.

Acknowledgments

The author wishes to express his sincere gratitude to Prof. J.M.J. Fréchet for the continuous inspiration and advice. This work would have not been possible without the enthusiasm and the hard work of several students - Kevin R. Lambrych, Richard Pracitto and Edward Gersten. Thanks are due to Dr. Y. Yuan and Prof. I. Cabasso for providing the optical microscope used in the investigation. The project was financially supported by a start-up fund from the College of Environmental Science and Forestry (SUNY) and grants from Cornell Center for Materials Research and Allied Signal Corporation.

Literature Cited

1. Fendler J.H.; *Catalysis in Micellar and Macromolecular Systems*, Academic Press, New York, 1975; Tuzar, Z., Kratochvil, P.; *Adv. Colloid Interface Sci.* **1975**, *6*, 201; Tuzar, Z., Kratochvil, P.; In *Surface and Colloid Science*, Matijevic, E. Ed., Plenum Press, New York, 1993, Vol. 1, p 1; Lawrence, D.S., Jiang, T., Levett, M.; *Chem. Rev.* **1995**, *95*, 2229; Moffitt, M., Khougaz, K., Eisenberg, A.; *Acc. Chem. Res.* **1996**, *29*, 95; Fendler, J.H.; *Chem. Mater.* **1996**, *8*, 1616; Tascioglu, S.; *Tetrahedron* **1996**, *52*, 11113

2. Price, C. in *Developments in Block Copolymers*; Goodman, I., Ed., Applied Science Publishers, London, 1982, Vol. 1, p 39; Remp, P.F., Lutz, P.J. in *Comprehensive Polymer Science*, Eastmond, G.C., Ledwith, A., Russo, S., Sigwalt, P. Eds., Pergamon Press, Oxford, 1989, Vol. 6, p 403; Kawakami, Y., *Progr. Polym. Sci.* **1994**, *19*, 203; Alexandridis, T.P., Hatton, A. in *Polymeric Materials Encyclopedia*, Salamone, J.S. Ed., CRC Press, Boca Raton, 1996, Vol. 1, p 743; Yokoyama, M. *ibid.*, p 754; Kabanov, A.V., Alakhov, V.Y., *ibid.*, p 757; Tuzar, Z., Kratochvil, P., Munk, P.; *ibid.*, p 761

3. Newkome, G.R., Yao, Z., Baker, G.R., Gupta, V.K., *J. Org. Chem.* **1985**, *50*, 2003; Tomalia, D.A., Berry, V., Hall, M., Hedstrand, D., *Macromolecules* **1987**, *20*, 1164; Newkome, G.R., Moorefield, C.N., Baker, G.R., Saunders, M.J., Grossman, S.H., *Angew. Chem.* **1991**, *103*, 1207; Hawker, C.J., Wooley, K.L., Fréchet, J.M.J., *J. Chem. Soc., Perkin Trans. I* **1993**, 1287; Newkome, G.R., Young, K.K., Baker, G.R., Potter, R.L., Audoli, L., Cooper, D., Weiss, C.D., *Macromolecules* **1993**, *26*, 2394; Lee, J.-J., Ford, W.T., Moore, J.A., Li, Y., *ibid.* **1994**, *27*, 4632; Frey, H., Lorenz, K., Mülhaupt, R., *Macromol. Symp.* **1996**, *102*, 19; Stevelmans, S., van Hest, J.C.M., Jansen, J.F.G.A., van Boxtel, D.A.F.J., de Brabander-van den Berg, E.M.M., Meijer, E.W., *J. Am. Chem. Soc.* **1996**, *118*, 7398

4. Tomalia, D.A., Naylor, A.M., Goddard, W.A., *Angew. Chem., Int. Ed. Engl.* **1990**, *29*, 138; Mekelburger, H.-B., Jaworek, W., Vögtle, F., *ibid.* **1992**, *31*, 1571; Fréchet, J.M.J., Hawker, C.J., Wooley, K.L., *J. Macromol. Sci. - Pure Appl. Chem.* **1994**, *A31*, 1627; Newkome, G.R., Moorefield, C.N., Vögtle, F., *Dendritic Molecules. Concepts, Syntheses, Perspectives*, VCH, Weinheim, 1996; Matthews, O.A., Shipway, A.N., Stoddart, J.F., *Prog. Polym. Sci.* **1998**, *23*, 1; Chow, H.F., Mong, T.K.K., Nongrum, M.F., Wan, C.W., *Tetrahedron* **1998**, *54*, 8543

5. Jansen, J.F.G.A., de Brabander-van den Berg, E.M.M., Meijer, E.W., *Science* **1994**, *266*, 1226; Jansen, J.F.G.A., Meijer, E.W., de Brabander-van den Berg, E.M.M., *J. Am. Chem. Soc.* **1995**, *117*, 4417

6. Newkome, G.R., Baker, G.R., Saunders, M.J., Russo, P.S., Gupta, V.K., Yao, Z.-Q., Miller, J.E., Boullon, K., *J. Chem. Soc., Chem. Commun.* **1986**, 752; Newkome, G.R., Baker, G.R., Arai, S., Saunders, M.J., Russo, P.S., Theriot, K.J., Moorefield, C.N., Rogers, L.E., Miller, J.E., Lieux, T.R., Murray, M.E., Phillips, B., Pascal, L., *J. Am. Chem. Soc.*, **1990**, *112*, 8458; Newkome, G.R., Moorefield, C.N., Baker, G.R., Behera, R.K., Saunders, M.J., *Angew. Chem. Int. Ed. Engl.* **1992**, *31*, 917

7. Gitsov, I., Wooley, K.L., Fréchet, J.M.J., *ibid.* **1992**, *31*, 1200

8. Gitsov, I., Fréchet, J.M.J., *Macromolecules*, **1993**, *26*, 6536

9. Halperin, A., *Macromolecules* **1991**, *24*, 1418; Balsara, N.P., Tirrell, M., Lodge, T.P., *ibid.* **1991**, *24*, 1975; Maechling-Strasser, C., Cloutet, F., Francois, J., *Polymer* **1992**, *33*, 1021; Zhang, K., Xu, B., Winnik, M.A., Macdonald, P.A., *J. Phys. Chem.*, **1996**, *100*, 9834

10. Hawker, C.J., Frechet, J.M.J., *J. Am. Chem. Soc.* **1990**, *112*, 7638

11. *Spectral Atlas of Polycyclic Aromatic Compounds*, Karcher, W., Fordham, R.J., Dubois, J.J., Glaude, P.G.J.M., Lighart, J.A.M., Eds., D. Riedel Publ. Co., Dordrecht, 1989, Vol.1

12. Hare, J.P., Kroto, H.W., Taylor, R., *Chem. Phys. Lett.* **1991**, *177*, 394

13. Gitsov, I., Wooley, K.L., Hawker, C.J., Ivanova, P.T., Fréchet, ., J.M.J., *Macromolecules* **1993**, *26*, 5621

14. Borch, R.F., Bernstein, M.D., Durst, H.D., *J. Am. Chem. Soc.* **1971**, *93*, 2897; Vulic, I., Loman, A.J.B, Feijen, J., Okano, J. T., Kim, S.W., *J. Polym. Sci., Part A: Polym. Chem.* **1990**, *28*, 1693

15. Gitsov, I., Fréchet, J.M.J., *to be published*

16. Yu, D., Fréchet, J.M.J., *PMSE Preprints* **1998**, *79*, 633

17. Gitsov, I., Ivanova, P.T., Fréchet, J.M.J., *Macromol. Rapid Commun.* **1994**, *15*, 387

18. Gitsov, I., *unpublished results*

19. Gitsov, I., *unpublished results*

20. Chapman, T.M., Hillyer, G.L., Mahan, E.J., Shaffer, K.A., *J. Am. Chem. Soc.* **1994**, *116*, 11195

21. Fréchet, J.M.J., Gitsov, I., Monteil, T. Rochat, S., Sassi, J.F., Vergelati, C., Yu, D., *Chemistry of Materials*, **1999**, *11*, 1267

22. Yekta, A., Duhamel, J., Brochard, P., Adiwidjaja, H., Winnik, M.A., *Macromolecules* **1993**, *26*, 1829; Wilhelm, M., Zhao, C.-L., Wang, Y., Xu, R., Winnik, M.A., *ibid.* **1991**, *24*, 7033

23. van Hest, J.C.M., Delnoye, D.A.P., Baars, M.W.P.L., Elissen-Román, C., van Genderen, M.H.P., Meijer, E.W., *Chem. Eur. J.* **1996**, *2*, 1616

24. Chen, X., Smid, J., *Langmuir* **1996**, *12*, 2207

25. Kalyanasundaram, K., Thomas, J.K., *J. Am. Chem. Soc.* **1977**, *99*, 2039

26. Wooley, K.L., Hawker, C.J., Pochan, J.M., Fréchet, ., J.M.J., *Macromolecules* **1993**, *26*, 1514; Hawker, C.J., Farrington, P.J., Mackay, M.E., Wooley, K.L., Fréchet, J.M.J., *J. Am. Chem. Soc.* **1995**, *117*, 4409

27. Hurter, P.N., Hatton, T.A., *Langmuir* **1995**, *8*, 1291

28. Hawker, C.J., Wooley, K.L., Fréchet, J.M.J., *J. Chem. Soc., Perkin Trans. I* **1993**, 1287

29. Kawaguchi, S., Akaike, K., Zhang, Z.M., Matsumoto, H., Ito, K. *Polym. J.* **1998**, *30*, 1004

30. Williams, R.M., Crielard, W., Hellingwerf, K.J., Verhoeven, J.W., *Rec. Trav. Chim. Pays-Bas* **1996**, *115*, 72 and refs therein; Beeby, A., Eastoe, J., Crooks, E.R., *J. Chem. Soc., Chem. Commun.* **1996**, 901

Surfactant-Modified Water-Soluble Polymers

HMPAM and HEUR Polymers

Chapter 6

Inter- and Intra-Molecular Aggregation of Associating Polymers in Water

Joseph Selb and Françoise Candau

Institut Charles Sadron/Centre de Recherches sur les Macromolécules (CNRS), 6, rue Boussingault, 67083 Strasbourg Cedex, France

Micellar polymerization processes have been used to prepare polyacrylamides hydrophobically modified with (I) N-alkyl-acrylamides or (II) the polymerizable surfactant (surfmer) n-hexadecyldimethyl-4-vinylbenzylammonium chloride. In process I, short hydrophobic blocks are formed whose length can be tuned by varying the hydrophobic monomer / surfactant ratio. In process II, long hydrophobic sequences are formed when pure surfmer micelles are used, whereas a random hydrophobe distribution can be obtained by adding the non-polymerizable analog to form mixed micelles. The competition between intra- and inter-molecular hydrophobic interactions and hence the thickening properties in aqueous solution depend both on the nature of the hydrophobe and on the blockiness of the copolymers.

Introduction

Polymerization in nanostructured media like micelles or microemulsions is presently the object of an increasing number of studies. Besides the fundamental interest of this type of process, potential and existing applications can be found in fields as diverse as paper manufacture, water-treatment, ultrafiltration, oil recovery, coatings, microencapsulation and drug delivery. The large interfacial area resulting from the small particles size and the particular microenvironment caused by the difference in polarity between bulk and dispersed phases can be taken advantage of to produce polymeric materials with interesting properties and/or morphologies (1, 2).

The present paper deals with hydrophobically modified water-soluble polymers prepared under two different radical micellar polymerization processes. We will show that it is possible to tune the aqueous solution properties of these copolymers, through inter- or intra-molecular hydrophobic associations according to the type of process used.

Copolymer Synthesis and Microstructure

The two following micellar routes were used for the synthesis of associating polymers. In the first process (I, Figure 1), the hydrophobic monomer, a N-alkylacrylamide (see Figure 3) which is insoluble in water, is solubilized within SDS (sodium dodecyl sulfate) micelles. In the second process (II, Figure 2), the hydrophobe is a water-soluble polymerizable surfactant (surfmer) in the micellar state: n-hexadecyldimethyl-4-vinylbenzylammonium chloride (N16, $cmc_{50°C} \approx 4 \times 10^{-4}$ M) (see Figure 6). In both cases, the water-soluble monomer, acrylamide (AM), is dissolved in the bulk aqueous phase together with the initiator: potassium persulfate (in process I) or 4.4'-azobis (4-cyano valeric acid), ACVA (in process II). The hydrophobe content in the monomer feed is low and varies from 1 to 3 mol%. The experimental procedures and the copolymer characterization have been described elsewhere (3-5).

Previous studies have shown that the high local concentration of the hydrophobic monomers in the micelles favors their incorporation in the hydrophilic backbone as random blocks rather than as isolated units (3, 6-8). In process I, it is assumed that the length of the blocks corresponds roughly to the number of hydrophobes contained in a micelle. The copolymer, after an appropriate recovery from the micellar solution, is free of surfactant. In process II, the hydrophobe block length should correspond to the micellar aggregation number of the surfmer. Therefore the block length of copolymers obtained by process II (~ 50 units) is much larger than that obtained in I (1 to 10 units).

Polyacrylamides Modified with an Alkylacrylamide

Heterogeneity in Composition

A problem encountered in this micellar process and discussed in some of our previous publications, is the compositional heterogeneity of the high-conversion samples (4, 9-12). This is illustrated in Figure 4 which shows the hydrophobe incorporation in the copolymer as a function of conversion for two polyacrylamides modified either with 5 mol% of N-hexylacrylamide (HexAM) or 5 mol% of N-methyl,N-hexylacrylamide (MeHexAM). A very severe drift in copolymer composition is found in the case of the N-monosubstituted acrylamide (HexAM) which is incorporated more rapidly at the beginning of the polymerization. On the other hand, a constant incorporation is observed for the N,N-disubstituted acrylamide (MeHexAM). Hence, the replacement of an N-H bond by an N-alkyl group is sufficient to suppress the drift in copolymer composition. As in both cases, AM and the alkylAM bear the same acrylamido polymerizable group, their reactivity should be rather similar (3). However, the reactivity of some monomers is known to strongly depend on the nature of the medium (13), and therefore, an interpretation based on microenvironment and polar effects in the micellar reaction medium can account fairly

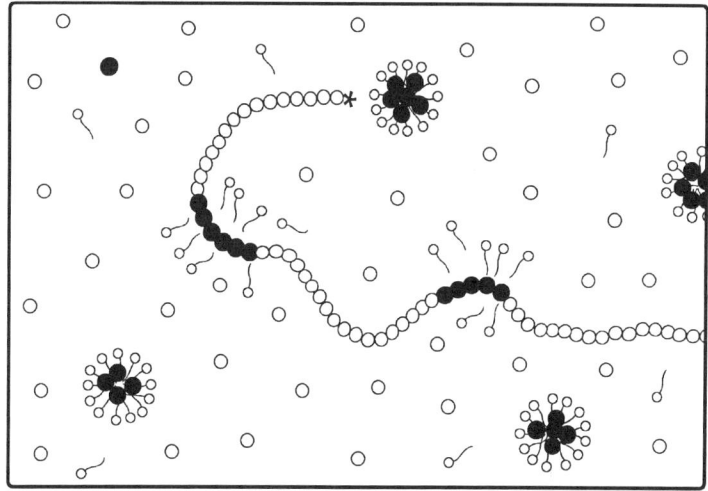

Figure 1. Schematic representation of the copolymerization process I:
○ : water-soluble monomer,   ● : hydrophobic monomer,   ⌒⌒ : surfactant

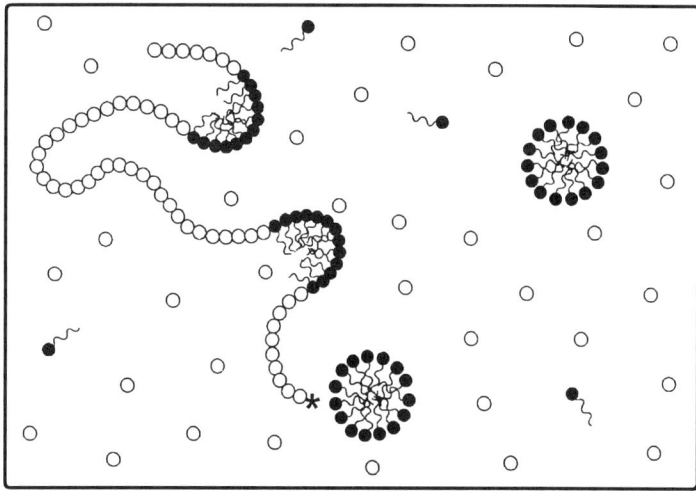

Figure 2. Schematic representation of the copolymerization process II:
○ : water-soluble monomer,   ●⌒⌒ : polymerizable surfactant

well for this result. The H atom present in monosubstituted AM is capable of forming intermolecular hydrogen bonding. This results in an enhanced reactivity for the hydrophobic monomer located within the micelle which is a medium of low polarity compared to the AM aqueous solution (11). This assumption is further supported by other studies performed on the copolymerization of AM derivatives in solvents of varying polarity (14-16). In the case of *N,N*-di-substituted AM, the absence of a hydrogen atom prevents the formation of hydrogen bonds and the reactivity of the monomer is unaffected by solvent polarity.

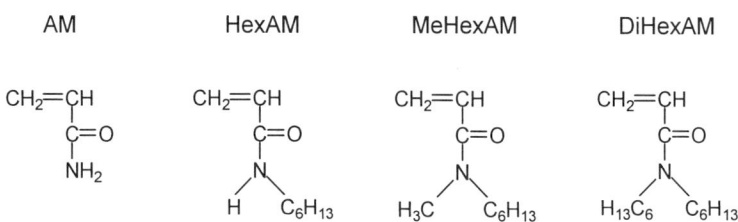

Figure 3. Structure and nomenclature of the monomers (copolymerization process I).

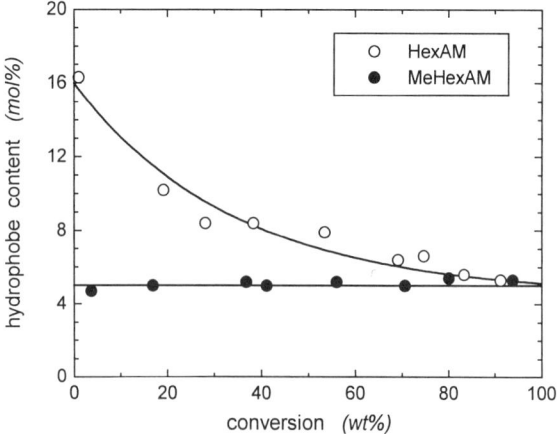

Figure 4. Variation of the copolymer composition as a function of conversion for the micellar copolymerization (process I) of acrylamide (AM) with N-hexylacrylamide (HexAM, O) or N-methyl,N-hexylacrylamide (MeHexAM, ●). Initial hydrophobe content in the monomer feed: 5 mol%.

The above result is a good illustration of the role played by interfacial and microenvironment effects on the copolymerization mechanism in micellar media. These effects can be capitalized upon to produce, by free-radical polymerization, copolymers homogeneous in composition and with a multiblock structure.

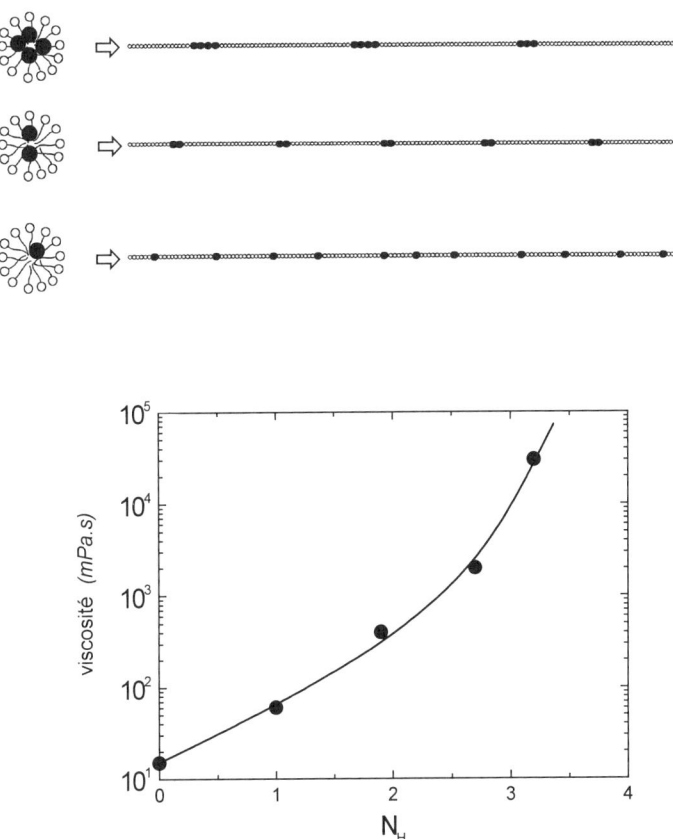

Figure 5. Variation of the zero-shear viscosity for various copolymers prepared by micellar copolymerization (process I) at a constant hydrophobe content (DiHexAM: 1 mol%) but at different surfactant concentrations i.e. at different N_H values (number of hydrophobes per micelle). $N_H = 0$ corresponds to homopolyacrylamide (PAM).

Relationship Between Molecular Structure and Properties in Solution

With process I, it is possible to tune the length of the hydrophobic blocks at constant hydrophobe level by varying the surfactant concentration used in the synthesis. Increasing the SDS concentration amounts to increasing the number of micelles in the medium. Therefore, the number of hydrophobes per micelle (N_H) is diminished and so the length of the block in the final copolymer, as schematically shown in Figure 5. In the same Figure is plotted the zero-shear viscosity of a series of homogeneous samples in aqueous solution versus N_H (i.e. ~ the block length in the copolymer). Viscosity measurements were carried out using a Haake RS100 controlled stress rheometer according to an experimental procedure described elsewhere (11, 12). The copolymer concentration, Cp, is 1 wt%; that is well above C*, where the intermolecular associations between the hydrophobic groups are predominant. The copolymers have same composition ([DiHexAM] = 1 mol%), same molecular weight (Mw ~ 2.10^6), but different blockiness. One observes a major effect of copolymer molecular structure on the macroscopic properties in aqueous solution. The viscosity increases by almost 3 decades upon increasing N_H from 1 (where the hydrophobic units are singly distributed) to ~ 3; the longer the hydrophobic blocks, the greater the thickening efficiency. The large increase in viscosity with N_H is likely due to the increase in the life-time of the associations. The rate of disengagement of a hydrophobic sequence from a cluster is an exponential function of the activation energy for disengagement and therefore it depends strongly on the block length.

Polyacrylamides Modified with a Polymerizable Surfactant

We have also attempted to tune, in process II, the molecular structure of the copolymers formed in order to determine its effect on the polymerization mechanism and on the macroscopic properties in solution. The method used in I to achieve this goal (variation of [hydrophobe] / [surfactant] ratio) cannot apply here. When a growing polyacrylamide radical encounters a N16 micelle, it would add in one step all the surfmer molecules forming the micelle (Nagg ~ 50). This difficulty can be overcome by using mixed micelles formed from N16 and the non-polymerizable analog, n-hexadecyldimethylbenzylammonium chloride (B16) (Figure 6). The molar ratio of [N16] / [B16] can be adjusted in such a way that one has about one N16 molecule per B16 micelle. This amounts to simulating a polymerization reaction in homogeneous solution and therefore the hydrophobic N16 units are expected to be singly and randomly distributed in the acrylamide backbone, as schematically represented in Figure 7.

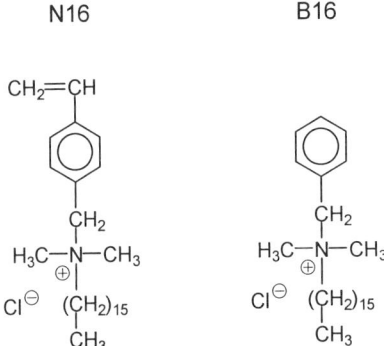

Figure 6. Polymerizable surfactant (N16) used in the copolymerization process II, and the non-polymerizable analog (B16).

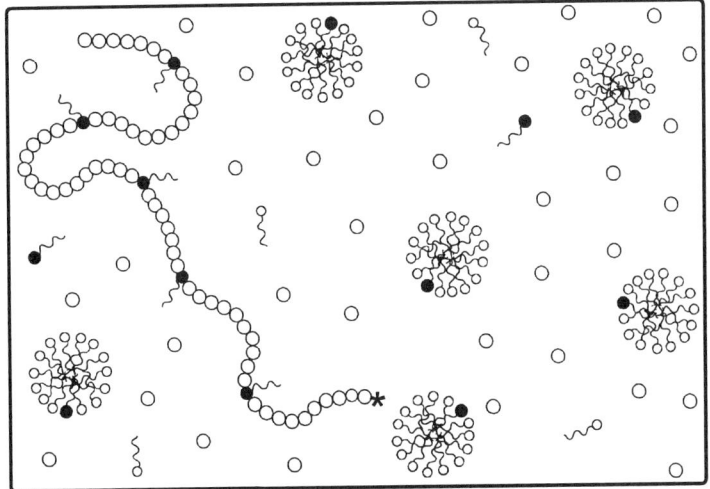

Figure 7. Schematic representation of the formation of a random copolymer from mixed micelles containing polymerizable and non-polymerizable surfactants.
O : water-soluble monomer,   ●⌒ : polymerizable surfactant,   ○⌒ : non-polymerizable surfactant.

Heterogeneity in composition

Figure 8a shows the evolution of the N16 hydrophobe incorporation in the copolymer as a function of conversion for two samples of same initial N16 content (1 mol%) but prepared under different experimental conditions: sample A was obtained by copolymerization of AM with pure N16 micelles and therefore has a multiblock structure. In sample B, which was obtained by copolymerization of AM with N16-B16 mixed micelles, the microstructure is random. As in the case of copolymers modified with N-monosubstituted AM (see Figure 4), one observes for both samples a more rapid hydrophobe incorporation at the beginning of the reaction. Samples A and B exhibit a similar drift in copolymer composition, taking into account the experimental error due to a lack of precision in the determination of chloride content by elemental analysis. Even though the samples have a different microstructure (blocky or random), no specific microenvironment effect should prevail here since the polarity sensed by the styryl groups remains unchanged when going from pure N16 micelles to mixed N16-B16 micelles. It is even surprising that the much higher local concentration of N16 in the former case does not lead to a faster consumption of this monomer. This result tends to indicate that the drift in copolymer composition observed for both samples is essentially caused by different values of the reactivity parameters of the AM-N16 monomer pair. To confirm this point, we have compared these incorporation curves to that which can be calculated from literature data relative to the copolymerization in water of AM and a monomer (N1) similar to N16, but in which the hexyl chain of N16 is replaced by a methyl group (17). The latter monomer is therefore molecularly dispersed in water and does not form micelles. As shown in Figure 8a, the calculated curve follows the same trends as those obtained from the micellar process. A similar agreement was obtained for an initial surfmer content of 3 mol% (Figure 8b).

From these results, we can conclude that the values of the reactivity ratios of the monomers control mainly the homogeneity of the sample composition. However, we cannot totally exclude a micellar effect (high local monomer concentration, residence time of the active end of the propagating radical within the micelle, and intermicellar exchange dynamics) as shown by a recent study performed in our group on the aqueous copolymerization of AM with a surfmer bearing the same acrylamido function (18). Further studies should provide additional information on the competition between the micellar effect and the monomer reactivity ratios.

A closer examination of Figure 8a-b shows that the composition/conversion curves for copolymers formed from pure N16 micelles (samples A) are always below the curves relative to copolymers formed from N16-B16 mixed micelles (samples B). This result can be ascribed to a difference in microstructure: samples A contain large N16 sequences connected by long AM sequences (blocky distribution), whereas samples B contain isolated N16 units separated by short AM sequences (random distribution); both types of samples contain copolymer chains with a high N16 content formed in the early stages of the polymerization. In the former case, the blocky copolymer chains with a very high N16 content (> 7-10 mol% i.e. > 30-40 wt%) can be lost during the polymer recovery process because of a weak precipitation efficiency

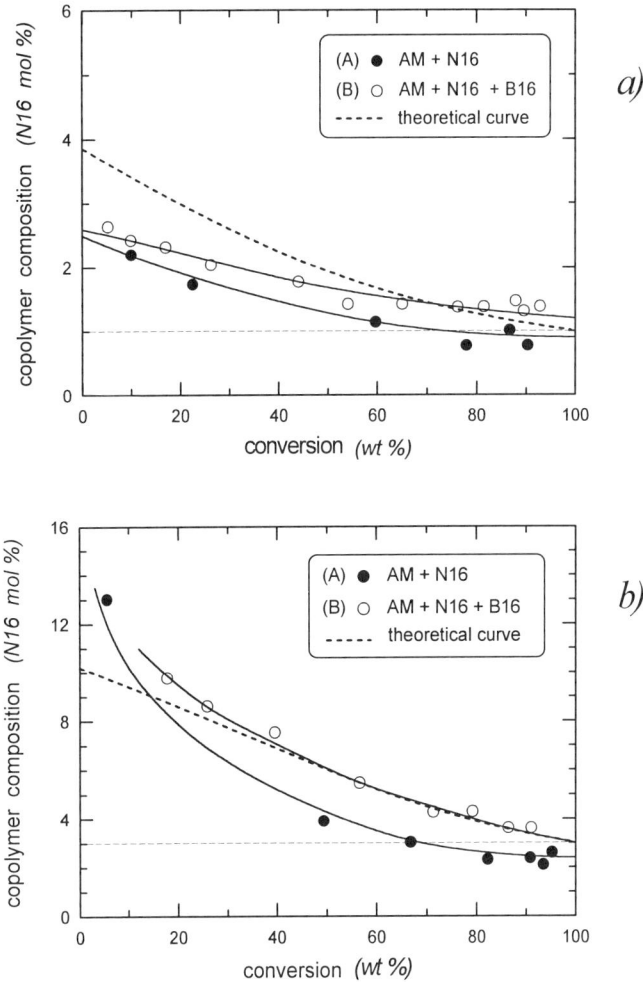

Figure 8. Variation of the copolymer composition as a function of conversion for the copolymerization of AM with the N16 polymerizable surfactant.
Samples A (●): synthesis with pure N16 micelles (see Figure 2);
Samples B (○): synthesis with N16-B16 mixed micelles (see Figure 7).
The dashed line is the theoretical composition/conversion curve calculated with $r_{AM} = 0.24$ and $r_{N16} = 0.046$ as reactivity ratios.
Initial surfmer content in the monomer feed: **a)** 1 mol%, **b)** 3 mol%.

of the nonsolvents used (acetone or acetone/methanol mixtures). In the latter case, the polymer chains with the N16 isolated units are likely more insoluble and are recovered more quantitatively upon precipitation. Such a difference in solubility properties between block and random copolymers with the same chemical composition is a well known phenomenon. (Note that a more important loss of some copolymer chains appeared when copolymers were recovered by precipitation in methanol, an usual precipitant for PAM and AM copolymers, but a solvent for polyN16. This led, in this case, to an apparent incomplete consumption of the surfmer (4)). As seen in the next section, this difference in microstructure was also confirmed by fluorescence experiments.

Relationship Between Molecular Structure and Properties in Solution

The correlation between molecular structure and macroscopic properties of samples A and B is illustrated in Table 1. Surprisingly, the viscosity in pure water ($C_p = 1$ wt%) of the blocky copolymer (A) is about one thousand fold less than that of the random one (B) in spite of an even higher molecular weight (sample A: Mw = 2×10^6; sample B: Mw = 0.9×10^6). This result clearly indicates that the thickening properties of these copolymers are much higher when the hydrophobic units are singly distributed along the backbone. This behavior is totally the opposite to that observed for AM-DiHexAM samples. In this case, the blocky copolymers ($N_H \sim 3$) exhibited a high association degree whereas almost no association was observed for the random copolymer ($N_H = 1$) (see Figure 5).

Table I. Influence of the Method of Synthesis on the Properties of AM-N16 Copolymers

experimental conditions	AM + N16 (see Figure 2) pure N16 micelles	AM + N16 + B16 (see Figure 7) mixed micelles
distribution of N16 units	blocky	random
Mw	2 000 000	900 000
viscosity (mPa.s) ($C_p = 1$ wt%)	~ 100	~ 100 000
I_3 / I_1 pyrene fluorescence ratio	0.75	0.66

Several possible explanations come into mind to account for the particular behavior of AM-N16 copolymers. A possible difference in compositional heterogeneity for samples A and B can be discarded in light of the results reported in the previous section (see Figure 8). Similarly, the eventual presence of residual surfactant cannot be responsible for this behavior since the interactions between AM-N16 copolymers and cationic surfactants were shown not to be very effective and not strongly modify the viscometric behavior of the copolymers (19).

The most plausible explanation is that the interactions which take place in the blocky AM-N16 copolymer are essentially of the intramolecular type, due to the particular polysoap-like structure. Such a structure, resulting from the presence of long sequences of amphiphilic units, differs from that of blocky AM-DiHexAM copolymers which contain short hydrophobic blocks with small alkyl chains.

This hypothesis was confirmed by fluorescence experiments using pyrene as a probe according to a previously described standard procedure (3). The I_3/I_1 ratio of the third to the first emission peaks of pyrene was found to be significantly higher for the blocky AM-N16 copolymer than for the random one (Table 1). This reveals a stronger degree of hydrophobic aggregation for the former sample, which is in apparent contradiction with the lower viscosity measured. In fact, a I_3/I_1 value similar to that obtained for the blocky copolymer ($I_3/I_1 = 0.75$) was also found in previous studies performed in our laboratory for both the pure N16 micelles solutions and for the N16 homopolymer obtained after polymerization (20). In the latter case, several complementary techniques allowed us to provide a good description of the microstructure of this polymer. The structure was that of a polysoap with intramolecular hydrophobic microdomains covalently linked and connected via linear spacers. The above result tends to confirm that the local conformation of the blocky copolymer is also that of a polysoap, i.e. a string of beads where the beads are large hydrophobic domains connected by hydrophilic polyacrylamide strings (Figure 9a). Such a structure is quite unfavorable to intermolecular associations due to steric hindrance, accounting thus for the low value of the viscosity.

In the case of the random copolymer, the hydrophobic microdomains are fewer and/or with less hydrophobes involved in a cluster (Figure 9b). The high viscosity observed is due to intermolecular associations. Note that the hydrophobic interactions are more efficient than for the random AM-DiHexAM copolymer, owing to the longer alkyl tail of the hydrophobe.

Examination of the full fluorescence emission spectra of the two AM-N16 copolymers provides another confirmation of their different behavior (Figure 10). In addition to the 'monomer' emission spectra, one observes for the blocky copolymer a broad peak at higher wavelength corresponding to excimer formation due to pyrene molecules in close vicinity. This result implies that some hydrophobic microdomains contain more than one pyrene probe. Such a peak is not observed for the random copolymer, which means that all pyrene molecules are singly distributed. As the pyrene concentration is similar in both cases, we are forced to conclude that there is, in the former case, a smaller number of large hydrophobic microdomains than in the latter case.

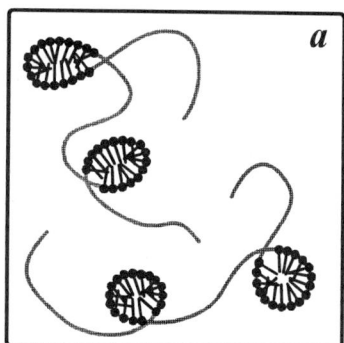 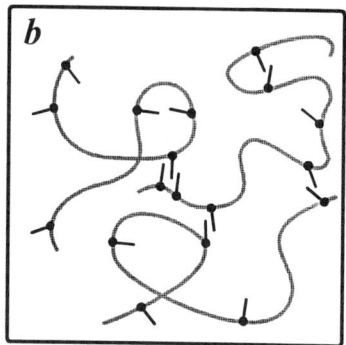

Figure 9. Schematic representation of the microstructure and conformation of AM-N16 copolymers prepared under different experimental conditions: **a)** from pure N16 micelles (see Figure 2); **b)** from N16-B16 mixed micelles (see Figure 7).

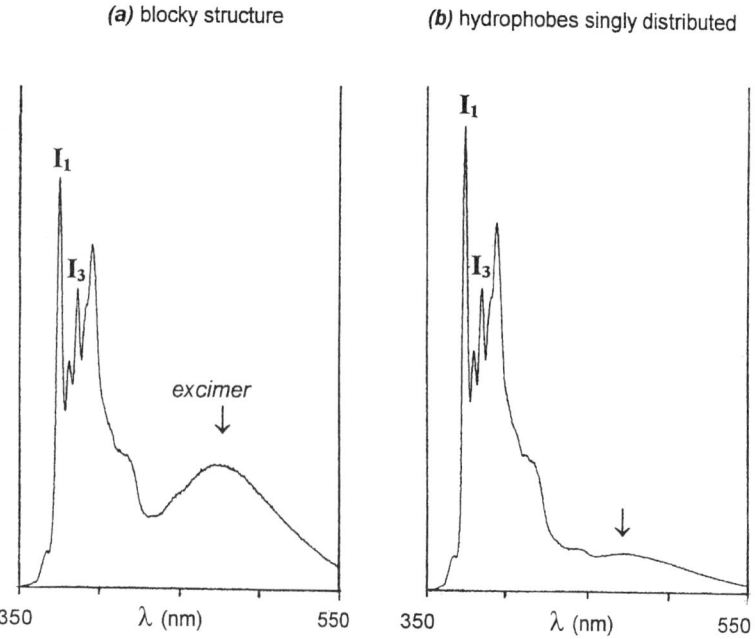

Figure 10. Pyrene fluorescence emission spectra for aqueous solutions of AM-N16 copolymers prepared from: **a)** pure N16 micelles, **b)** N16-B16 mixed micelles. ($Cp = 0.5$ wt%).

Conclusion

The results presented here give an illustration of the possibilities the chemist has at his disposal for tuning the properties of amphiphilic copolymers by playing with their microstructures. By using some 'tricks' to disperse the hydrophobic compounds, it is possible to build up various supramolecular architectures.

When the hydrophobe is solubilized within direct micelles, the polymer chains formed contain the hydrophobic units distributed as small blocks. In aqueous solution, these blocks associate **inter**molecularly and form a transient physical network with improved thickening properties.

When the hydrophobe is a polymerizable surfactant, the microstructure of the copolymer formed is that of a polysoap essentially controlled by **intra**molecular associations. Potential applications can be envisioned in controlled drug delivery due to the possibility of the microdomains to encapsulate hydrophobic compounds.

Acknowledgments

We thank F. Vadeboin for her technical assistance in the synthesis of some samples and in rheological experiments.

References

1. Candau, F. In *Microemulsions: Fundamental and Applied Aspects*; Mittal, K. L., Kumar, P., Eds.; Dekker: New York, in press.
2. Candau, F.; Selb, J. *Adv. Colloid Interface Sci.* **1999**, *79*, 149.
3. Hill, A.; Candau, F.; Selb, J. *Macromolecules* **1993**, *26*, 4521.
4. Selb, J.; Biggs, S.; Renoux, D.; Candau, F. In *Hydrophilic Polymers: Performance with environmental acceptability*; Glass, J. E., Ed.; Advances in Chemistry Series: American Chemical Society: Washington, DC, 1996; Vol. 248, Chapter 15, p 251.
5. Renoux, D.; Selb, J.; Candau, F. *Prog. Colloid Polym. Sci.* **1994**, *97*, 213.
6. Dowling, K. C.; Thomas, J. K. *Macromolecules* **1990**, *23*, 1059.
7. Ezzell, S. A.; Hoyle, C. E.; Creed, D.; McCormick, C. L. *Macromolecules* **1992**, *25*, 1887.
8. Branham, K. D.; Shafer, G. S.; Hoyle, C. E.; McCormick, C. L. *Macromolecules* **1995**, *28*, 6175.
9. Biggs, S.; Hill, A.; Selb, J.; Candau, F. *J. Phys. Chem.* **1992**, *96*, 1505.
10. Lacik, I.; Selb, J.; Candau, F. *Polymer* **1995**, *36*, 3197.
11. Volpert, E.; Selb, J.; Candau, F. *Macromolecules* **1996**, *29*, 1452.

12. Volpert, E.; Selb, J.; Candau, F. *Polymer* **1998**, *39*, 1025.
13. Plochocka, K. *J. Macromol. Sci., Rev. Macromol. Chem.* **1981**, *C20*, 67.
14. Saini, G.; Leoni, A.; Franco, S. *Makromol. Chem.* **1971**, *144*, 235.
15. Saini, G.; Leoni, A.; Franco, S. *Makromol. Chem.* **1971**, *146*, 165.
16. Leoni, A.; Franco, S.; Saini, G. *Makromol. Chem.* **1973**, *165*, 97.
17. Jones, G. D.; Goetz, S. J. *J. Polym. Sci.* **1957**, *25*, 201.
18. Stähler, K.; Selb, J.; Candau, F. *Colloid Polym. Sci.* **1998**, *276*, 860.
19. Renoux, D. Thesis, Université Louis Pasteur, Strasbourg, France, 1995.
20. Cochin, D.; Candau, F.; Zana, R. *Macromolecules* **1993**, *26*, 5755.

Chapter 7

Collapsed and Extended Polysoaps

O. V. Borisov[1] and A. Halperin[2]

[1]BASF AG ZOI/ZC-C13, 67056 Ludwigshafen, Germany
[2]UMR 5819 (CEA-CNRS-UJF), DRFMC-SI3M, CEA-Grenoble, 38054 Grenoble Cedex 9, France

Polysoaps are flexible hydrophilic chains that incorporate, at intervals, amphiphilic monomers. The amphiphilic monomers can self assemble into intrachain micelles. The resulting hierarchical self organization strongly affects their configurations, dimensions and elasticity. In solutions of high ionic strength the equilibrium state is a globule of closely packed intrachain micelles. The static deformation behavior of such globules, as well as their response to flow fields, exhibit novel features due to the coupling to their internal degrees of freedom. The formation of interchain micelles can give rise to phase separation that is simply described in terms of the p-cluster model.

The molecular architectures of waterborne associating polymers are highly diversified. Such polymers consist of a water soluble backbone and covalently bound associating groups, "stickers". However, the molecular architecture as well as the characteristics of the hydrophilic chain and of the stickers can vary widely [1][2]. For example, the distinguishing features of the hydrophilic chain include its rigidity and the origin of its solubility in water. In particular, the chain can be a polyelectrolyte or it may be water soluble because of the formation of hydrogen bonds. The nature of stickers, their number and their distribution along the chain also vary widely. The architectural richness of associating polymers impedes the formulation of a general theory of these systems [3][4][5]. In the following we discuss a highly simplified model for one particular family of associating polymers: Polysoaps [2][5][6][7]. Strictly speaking, we use this term to describe associating polymers comprising of a flexible, neutral chain incorporating, at intervals, many amphiphilic monomers (Figure 1). However, as we shall discuss in the final section, much of the analysis applies to a wider class of polymers. For example, it describes flexible polyelectrolytes carrying many hydrophobic side chains in the high salt limit, when electrostatic interactions are screened.

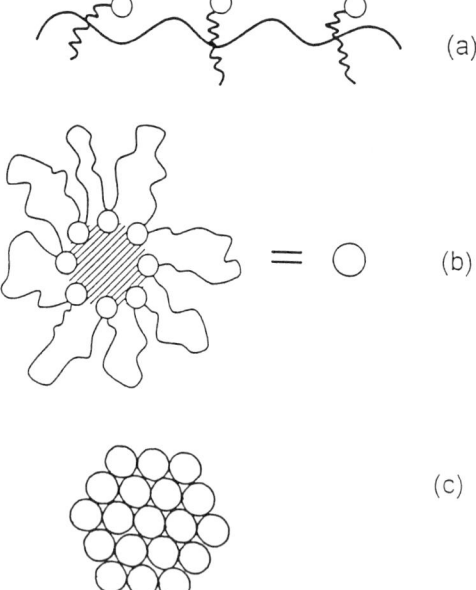

Figure 1. A polysoap consists of amphiphilic monomers joined by flexible spacer chains (a) self assembles into intrachain micelles endowed with a corona of loops formed by the spacer chains (b). The equilibrium state of long polysoaps is a globule of closely packed intrachain micelles (c). The headgroups of the amphiphiles are represented by small circles while intrachain micelles are depicted as large ones.

An important distinguishing feature of polysoaps is the occurrence of both interchain and intrachain self assembly. In other words, the polymerized amphiphiles can micellize within a single chain or form interchain micelles incorporating amphiphiles that belong to different chains. The intrachain self assembly, within an isolated polysoap, results in qualitative modification of the single chain characteristics: dimensions, configurations [8][9] and elasticity [10][11][12]. We discuss these issues because an understanding of the single chain behavior is a necessary step for the formulation of a theory of solutions of polysoaps. In addition to these topics, we consider the effect of flow fields on the configurations of isolated chains [12]. Finally we discuss the implementation of the p-cluster model to describe the phase diagram of polysoap solutions [13]. While most of the results discussed in this review have been reported previously, the behaviour of globular polysoaps in uniform flow was not presented elsewhere.

Within this model, a polysoap is a copolymer consisting of $m \gg 1$ amphiphilic monomers joined by flexible, neutral, hydrophilic spacer chains. An amphiphilic monomer is assumed to consist of an ionic head-group and a hydrophobic tail. Each spacer chain comprises of n monomers of size b and the overall polymerization degree is thus $N \approx mn$. For simplicity we treat the hydrophilic backbone in water as a flexible polymer immersed in a good solvent. We focus on the high salt limit that is, solutions of high ionic strength, when the long range electrostatic repulsion between the ionic head-groups are screened. The polymerized amphiphiles can self assemble into intrachain and interchain micelles. The intrachain self assembly gives rise to a hierarchical self organization within the chain which is somewhat reminiscent of protein folding. The primary structure, the sequence of monomers, is determined by the synthesis. The intrachain micelles may be viewed as a secondary structure. When m is small, there is no higher level of self organization. However, when m is large enough so as to allow the formation of many intrachain micelles there appears a higher level of self organization: A tertiary structure associated with the configurations of the string of intrachain micelles. This hierarchical intrachain self assembly gives rise to two important features of the isolated polysoaps: (i) The resulting configurations are compact in comparison to the corresponding homopolymer (ii) The force law associated with the chain deformation is modified due to the coupling between the strain and the "internal" structure of the chain. An additional feature, not considered here, is the possibility of using free, unpolymerized amphiphiles in order to disrupt the intrachain self assembly [14]. The micellization of the polymerized amphiphiles gives rise to an effective attraction between the polysoap chains. In turn, this can result in phase separation involving a dense gel phase and a dilute solution of free chains. A simple description of the phase behavior is made possible by the p-cluster model [13], as proposed by de Gennes in a different context [15] [16].

Intramolecular Aggregation in Polysoaps

In many respects, intrachain micelles formed by polysoaps are similar to free micelles formed by unpolymerized, monomeric amphiphiles [17]. However, the formation of intrachain micelles is not associated with a loss of translational entropy since the polymerized amphiphiles can not move independently. Consequently, the intrachain micellization is not associated with a critical micelle concentration (CMC). As we shall discuss, penalty due to the translational entropy is replaced by a penalty associated with the configurational entropy. As a first step we consider the equilibrium structure and the free energy of a single intrachain micelle. We then discuss the configuration of a long polysoap capable of forming many intrachain micelles. At this stage we focus on the case of strong amphiphiles when all the polymerized amphiphiles aggregate. This corresponds to the case of monomeric amphiphiles characterized by a very low CMC in the unpolymerized state.

Intramolecular micelles

An intrachain micelle comprises of two regions. The inner region is reminiscent of a micelle formed by free surfactants. In other words, it consists of a dense hydrophobic core formed by the hydrocarbon tails of the surfactants with the ionic head-group straddling the core-water interface. This inner region is surrounded by a swollen corona of loops formed by the spacer chains attached to the polymerized amphiphiles (Figure 1) [8]. The corona is responsible for two distinctive features of the intrachain micelles. First, their contribution may dominate the overall size of the micelles. Second, the crowding of the coronal loops gives rise to a free energy penalty that affects both their shape and size. Since the micellar corona is reminiscent of the corona of a star polymer we base our analysis on the combination of two models: The model of Israelachvili *et al* [17] for free micelles and the Daoud-Cotton model [18][19] [20] of star polymers.

Within the model of Israelachvili *et al* the aggregated amphiphile is characterized by three attributes: The volume v and maximal extension of the hydrophobic tail, l as well as the area per head-group at the core-water interface, a. The packing parameter, v/al determines the shape of the aggregate. v and l are determined directly by the molecular structure of the amphiphile. On the other hand, a is set by the equilibrium condition as determined by the free energy per aggregated amphiphile. It reflects a balance between the screened electrostatic repulsion between the head-groups and the surface energy per amphiphile. In the case of intrachain micelles it is necessary to supplement this two terms by a free energy penalty reflecting the crowding of the coronal loops. In the following we limit the discussion to spherical intrachain micelles for which the number of amphiphiles per micelle is $p \approx v^2/a^3$ and the corona is described by the Daoud-Cotton model. The equilibrium values of the aggregation number, p, and of a correspond to the minimum of the free energy per surfactant, ϵ. ϵ is expressed in terms of the dimensionless variable $a/a_0 \equiv (p_0/p)^{1/3}$ where a_0 is the optimal area per head group in a micelle formed by unpolymerized amphiphiles and $p_0 \approx v^2/a_0^3$ is the corresponding aggregation number as

$$\epsilon/kT \approx -\delta + \gamma a_0[a/a_0 + a_0/a + \kappa(a_0/a)^{1/2}] \quad (1)$$

Here k is the Boltzmann constant and T is temperature. The first term allows for the transfer free energy of the hydrophobic tail from water into the core, $-\delta kT$. As in the case of free micelles, this is the driving force for the self assembly. The second term is the free energy of the core-water interface, where γkT is the surface tension of the core. This term favors micellar growth while the next two oppose it. The repulsion between the ionic head-groups gives rise to the third term. The fourth term reflects the crowding of the coronal loops as specified by the Daoud-Cotton model. In particular, the micellar corona is considered as indistinguishable from the corona of star polymer with $2p$ arms of length $n/2$. Within this model the overall radius of the corona in a good solvent is

$$R_{corona} \approx n^{3/5} p^{1/5} b \qquad (2)$$

while the coronal free energy is $p^{3/2} \ln R_{corona}$. The logarithmic n dependence is neglected in our expression for ϵ. The perturbation parameter $\kappa \cong p_0^{1/2}(\gamma a_0)^{-1}$ is the ratio of the coronal and head group repulsion penalties when $p = p_0$. It determines the relative importance of these two contributions in determining the equilibrium state of the intrachain micelles.

The corona affects the structure of the intrachain micelles in two respects. First, the radius of the micelle is always larger than that of micelles formed by free amphiphiles and may be approximated by R_{corona}. Second, even though the span of the intrachain micelles is larger, the aggregation number, p, is smaller than p_0 because of the coronal penalty. The effect on p is determined by κ. This in turn is tuned by controlling n. In the limit of $\kappa \ll 1$, when n is relatively small, the coronal penalty is negligible thus leading to $p \approx p_0$ and $a \approx a_0$. In the opposite limit, of large n, when $\kappa \gg 1$, the coronal term is dominant and $p \ll p_0$. Eventually, for very large κ the intrachain micellization is repressed altogether. For brevity we focus in the following on the case of $\kappa \ll 1$. The generalization to the case of $\kappa \gg 1$ is straightforward [8].

Large-scale configurations of polysoaps: the role of exchange attraction

For short polysoaps, when $m \approx p$, only the secondary structure can develop. In other words, the chain will self assemble into a single intrachain micelle. In the opposite limit, of $m \gg p$, it is necessary to consider the tertiary structure of a string of m/p intrachain micelles. As we shall discuss, the equilibrium state is a spherical globule of closely packed intrachain micelles. This arises because the exchange of amphiphiles between the micelles gives rise to an effective "exchange attraction" between the micelles. However, many other configurations such as linear and branched strings of intrachain micelles may appear as metastable states. This is because the reorganization of the configurations of the micellar string is an activation process involving, as intermediate step, the expulsion of micellized amphiphiles into the aqueous medium [9]. In the following we will consider two extreme configurations that play a role in our subsequent discussion. In particular, a linear string of micelles and a dense globule (Figure 1).

In our subsequent discussion we will encounter a linear string of micelles as a configuration of an extended chain. It may also occur, depending on the sample history, as a metastable state. In this situation the exchange interactions are not operative on the relevant time scales and the micelle-micelle interactions are purely repulsive. The repulsion results from the overlap of the two coronas and the subsequent increase in unfavorable monomer-monomer contacts. This situation is similar to that of two star polymers at close proximity. The interaction free energy then varies as [21]

$$U(r)/kT \approx p^{3/2}\ln(R_{corona}/r), \qquad r \leq R_{corona} \qquad (3)$$

where $2r$ is distance between the centers of the two micelles. In this situation the chain may be viewed as a self avoiding random walk of intrachain micelles *i.e.*, the effective monomer size is $R_{micelle} \approx R_{corona}$. Consequently, the span a linear string of intrachain micelles is

$$R \approx (m/p)^{3/5} R_{micelle} \qquad (4)$$

The interaction between the micelles is purely repulsive only when there is no exchange of amphiphiles between them. When exchange is possible, it gives rise to attractive interaction that favors grazing contact between the coronas. The origin of the attraction is entropic. When the micelles are at grazing contact and exchange is allowed, each amphiphile can reside in two micelles thus lowering the free energy by $kT\ln 2$ per amphiphile [22]. This rough argument overestimates the attraction since not all the amphiphiles are equally likely to exchange. A more detailed analysis leads to [9][23]

$$U_{att}/kT \approx -p^{1/2}. \qquad (5)$$

at grazing contact. This, in turn, gives rise to a negative second virial coefficient of micelle-micelle interactions. In other words, a string of micelles experiences poor solvent conditions even though the backbone is immersed in a good solvent. As a result, the equilibrium configuration of the micellar string is collapsed [9]: it forms a spherical globule of closed packed intrachain micelles with an overall radius of

$$R_{globule} \approx (m/p)^{1/3} R_{micelle} \qquad (6)$$

The boundary of the globule is associated with a surface tension $kT\gamma_g$. It arises because the micelles at the interface experience exchange interactions only with the interior. Consequently, $kT\gamma_g \approx |U_{att}|/R_{micelle}^2$ or

$$\gamma_g \approx p^{1/2}/R_{micelle}^2 \qquad (7)$$

It is important to note two points that distinguish globular polysoaps from collapsed homopolymers in a poor solvent. First, while the large scale configuration of a long polysoap is collapsed, the coronas of the individual intrachain micelles remain swollen. Second, the collapse of homopolymers in poor solvents does not involve metastable states. In marked distinction, the exchange attraction in polysoaps is an activation process and non-globular metastable states are possible.

The Deformation Behavior of a Globular Polysoap

The familiar elasticity of flexible polymers involves no internal degrees of freedom. The chain is envisioned as a structureless random coil, modeled as a random walk or self avoiding random walk. A deformation decreases the number of attainable configurations thus lowering the entropy of the chain and giving rise to a restoring force. In the case of polysoaps it is necessary to allow for the effect of the deformation on the hierarchical self organization within the chain. The deformation couples to the self assembled structures within the chain. The existence of these internal degrees of freedom modifies the elastic behavior of the chain according to Le Chatelier principle [10][11]. To illustrate this point we consider the extension behavior of a globular polysoap [12]. As we shall see, weak deformation couple to the tertiary structure while stronger extensions affect the secondary structure (Figure 2). Eventually, when the chain is highly extended, all traces of the intrachain self assembly disappear. The weak deformation behavior is reminiscent of that of collapsed homopolymers [24][25] while the strong deformation scenario is identical to that encountered for a linear string of micelles [10][11].

Weak strains deform the globule from a sphere into an ellipsoid. The restoring force within this linear response regime is due to the increase in the surface free energy of the globule

$$\Delta F_s/kT \approx \gamma_g \Delta A \approx p^{1/2}(R - R_{globule})^2/R_{micelle}^2 \qquad (8)$$

where $\Delta A \approx (R - R_{globule})^2$ is the change in surface area of the globule, γ_g is the surface tension of the globule defined by equation (7) and $R_{globule}$ is the radius of the unperturbed globule. The associated restoring force is

$$\frac{f}{kT} \approx \frac{\partial}{\partial R}\frac{\Delta F_s}{kT} \approx \frac{p^{1/2}}{R_{micelle}^2}(R - R_{globule}), \quad (R - R_{globule}) \ll R_{micelle} \qquad (9)$$

This process cannot proceed indefinitely. Once the ellipsoid approaches cylindrical form the fR diagram exhibits a van der Waals loop. This indicates that the cylindrical form is instable with respect to coexistence of weakly deformed globule and an extended string of micelles. The onset of the instable regime occurs at f_c that can be estimated from the condition $f_c R_{micelle} \approx kTp^{1/2}$, where $kTp^{1/2}$ is the exchange attraction per micelle. When the end to end distance in imposed and finite size corrections are ignored the globule string coexistence occurs in the range $R_{globule} + R_{micelle} \ll R \ll R_{max}$ where $R_{max} \approx (m/p)R_{micelle}$ is the length of the fully extended string of micelles. It is associated with a constant tension

$$f_c/kT \approx p^{1/2}R_{micelle}^{-1} \qquad (10)$$

Note that the tension is actually a decreasing function of R because of finite size effects due to the surface free energy of the globule. As a result the globule will unravel fully when subjected to a constant tension of $f = f_c$

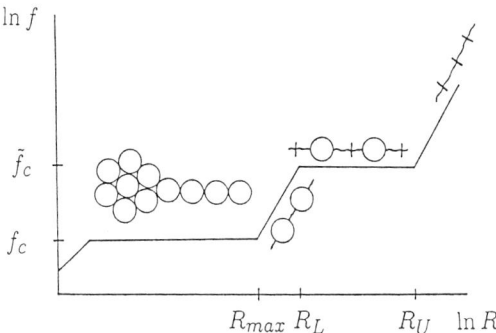

Figure 2. The force law, f/kT vs R plot, characterizing the extension of a globular polysoap. The insets depicts the chain configurations in the various regimes. Intrachain micelles are represented by circles and dissociated amphiphiles by short vertical lines.

While $R < R_\text{max}$ the individual micelles are only weakly perturbed. Only the tertiary structure is affected. Stronger extensions can only be accommodated by modifying the secondary, micellar structure. This scenario is identical that of strong stretching of a linear string of micelles. Three regimes are involved. In the first regime, the extension results in stretching the bridges between the micelles with little effect on p and the overall micellar structure. This occurs while the imposed tension is in the range $f_c < f < f_{co}$ where f_{co} is the tension associated with the onset of micellar dissociation. f_{co} can be estimated by equating the free energy of an amphiphile incorporated into unperturbed micelle $kT|\epsilon(a_0)|$ to the elastic energy of a stretched bridge. The elasticity of the stretched bridges is described by the Pincus law [26][27] which relates the extension of a bridge r to the applied tension f as

$$r \approx n(fb/kT)^{2/3}b \qquad (11)$$

The elastic energy is $f_{co}r_{co}$ leading to the criterion for onset of micellar dissociation

$$f_{co}r_{co} \approx kT|\epsilon(a_0)| \approx kT|-\delta + 2\gamma a_0| \qquad (12)$$

or

$$f_{co}b/kT \approx (|\epsilon(a_0)|/n)^{3/5} \qquad (13)$$

The range $f_c < f < f_{co}$ corresponds to end to end distances such that $R_\text{max} < R < R_L$ where R_L, the upper boundary of the stretched bridges regime, will be specified below. In this range, the chain span is given by the Pincus law

$$R/b \approx \frac{m}{p}n(fb/kT)^{2/3} \qquad (14)$$

The upper limit of this domain, R_L can be estimated by inserting f_{co} as given by (13) into (14). Stronger extension, $f > f_{co}$ or, equivalently, $R > R_L$, can not be sustained without major effect on the micellar structure. It is accommodated by partial dissociation of the micelles. In particular, it gives rise to a coexistence of weakly perturbed micelles and fully dissociated amphiphiles. The coexistence regime is associated with a plateau in the fR diagram. The tension at the coexistence would have been independent of R, $f \approx f_{co} \sim R^0$, if the coexistence would have been associated with a first order phase transition. However, since the stretched polysoap is a one dimensional system with no long range interaction it can not undergo a first order phase transition. The mixing entropy of the dissociated amphiphiles and the micelles smoothes out the transition and gives rise to $f \sim \ln R$ dependence. Eventually, when the end to end distance reaches R_U to be specified below, all the intrachain micelles dissociate and there is no trace of the intrachain self assembly. In the coexistence range, $R_L \leq R \leq R_U$, the fraction of micellized amphiphiles, ψ increase continuously with R according to the lever rule

$$\psi \approx \frac{R - R_L}{R_U - R_L} \qquad (15)$$

A rough approximation of R_U is possible if we ignore the logarithmic R dependence of f and approximate the tension at the upper boundary of the coexistence regime, $R = R_U$, by f_{co} as given above. When $R \geq R_U$ the polysoap is fully dissociated and the elasticity of the chain is identical to that a flexible chain comprising mn monomers as given by the Pincus force law

$$R/b \approx mn(fb/kT)^{2/3} \qquad (16)$$

Inserting $f \approx f_{co}$ into (16) determines

$$R_U \approx pR_L \qquad (17)$$

It is of interest to note that force laws similar to that of a linear string of intrachain micelles were observed for certain biopolymers [28]. The similarity arises because of the involvement of internal degrees of freedom in both types of systems.

Extension of Globular Polysoaps by Flow Fields

In our preceding discussion of the extension of polysoaps either the tension on the chain, f, or the end to end distance, R, were externally imposed. When the extension is accomplished by a flow field, this is no longer the case. In such situation, one controls the characteristics of the flow field, shear rate or velocity, rather than f or R. The chain stretching is due the Stokes drag force. In turn, the drag force varies with the configuration of the chain. The resulting scenarios differ accordingly from those considered in the previous section. In the following we briefly consider two situations: longitudinal shear flow and uniform flow.

Extensional flow

Longitudinal shear flow is defined by the Cartesian components of the velocities

$$V_x = sx, \quad V_y = -sy/2, \quad V_z = -sz/2. \qquad (18)$$

Such flow fields are realized, for example, in convergent ducts or at the entry of a capillary. Our analysis concerns the behavior of a polymer placed at the origin, the stagnation point. The residence time of the chain in this region is assumed to be large in comparison to its relaxation time. Flexible homopolymers placed in this situation undergo an abrupt coil-stretch transition [29][30]. This is due to the sharp increase in the drag force with the end to end distance of the stretched chain. The drag force increases as R^3 while the elastic restoring force increases only as $R^{3/2}$. As a result the free energy curve of the chain, allowing for the stationary hydrodynamic force field, exhibits two distinct minima corresponding, respectively, to weakly and strongly stretched configurations [31][32]. This situation differs qualitatively from the behavior of a chain subjected to given strain, where the free energy exhibits a single minimum. The case of globular polysoaps [12], as of collapsed homopolymers

[37][38], is simpler to analyze because a coexistence occurs even in the absence of a hydrodynamic flow field. As a result, force balance consideration suffice and it is not necessary to consider the free energy curve. In particular, one balances the critical tension f_c corresponding to the unraveling of a globular polysoap, as give by (10) with the drag force acting on the globule subjected to the extensional flow field specified by (18) thus leading to

$$f_c \approx s_c \eta R_{globule}^2 \qquad (19)$$

Here η is the solvent viscosity and $R_{globule}$ is the size of an unperturbed globule. This specifies the critical shear rate, s_c, for the unfolding of the globule

$$s_c \eta / kT \approx p^{29/30} m^{-2/3} n^{-3/5} \qquad (20)$$

At first sight one may have expected two critical shear rates corresponding to the two plateaus characterizing the "static" extension of globular polysoaps. However, because of the strong increase of the drag force with the end to end distance, only a single transition occurs. s_c is sufficient to cause complete unfolding of the globular polysoap into a fully dissociated chain.

Uniform flow

Chain extension can also be induced by uniform flow specified by

$$V_x = -V, \quad V_y = V_z = 0 \qquad (21)$$

This flow field obtains for a chain which is terminally grafted to a small bead moving at constant velocity V [33][34]. The extension of swollen and collapsed homopolymers in this situation gives rise to "stem-flower" configurations [35][36]. The tension in the chain decreases with the distance from the anchoring site. Consequently, the chain segment attached directly to the bead can be fully extended while the more distant parts of the chain are progressively less stretched. A similar picture emerges for globular polysoaps. Because the tension along the chain varies the different "phases" can coexist, in marked distinction to the behavior encountered in the longitudinal shear flow.

Two main regimes emerge as V increases. For small V the globular polysoap is weakly deformed into an ellipsoid. The span of the ellipsoid along the major axis, R, is specified by equating the restoring force due to the surface free energy of the globule, as given by (9), with the drag force

$$f \approx \eta V R_{globule} \qquad (22)$$

This regime ends when the drag force becomes equal to the critical force f_c defined by (10). This condition sets a critical velocity, V_c, for the onset of the unraveling of the globule

$$V_c \approx f_c / \eta R_{globule} \approx p^{1/2} kT / \eta R_{micelle} R_{globule} \qquad (23)$$

A stem-globule configuration develops when V exceeds V_c. It consists of a depleted globule of radius $r_g < R_{globule}$ attached to the bead by an extended stem. At the boundary between the globule and the stem the tension is equal to f_c. This tension is balanced by the hydrodynamic drag $\eta V r_g$ on the globule thus leading to

$$r_g \approx \frac{p^{1/2}}{R_{micelle}} \frac{kT}{\eta V} \approx R_{globule} \frac{V_c}{V}, \quad V \geq V_c \qquad (24)$$

The tension along the stem decreases with the distance from the bead. As a result, in the general case the stem consists of three regions (Figure 3): (i) A stretched string of micelles, of length L_{mic}, is joined directly to the globule. The tension at the junction with the globule is f_c. (ii) The next region, closer to the bead, comprises of an extended, fully dissociated chain of length L_{dis} The tension at the junction with the micellar string is f_{co}. (iii) Finally, the chain segment attached to the bead is dissociated and fully stretched. The tension at the boundary with the adjacent section is kT/b where b is the monomer size.

Simple force balance considerations determine L_{mic} and L_{dis}. In the non draining limit, the tension at the junction of the micellar string and the dissociated region is $f_c + \eta V L_{mic}$ thus leading to

$$L_{mic} = (f_{co} - f_c)/\eta V \qquad (25)$$

similarly, the tension at the junction between the two dissociated regions is $f_{co} + \eta V L_{dis}$ giving rise to

$$L_{dis} = (kT/b - f_{co})/\eta V \qquad (26)$$

A complete description of the stem specifies the variation of tension along the various regions as well as the number of monomers in each. To achieve this, it is necessary to allow for the local tension. The type of argument involved is easiest to illustrate in the case of the dissociated region. One considers a chain element with an end to end distance dx incorporating $dN(x)$ monomers where x measures the distance from the junction point with the micellar region. Since the chain is strongly stretched, every chain element obeys the Pincus law, $dx/dN(x) \approx (fb/kT)^{2/3}b$ or $(dx/dN)^{3/2} \approx fb^{5/2}/kT$. Accordingly, at any point x along the dissociated segment

$$\frac{kT}{b^{5/2}} \left(\frac{dx}{dN}\right)^{3/2} = \eta V x + f_{co} \qquad (27)$$

Integrating this expression between the two junctions, where the tension is known, yields the number of monomers in the region and the position dependence of the tension. Similar argument can be applied to the micellar region, bearing in mind that the elementary unit in this region is a micelle. These results, together with the monomer conservation and (24) specify the state of the chain. The details of this analysis are beyond the scope of this arti-

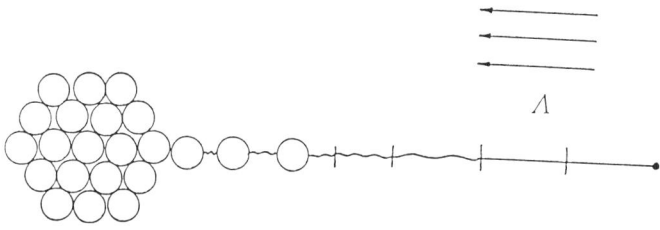

Figure 3. A coexistence of different configurational states is possible in a polysoap unfolded by uniform flow.

cle. The main point is that by increasing V beyond V_c the chain undergoes progressive stretching. Initially the extended chain consists of a globule, a stretched string of micelles, an extended dissociated segment and a fully extended dissociated region. Eventually, as V increases further, the chain as a whole becomes fully dissociated.

Phase separation in Solutions of Polysoaps

Thus far, our discussion focused on the configurations of isolated polysoaps immersed in a good solvent for the backbone. It was further limited to the case of "strong" amphiphiles, where all the polymerized amphiphiles self assemble into micelles *i.e.*, the unperturbed chain does not contain dissociated amphiphiles. This limit corresponds to the case of large δ and relatively short spacer chains. In this situation, the exchange interactions that favor the globular configuration also give rise to phase separation involving a dense gel phase of closely packed micelles and a dilute phase of isolated globules. The volume fraction of polymer in the dilute phase is roughly

$$\phi_{dilute} \approx N \exp(-F_s/kT) \tag{28}$$

where F_s is the surface free energy of the globule, $F_s/kT \approx \gamma_m R_{globule}^2 \approx m^{2/3} p^{-1/6}$. This situation is reminiscent of the phase separation of homopolymers in a poor solvent [27]. It is however important to note two differences. First, both the isolated globules and the dense phase are locally swollen by the solvent. Second, the dense phase forms a physical gel.

The consideration described above are limited to the case of fully assembled polysoaps. It is possible to extend this picture to the general case by utilizing a simple theoretical approach, the p-cluster model [13], proposed by de Gennes in a different context [15, 16]. Within this approach the familiar Flory free energy per unit volume is supplemented by a correction term to the pth term in the virial expansion

$$F_{pc}/kT \approx -\rho(T)\phi^p \tag{29}$$

F_{pc} allows for the free energy gain due to formation of stable clusters, micelles, consisting of p monomers. In the simplest situation, of $N \to \infty$ and an athermal solvent, $\chi = 0$, the augmented Flory free energy is

$$F/kT = (1-\phi)\ln(1-\phi) + \rho(\phi - \phi^p) \tag{30}$$

where the $\rho\phi$ term ensures that $F = 0$ for $\phi = 0$, as required from a mixing free energy. Three regimes are possible within this picture. When $0 < \rho < \rho_c$ the uniform solution is stable for any ϕ. The poor solvent scenario considered above is recovered when $\rho > \overline{\rho}$. In the limit of $N \to \infty$ this involves a coexistence of a dense phase, ϕ_+ with a neat solvent $\phi_- \approx 0$. At the intermediate range, $\rho_c < \rho < \overline{\rho}$ the p-cluster model predicts a novel coexistence of a dense phase, ϕ_+ and a semidilute phase $\phi_- > 0$. In the high p case, of interest here, $\rho_c \approx e/p$ and $\overline{\rho} \approx 1 - \exp(-p)$. In its original version, the p-cluster model does not specify the phenomenological parameters p and ρ. These constants can be however determined within the framework described in the earlier sections leading to

$$\rho = \frac{\exp(-p\epsilon)}{n^p} \qquad (31)$$

When the spacer chains are short enough this reduces to

$$\rho = \left(\frac{1}{n\phi_{cmc}}\right)^p \qquad (32)$$

where ϕ_{cmc} is the critical micelle concentration of the unpolymerized amphiphiles. Not surprisingly, ρ increases as ϕ_{cmc} decreases, since ϕ_{cmc} is a measure of the strength of the amphiphilic stickers. ρ decreases as n, and the coronal penalty, increase. Note that (32) should be modified for high n so as to allow for the $\kappa \gg 1$ behavior of the intrachain micelles. This does not however modify the trends described above. It is also important to notice that in the limit of $N \to \infty$ considered above there is no distinction between a dense phase and a gel.

Discussion

A number of assumptions underlie the theoretical model presented above. The consequences of these assumptions, the resulting constraints on the applicability of the model, are not of equal importance. Three assumptions are of significance: First, both the hydrophilic backbone and the hydrophobic tails were considered as flexible. Second, long range electrostatic interactions were assumed to be screened. Third, the length of the spacers between the stickers was taken to be long and monodispersed. The flexibility of the backbone enables intrachain micellization. Clearly, intrachain self assembly is impossible when the backbone is perfectly rigid and the distance between the hydrophobic moieties is large. The assumed flexibility of the hydrophobic tails is crucial for the utilization of the micellar model of "simple" amphiphiles . The packing constraints of rigid hydrophobic moieties are different. The presence of long ranged electrostatic interactions is expected to modify the configurations of the micellar string. It favors extended linear configurations [39]. The model of intrachain micelles having a star like corona is reasonable only when the spacer chains are long. Finally, when the stickers are placed in blocks along the chain, the intrachain self assembly will be modified. In such cases it is necessary to allow for blocks that do not undergo micellization.

In marked contrast, the limitations due the assumed molecular architecture are less stringent. Our discussion begun with the case of monomeric amphiphiles whose ionic head-groups are joined by flexible and neutral spacer chains. However, the case of graft copolymers with neutral flexible backbone and simple hydrophobic side chains, with no head groups, is described by the $\kappa \gg 1$ limit of our model. In this limit, the micellar structure reflects only the balance of the surface tension of the core and the coronal penalty. This also corresponds to the case of polyelectrolyte backbone carrying flexible hydrophobic grafts in solutions of high ionic strength when long range electrostatic repulsions are screened. Finally, it is straightforward to apply this model to the case of multiblock copolymers comprising of hydrophilic and hydrophobic blocks. With the important caveats noted earlier, concerning flexibility *etc*, our considerations are generic to a wide class of polymeric architectures leading to the formation of intrachain micelles.

While our consideration are aimed at waterborne systems, it should be stressed that our model does not explicitly allow for the distinctive aspects of aqueous solutions. Two main issues are involved. The first concerns the modeling of neutral, water soluble polymers, such as poly(ethylene oxide) (PEO), that owe their solubility to the formation of hydrogen bonds. A full description of aqueous solutions of PEO typically allows the monomers to exist in two different states [15][40][41][42][43]. In our model the hydrophilic backbone behaves as a simple polymer in a good solvent. This assumption in not valid at elevated temperatures and pressures. It is an uncontrolled but convenient approximation that captures the role of the configurational entropy of the chain. Second, the hydrophobic effect plays a crucial role in the micellization. It is involved in determining δ, γ and a_0. Clearly, our approach does not explicitly allow for the detailed aspects of the hydrophobic effect. Rather, δ, γ and a_0 are treated as phenomenological parameters. However, these parameters can be independently determined in the case of polysoaps, as defined in the introduction, from the experimental data on the aggregation behavior of the unpolymerized amphiphiles. For example, it is possible to consider on this basis the effect of temperature on the stability of intrachain micelles [44].

References

[1] Glass, J.E. Ed. *Polymers in Aqueous Media: Performance Through Association*, ACS press, Washington, **1989**.

[2] Glass, J.E. Ed. *Hydrophilic Polymers: Performance with Environmental Acceptance*, ACS press, Washington, **1996**.

[3] Tanaka, F. *Adv. Colloid Interface Sci.* **1996**, 63, 23

[4] Rubinstein, M.; Dobrynin, A.V. *Trends Polym. Sci.* **1997**, 5, 181

[5] Borisov, O.V.; Halperin, A., *Curr. Opin.Colloid Interface Sci.*, **1998**, *3*, 415.

[6] Laschewsky, A., *Adv. Polym. Sci.*, **1995**, *124*, 1.

[7] Ringsdorf, H.; Schlarb, B.; Venzmer, J., *Angew. Chem. Int. Ed. Engl.* **1988**, *27*, 113.

[8] Borisov, O.V.; Halperin, A., *Langmuir*, **1995**, *11*, 2911.

[9] Borisov, O.V.; Halperin, A., *Macromolecules*, **1996**, *29* 1996.

[10] Borisov, O.V.; Halperin, A., *Europhys. Lett.*, **1996**, *34*, 657.

[11] Borisov, O.V.; Halperin, A., *Macromolecules*, **1997**, *30*, 4432.

[12] Borisov, O.V.; Halperin, A. *Euro. Phys. J. B* (in press)

[13] Borisov, O.V.; Halperin, A. *Macromolecules* (in press)

[14] Borisov, O.V.; Halperin, A *Phys. Rev. E* **1998**, 30, 812,

[15] de Gennes, P.-G. *CR Acad. Sci, Paris II*, **1991**, *117*, 313.

[16] de Gennes, P.-G. *Simple Views on Condensed Matter*, World Scientific, Singapore, **1992**.

[17] Israelachvili, J.N., *Intermolecular and Surface Forces*, 2nd edition, Academic Press, London, **1992**

[18] Daoud, M.; Cotton, J.P., *J.Phys. France*, **1982**, *43*, 531.

[19] Halperin, A.; Tirrell, M.; Lodge, T.P. *Adv Polym. Sci.* **1992**, 100, 31.

[20] Zhulina, E.B., *Polymer Science USSR*, **1984**, *26*, 885.

[21] Witten, T.A.; Pincus, P.A., *Macromolecules*, **1986**, *19*, 2509.

[22] Witten, T.A. *J. Phys. France* **1988**, 49, 1056.

[23] Semenov, A.N.; Joanny, J.-F.; Khokhlov, A.R., *Macromolecules*, **1995**, *28*, 1066.

[24] Halperin, A.; Zhulina, E.B., *Europhys. Lett.*, **1991**, *15*, 417.

[25] Halperin, A.; Zhulina, E.B., *Macromolecules*, **1991** *24*, 5393.

[26] Pincus, P.A., *Macromolecules*, **1977**, *10*, 210.

[27] de Gennes, P.-G., *Scaling Concepts in Polymer Physics*, Cornell Univ.Press, Ithaca, **1985**.

[28] Rief, M.; Fernandez, J.M.; Gaub, H.E. *Phys. Rev. Lett.* **1998**, 81, 4764.

[29] Frank, F.C.; Keller, A.; Mackley, M.R., *Polymer*, **1971**, *12*, 476.

[30] Odell, J.A.; Keller, A.; Muller, A.J in Glass, J.E. Ed. *Polymers in Aqueous Media: Performance Through Association*, ACS press, Washington, **1989**.

[31] de Gennes, P.-G., *J. Chem. Phys.*, **1967**, *87*, 962.

[32] de Gennes, P.G., *J. Chem. Phys.*, **1974**, *60*, 5030.

[33] Smith, S.B.; Finzi, L.; Bustamante, C., *Science* **1992**, *258*, 1122.

[34] Perkins, T.T.; Quake, S.R.; Smith, D.E.; Chu, S., *Science* **1994**, *264*, 819

[35] Brochard-Wyart, F., *Europhys. Lett.*, **1993**, *23*, 105.

[36] Buguin, A.; Brochard-Wyart, F., *Macromolecules*, **1996**, *29*, 4937.

[37] Borisov, O.V.; Darinskii, A.A.; Zhulina, E.B., *Macromolecules*, **1995**, *28*, 7180.

[38] Borisov, O.V.; Darinskii, A.A., in *Flexible Polymers in Elongational Flow: Theories and Experiments*; Nguyen, T.Q.; Kausch, H.H., Ed., Springer Verlag, Heidelberg, **1999**

[39] Turner, M.S.; Joanny, J.-F., *J.Phys.Chem.*, **1993**, *97*, 482.

[40] Karlstrom, G. *J. Phys.Chem.* **1985**, 89, 4962.

[41] Matsuyama, A; Tanaka, F. *Phys. Rev. Lett.* **1990**, 65, 341.

[42] Bekiranov, S.; Bruinsma, R.; Pincus, P. *Europhys. Lett.* **1993**, 24, 183.

[43] Bekiranov, S.; Bruinsma, R.; Pincus, P. *Phys. Rev. E* **1997**, 55, 577.

[44] Leckband, D.; Borisov, O.V.; Halperin, A. *Macromolecules*, **1998**, 31, 2368.

Chapter 8

Solution Structure and Shear Thickening Behavior of Ionomers and Hydrophobically Associating Polymers

Srinivas Nomula[1], Sharon Ma[1], and Stuart L. Cooper[2,3]

[1]Department of Chemical Engineering,
University of Delaware, Newark, DE 19716
[2]Department of Chemical Engineering,
Illinois Institute of Technology, Chicago, IL 60616

Solution properties of ionomers as well as hydrophobically modified associating polymers are investigated. Model polyurethane ionomers were synthesized with regularly spaced ionic groups along the polymeric backbone. Light scattering and fluorescence applied to these ionomers in a variety of solvents revealed that in addition to the polarity of the solvent, polymer-solvent interactions and polymer ionic content are critical factors in determining the solution behavior of ionomers. The results provide insight into the controversial origin of polyelectrolyte behavior. The effect of deformation on the structure and chain stretching of hydrophobically terminated associating polymers is determined through rheological studies. It is found that both association strength and polymer chain length play a key role in shear thickening behavior.

Ionomers contain a small number of ionic groups, up to 15 mol %, attached to nonionic backbone chains. Ionomer solutions are commercially very important, for instance as stabilizers for oil based paint suspensions and as coating materials. Though extensive studies have been performed in the bulk, solution properties of ionomers have only recently been investigated (*1-3*). It has been well established that ionomer solutions show two types of behavior depending on the polarity of the solvents used (*4*), namely 1. aggregation due to dipolar attractions between chain segments in non-polar or low-polarity solvents, and 2. polyelectrolyte behavior due to Coulombic interactions in high-polarity solvents. Comparison of the solution behavior

---

[3]Corresponding author.

exhibited by the ionomer solutions to that observed by the corresponding nonionic form of the polymer provides evidence of the unique structure of the ionomer solutions.

The association of ionomer molecules in low-polarity solvents results from strong dipole-dipole interactions among the ionic groups. Evidence of intermolecular association in ionomer solutions in low-polarity solvents has been most dramatically demonstrated by a significant enhancement in the viscosity of the ionomer in solution compared to the parent polymer (*4-8*). Light and neutron scattering data obtained on these solutions indicated the formation of multimers and also showed that the extent and average size of the multimers increases with concentration (*9, 10*). In highly polar solvents, ionomers behave identical to aqueous polyelectrolyte solutions (*4, 11-13*) and exhibit a dramatic increase in reduced viscosity with decreasing polymer concentration. This is quite surprising since ionomers contain relatively low ionic content while polyelectrolytes contain very high ionic content, usually one ionic group per repeat unit. The identification of polyelectrolyte behavior in high polarity ionomer solutions has also been supported by static and dynamic light scattering (*9, 12-14*) and small angle neutron scattering (SANS) experiments (*10, 15*) that show features that are characteristic of polyelectrolyte solutions. Hara recently reviewed the behavior of polyelectrolytes and ionomers in non-aqueous solution and demonstrated that ionomers offer good model systems to investigate the characteristic behavior of salt-free polyelectrolyte solutions (*16*).

Previous studies on the solution properties of ionomers have focused primarily on random copolymer ionomers and model telechelic ionomers. Because of the imposition of ionic interactions over polymer-solvent interactions, a model ionomer system would be preferable for these studies. Telechelic ionomers have ionic groups located at the ends of the polymer chains and are typically of low molecular weight in order to achieve a sufficiently high ionic content. Also, the influence of chain length and ionic content on the morphology of ionomers cannot be studied independently using these systems. Model polyurethane ionomers, developed based on polyurethane chemistry, have regularly spaced ionic groups along the polymer backbone and allow a direct correlation of ionomer structure and properties (*17*). In the first part of the paper, solution structure of ionomers is investigated through studies of model polyurethane ionomers in a variety of solvents.

The random association of macromolecules observed in solutions of ionomers in low polarity solvents also takes place in solutions of multiblock copolymers in selective solvents and hydrophobically modified water-soluble polymers in water. Water-soluble associating polymers containing hydrophobic groups show enhanced viscosity and more elastic properties in aqueous solution, similar to the behavior of ionomers in low polarity solvents. It is generally accepted that such unique solution behavior is due to the formation of a three dimensional network, with the chains connected through micelles formed from the association of the hydrophobic groups. A manifestation of such a complex structure is demonstrated in the concentrated solution regime where the extent of association is large. At concentrations above the overlap concentration, these solutions have been shown to display shear thickening (i.e., an increase in viscosity with shear rate), a phenomenon that is not observed in the parent polymer. Hexadecyl-terminated poly(ethylene oxide) displays shear-thickening behavior similar to ionomer solutions. At low shear rates the solutions exhibit Newtonian behavior, as shear rate increases shear thickening takes place, as evidenced by an increase in viscosity with shear in the shear rate regime 1 to 100 sec^{-1}, followed by shear thinning at higher shear rates (*18*). With increasing concentration the onset of shear thickening occurs at lower shear rates, accompanied by enhancement of the solution viscosity. It has been noted that the maximum in viscosity occurs in a narrow shear stress range, indicating that the structure that causes shear thickening may be lost at a critical shear stress.

Due to the importance of this behavior for potential applications, there have been many studies on shear thickening of associating polymers both theoretically (*19-28*) and experimentally (*18, 29-35*). However, most studies qualitatively explain shear thickening as arising due to a transition from intra- to inter-molecular association. There is also a lack of quantitative comparison between experimental results and theoretical predictions. Telechelic polymers (polymers with associating groups at each chain end) make useful model systems for studying shear thickening due to simplicity of the architecture and known location of the associating groups. In the second part of this paper shear-thickening behavior is investigated through studies of telechelic ionomers in toluene and hydrophobically end-capped poly(ethylene oxide) (PEO) in aqueous solution.

Experimental Details

Light Scattering. Static and dynamic light scattering measurements were conducted with a BI-200SM goniometer (Brookhaven) and a BI-9000AT digital correlator (Brookhaven) at a wavelength of 488 nm from a 300 mW Ar Lexel laser. Measurements were conducted at 35±0.1 °C. Angular measurements were made using a goniometer in the range of scattering angles from 30° to 150°. Solutions were filtered using 0.45-µm and 1.0-µm syringe filters (Gelman Sciences, Ann Arbor, MI). In dynamic light scattering measurements, the channel numbers were 256, of which 6 data channels were used to determine the baseline. To eliminate dust effects, the data were discarded whenever the difference between measured and calculated baselines was greater than 0.1%. Sample times from 0.1 µsec to 1 sec were used.

Fluorescence. Fluorescence spectra were measured on a SLM Aminco 8100 spectrofluorometer. Quartz cuvettes filled with 2.5 mL aliquots of the ionomer solution were used. The excitation wavelength was 385 nm and emission intensities were measured between 400 and 650 nm.

Rheology. A Bohlin VOR with a C-14 couette geometry was used to obtain the steady-shear and dynamic data at 25 °C. Prior to the oscillatory measurement, a strain sweep test was run to ensure the dynamic test was in the linear viscoelastic region. For ionomer studies, the rheometer was placed in an environmental chamber with less than 5% relative humidity.

Results and Discussion

Structure of Polyurethane Ionomers in Solvents of Varying Polarity

The synthesis and characterization of model sulfonated polyurethane ionomers is described in detail elsewhere (*17*). The strucure is shown in Figure 1. Poly(tetramethylene oxide) (PTMO) of molecular weights 1000 and 4500 were used and the ionomers are designated PU1 and PU4 respectively.

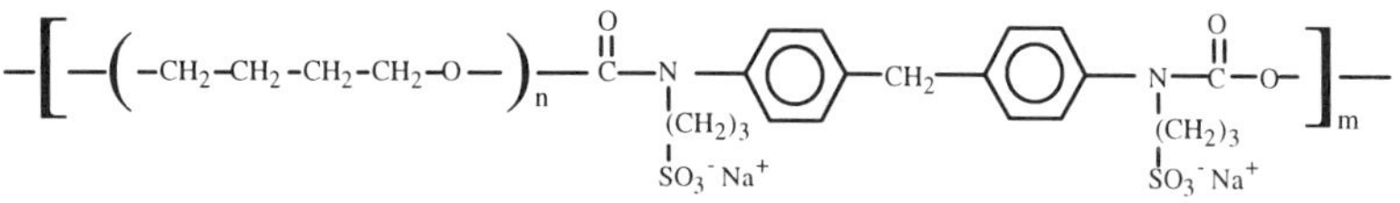

Figure 1. Structure of model polyurethane ionomer.

The influence of the solvent characteristics on the ionomer solution behavior has not received serious consideration since the solvents were limited to those that can dissolve the ionomers. Hence, ionomer solution studies have been mainly carried out either only in nonpolar solvents or only in polar solvents. By varying the distance between the ionic groups model polyurethane ionomers can be dissolved in a wide variety of solvents. Thus PU4 was studied in toluene, tetrahydrofuran (THF), propanol, 1-methyl-2-pyrrolidinone, dimethylacetamide (DMAc), formamide, N-methyl formamide (NMF), and mixtures of toluene/dimethylacetamide. Three distinct regimes have been observed in the solution behavior. In low-polarity solvents (toluene and THF with dielectric constant, $\varepsilon = 2.5$ and 8 respectively), single chains as well as aggregates consisting of physically crosslinked chains are present (*36*). In high-polarity solvents (higher dielectric constant than that of water, formamide and N-methylformamide with dielectric constant, $\varepsilon = 109$ and 182 respectively) which are usually poor solvents for the backbone, hydrophobic aggregation takes place.

Ionomers in polar solvents show polyelectrolyte behavior. The behavior of polyelectrolytes with added salt is well understood in terms of the screening effect by simple ions of electrostatic interactions among fixed ions and can be described by scaling theory developed for neutral polymer solutions (*37*). Although the polyelectrolyte behavior of salt free aqueous polyelectrolyte solutions has been extensively studied, its nature is not clear yet. Several models have been proposed to explain the formation and structure of some kind of ordered clusters in their solutions; however, the molecular structure and thermodynamic driving force of such ordered structures remain as unsolved issues.

Viscometric and DLS measurements of PU1 as well as PU4 were used to demonstrate that sulfonated polyurethane ionomers in DMAc, which is a polar solvent, show polyelectrolyte behavior (*17*). Viscometric measurements showed an upturn in reduced viscosity at very low concentrations and CONTIN analysis of the

DLS data revealed two diffusive modes for the ionomer solutions. The fast mode increases with concentration and it is well accepted that it arises due to intermolecular electrostatic interactions. The slow mode is independent of concentration within the concentration range investigated and corresponds to the existence of large scale "heterogeneities", whose origin, nature and structures are currently controversial. The slow mode has been attributed to filterable aggregates (38, 39), clusters of higher ionic content polymer chains, particles formed due to hydrophobic aggregation, and aggregates which take considerable time to dissolve. It has been shown that polar solutions of ionomers contain free polyions as well as loose aggregates, aggregates which consist of polyions and counterions held together due to electrostatic interactions (17).

The polyurethane ionomer, PU4, is studied in different polar solvents: 1-propanol, 1-methyl-2-pyrrolidinone, and DMAc. Diffusion coefficients extrapolated to zero angle, corresponding to the fast mode in different polarity solvents, are shown in Table I. The fast mode diffusion coefficient can be seen to increase with the polarity of the solvent showing that the diffusion coefficient of single polyions is enhanced by electrostatic interactions (so, the fast mode is dominated by coupled polyion-polyion dynamics). Since as the solvent changes, specific solvent properties like viscosity change, a parameter *PE*, defined as the ratio of the fast mode diffusion coefficient (D_f) to that of the diffusion coefficient of the unionized polymer (D_{un}) is used to examine the effect of solvent polarity. *PE* is a measure of the increase in D_f due to ionic interactions. From Table I it can be seen that *PE* increases with the polarity of the solvent showing that the diffusion coefficient of single polyions is enhanced by electrostatic interactions and ionomer solution behavior is influenced by a change in the dielectric constant of the solvent.

Table I. Diffusion coefficients corresponding to the fast mode (extrapolated to zero angle) for PU4 at 0.2 g/dl in different polar solvents.

Solvent	*Dielectric constant*	D_f (cm²/s)	D_{un} (cm²/s)	*PE*
1-Propanol	18	3×10^{-7}	1.8×10^{-7}	1.67
1-Methyl-2-Pyrrolidinone	32	6×10^{-7}		
N,N'-Dimethylacetamide	37	8×10^{-7}	2×10^{-7}	4.0

The change in the loose aggregate as a function of solvent polarity can be monitored in terms of hydrodynamic radius. The hydrodynamic radii calculated from the slow mode diffusion coefficients are shown in Table II. The size of the loose aggregate first decreases and then increases with dielectric constant. The results can be understood in terms of the interaction of the solvent with the ionomer backbone. 1-

Methyl-2-pyrrolidinone is a poor solvent for the ionomer backbone and hence the aggregate is smaller. Considering propanol and DMAc, both of which are good solvents for the backbone, it can be seen that the loose aggregate size increases with dielectric constant. Thus the size corresponding to the slow mode responds to the solvent-backbone interactions and provides evidence for the presence of loose aggregates in polar solutions of ionomers.

Table II. Hydrodynamic radii corresponding to the slow mode for PU4 at 0.2 g/dl in different polar solvents.

Solvent	Dielectric constant	R_h (nm)
1-Propanol	18	75
1-Methyl-2-Pyrrilidinone	32	50
N,N'-Dimethylacetamide	37	100

Viscometric measurements of PU4 solutions in NMF indicated absence of polyelectrolyte behavior. Static and dynamic light scattering applied to these solutions revealed that spherical particles of radius 30 nm with a polymer core and outer ionic shell are formed (40). The particle formation is driven by hydrophobic interactions since the solvent NMF is a poor solvent for the PU4 backbone. Absence of polyelectrolyte behavior was explained in terms of the screening length for electrostatic interactions being less than the interparticle distances in the solution. Aqueous solutions of Nafion show similar behavior. Fluorescence measurements provide valuable complementary information to that obtained from scattering studies (41-43). Hydrophobic aggregation in high-polarity solvents (40) is studied using 8-anilino-1-naphthalene sulfonic acid (ANS). Fluorescence intensities of ANS, for solutions of PU1 and PU4 in DMAc as well as NMF, normalized by the intensities from the solvent containing only the probe are shown as a function of concentration in Figure 2. The intensities plotted are emission intensities at 463 nm and 484 nm for DMAc and NMF solutions, respectively. The emission intensities of PU4 solutions in NMF rapidly increase with increase in concentration indicating hydrophobic aggregation. For DMAc solutions, addition of ionomer does not make any difference to the emission intensities providing evidence for the absence of hydrophobic aggregation. For PU1 solutions in NMF the emission intensity increases very slowly with concentration indicating only a small amount of hydrophobic aggregation. The results confirm how the nature of the aggregate changes with solvent polarity. In perfluorosulfonated ionomer solutions electrostatic persistence length arguments were used to explain both the increase in polyelectrolyte effect with dielectric constant of alcohols and the disappearance of the effect in high polarity solvents (44, 45). From these results it can be seen that change in solution structure is responsible for the disappearance of polyelectrolyte behavior in highly polar solvents.

Recent studies with higher ionic content ionomer PU1 in NMF have shown that at sufficiently high ionic content, ionomer solutions exhibit polyelectrolyte behavior even if the solvent is a poor solvent for the backbone (46). Previously, ionomer solution behavior has been classified based solely on the polarity of the solvent. (4) From these results it can be seen that, in addition to the polarity of the solvent, polymer-solvent interactions (interactions between the ionomer backbone (the non-ionic part of the ionomer) and the solvent) and the ionic content of the ionomer are critical in determining the solution behavior of ionomers.

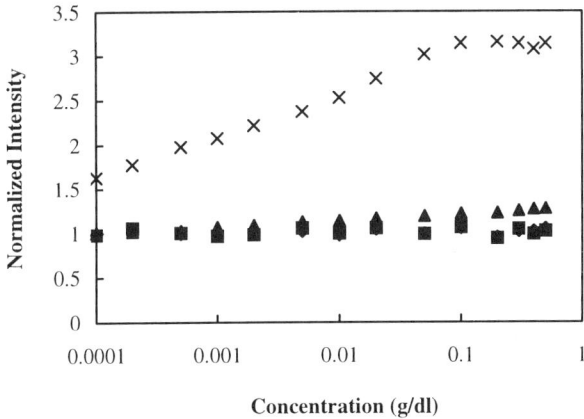

Figure 2. Normalized fluorescence intensities as a function of concentration PU1 in DMAc (♦) PU4 in DMAc (■) PU1 in NMF (▲) and PU4 in NMF (✗).

Shear Thickening of Associating Polymers

Although rarely observed in common polymer melts or solutions, shear thickening effects have been observed in complex fluids including dense suspensions, worm-like micelles and associating polymers solutions. Several theoretical models have been proposed to describe the steady shear or dynamic behavior of associating polymer solutions (19, 23-25, 47-54). These theories differ in the mechanisms they propose for chain conformation or change in aggregation under deformation. Witten and Cohen first proposed a mechanism of shear thickening in the framework of a mean-field approximation (19). They showed in their calculation that the shear flow can increase the probability of interchain association over that of intrachain association, thus leading to the viscosity increase. Ballard et al. extended Flory's theory of the expansion parameter to the self-associating polymers, and predicted that shear thickening will occur in a flow field of sufficiently high strain rate (22). Their proposed mechanism for shear thickening is similar to that of Witten and Cohen. Other models are based on the existence of a transient network in solution. Tanaka and Edwards developed a transient network theory to describe the dynamic properties of physical gels (23). However, it did not predict shear thickening behavior. Wang modified TE model by introducing free chains into the system, and predicted that shear thickening is the result of more chains participating in the network (24). Marrucci et al. explained shear thickening as arising due to a non-Gaussian chain stretching effect (25). They also showed that fast recapture of the end-group by the network can produce pronounced shear thickening. The network relaxation time depends on both the average life time of an association and the proximity of the neighboring aggregates. The experimental data that could quantitatively support various theories are scarce and therefore the actual mechanisms that lead to the observed behavior are still unknown. Success in the development of an accurate analytical description of the shear-thickening phenomenon lies in closer examination of the existing experimental data and further experimental data collection. In the following sections, studies of telechelic ionomers in toluene and hydrophobically end-capped poly(ethylene oxide) (PEO) in aqueous solution will be discussed.

It has been shown that solution behavior of ionomers is very sensitive to the addition of small amount of polar solvents. Quantitative effects of water on the viscosity and rheology of magnesium-and sodium-neutralized dicarboxy-terminated polybutadiene in toluene were reported (55). A change in the water concentration from 50 to 250 ppm led to a viscosity decrease of 2 orders of magnitude for solutions near the gel concentration (See Figure 3). The dynamic measurements showed that the viscosity reduction was related to a reduction in both the plateau modulus and the terminal relaxation time. The magnitude of shear-thickening decreased and the onset of shear-thickening shifted to higher shear rates with increasing water content (Figure 4). Because of the hygroscopic nature of ionomer solutions, rheological experiments were conducted in a humidity-free environment. Rheological studies on telechelic ionomers on effects of concentration show an increase in relaxation time and an increase in the magnitude of shear thickening with concentration (56), which are in qualitative agreement of the prediction of theory of Marrucci et al. (25)

Above study on telechelic ionomers showed that moisture readily solvates the ionic groups and greatly influences the solution behavior. Hydrophobically end-capped poly(ethylene oxide) (PEO) in aqueous solution is thus a good candidate for a more quantitative study. Furthermore, PEO can be synthesized through anionic polymerization, enabling study of narrow molecular weight distribution polymers. Since most of the network models assume that all polymer chains in the network have the same length, hydrophobically end-capped monodisperse PEO will provide a better system to compare with theoretical predications. PEOs used in this study have a polydispersity of 2 (Polysciences Inc.). The effect of polydispersity of polymeric backbone will be discussed in detail in a future publication.

Synthesis of hydrophobically modified PEO is reviewed elsewhere. (58). It can be synthesized in mainly two ways. One is to react hydroxy-terminated poly(ethylene oxide) with a large excess (200-fold molar excess) of a diisocyanate such as isophorone diisocyanate (IDPI), followed by purification and reaction with a long chain alcohol (34). Another is to utilize urethane chemistry through a one-step reaction as shown in figure 5. In this study, we chose the second method as the direct addition of monoisocyanate to PEO guarantees the telechelic structure. The end-

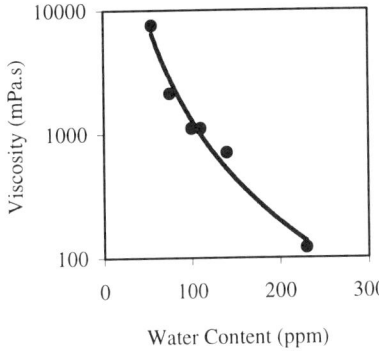
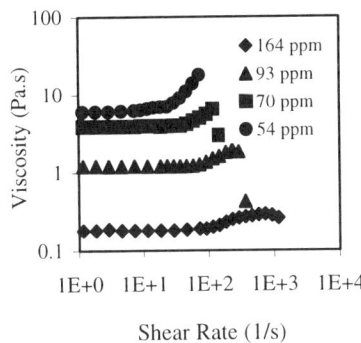

Figure 3. Effect of water content on zero-shear viscosity of 4.91 g/dl Mg(110)-neutralized dicarboxy-polybutadiene in toluene.

Figure 4. Effect of water content on shear-thickening of 4.91 g/dl Mg(110)-neutralized dicarboxy-polybutadiene in toluene.

capping efficiency, determined by ^{1}H NMR was close to 100% within experimental error.

$$HO-PEO-OH + 2\, C_nH_{2n+1}NCO \longrightarrow$$

$$C_nH_{2n+1}NHCOO-PEO-OOCNC_nH_{2n+1}$$

Figure 5. Synthesis of hydrocarbon-terminated poly(ethylene oxide)

Polymer solution at rest

Compared to the parent polymers, the associating polymers show large viscosity enhancement. It is believed that this large enhancement in viscosity can be attributed to the formation of a network. Telechelic polymers with sufficiently long hydrophobic end groups associate to form micelles even at very low concentration. As concentration approaches the overlap concentration (ϕ^*), micelles begin to connect to each other through bridge chains. Above ϕ^*, a transient network is formed with the fraction of bridge chains increasing with an increase in polymer concentration. Various spectroscopic studies, such as fluorescence (59-61) and dynamic light scattering (62, 63) have provided evidence for this gradual structure change with concentration.

Not only is the viscosity much larger than that of parent polymers, these associating polymers also show a much stronger concentration dependence of zero-shear viscosity. At high enough concentration, where the hydrodynamic relaxation time of the network is longer than the relaxation time of the association, transient network theory can be used to describe the behavior of these associating polymers. In the transient network theory they developed, Tanaka and Edwards calculated the stationary properties in terms of two parameters -- the chain breakage function $\beta(r)$ and the recombination probability p of the sticky dangling ends (23). They predicted that the zero shear viscosity is proportional to the equilibrium number of the active chains, which is, in turn, linearly proportional to total number of chains, or concentration.

$$\eta_0 = \frac{v_0}{\zeta_0}\sigma_1 = \frac{p\zeta_0}{1+p\zeta_0}n \qquad (1)$$

where v_0 is the equilibrium number of the active chains, ζ_0 is the average of the reciprocal of the chain breakage function $\beta(r)$ over the Gaussian distribution of polymer chain configuration. Marrucci et al. (25) have predicted the augmented concentration dependence when including the effect of fast recapture of end-groups by the neighboring aggregates. As concentration increases, not only the number of active chains increases, but also the average distance between the aggregates decreases, which leads to an increase in the relaxation time, resulting in a stronger concentration dependence.

$$\eta_0 = vkT\tau = \frac{vkT}{\beta_0}\frac{Nb^2}{a^2} \propto v^{5/3} \qquad (2)$$

where v is the number of elastically active chains, τ is the relaxation time of the network, and a and b are the average distance between the aggregates and persistance length, respectively.

According to the simple rubber elasticity theory (64), the plateau modulus, to the first approximation, is proportional to the number of effectively elastic chains ($G_\infty = vRT$). Thus, zero-shear viscosity should scale with $G_\infty^{5/3}$ as well according to the prediction of Marrucci's model. The plateau modulus was determined by fitting the Maxwell model with a single relaxation time. Despite the complexity of the system, the simple viscoelastic model fits experimental data fairly well over two decades of frequency (see Figure 6). This relatively simple oscillatory shear behavior was also observed by other researchers (18, 65). Figure 7 shows the log-log plot of zero-shear viscosity as a function of plateau modulus at different concentrations of polymer C_{16}-PEO8k-C_{16}. The slope 1.5 is in reasonable agreement with the prediction of Marrucci's model (slope of 1.67).

Terminal relaxation times of the network (τ) were also obtained at different concentrations from the oscillatory shear experiment. τ increases slowly at low concentrations and more rapidly at higher concentrations. As predicted in the theory of Marrucci et al., the terminal relaxation time of the network depends on the proximity of the aggregates and increases with an increase in concentration. At low concentration, the aggregates are isolated and their presence has a relatively small effect on network relaxation time. At higher concentration, the aggregates are packed close to each other so that the "dangling" end-groups are captured into the neighboring aggregates before they get a chance to completely relax, thus increasing the relaxation time.

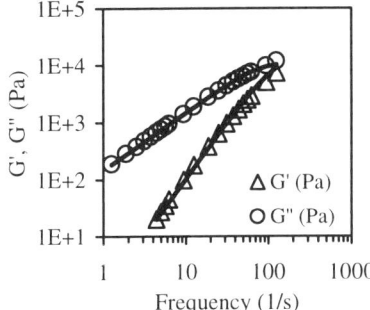
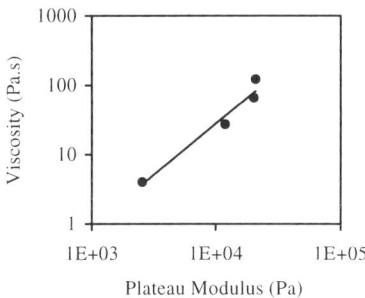

Figure 6. Shear moduli of hexadecyl-terminated PEO with MW of 8000 at concentration of 12 wt%.

Figure 7. Zero-shear viscosity of hexadecyl-terminated PEO with MW of 8000 as a function of plateau modulus at different concentrations.

Polymer solution under shear

As mentioned above, viscosity depends on the number and lifetime of the elastically active chains (Equation 2). At rest, the lifetime of an elastically active chain is governed solely by the association strength of the end group. However, when polymer chain is under shear, shear flow moves the junctions affinely, causing the elastically active chain to stretch. When the tension in the stretched chain reaches a critical value, the chain end dissociates from the junction, decreasing both the overall lifetime and number of elastically active chains. As a result, shear thinning occurs. This is often the case for the telechelic associating polymer with relatively low molecular weight (much lower than the entanglement molecular weight), where the dynamics are dominated by junction formation and destruction. When molecular weight of associating polymer is larger than the entanglement molecular weight, complications arise due to topological entanglements. Reduction in the density of topological entanglements caused by shear flow also leads to shear thinning. At very high shear rate, a catastrophic shear thinning is the result of both the rupture of network and depletion of topological entanglements density. Studies on hydrophobically modified alkali soluble emulsion (HASE) polymers, usually with molecular weight of 10^5 or higher, have shown this effect of topological entanglements (35, 66).

Although shear thinning is universally observed in all systems, shear thickening only takes place in certain system. Figure 8 shows the response of hexadecyl end-capped PEO with various polymer chain lengths under steady shear. Systems with longer chains (20k, 35k) show only shear thinning behavior, while shear thickening was observed in shorter polymer chain (8k). This is consistent with the prediction of non-Gaussian chain stretching. As the elastically active chain is being progressively stretched, a point will be reached where the linear force law no longer holds. Further stretching the chain requires an exceedingly large amount of work (25), and the system exhibits shear-thickening. To allow this non-Gaussian chain stretching, the association has to be strong enough. Even though the end group is the same for three different molecular weights studied, the longer the polymer chain, the less effective the hydrophobic association is. For the polymer with the highest molecular weight, the chain end is pulled out of the junction before the chain is stretched into the non-

Gaussian regime, resulting in the absence of shear-thickening. The same effect of polymer chain length was observed in similar systems (octadecyl (C_{18}) end-capped PEOs) at relatively low concentrations (*34*). It was argued that shear thickening is caused by the transition from intra- to inter-molecular association, and short EtO spacing is in favor of the transition. It has been shown that such transition is more pronounced at low polymer concentration near overlap concentration (*35*). When the polymer concentration is in semi-dilute or concentrated region, intermolecular association dominates. Since the polymer concentration in this study is well above the overlap concentration, the transition from intra- to inter-molecular association is not expected to be a dominant mechanism for shear thickening.

For hydrophobic C_{16} end-group, shear thickening was observed in the system containing PEO with molecular weight of 8000. As shown in Figure 9, the critical shear rate (the onset of shear thickening) shifts to lower shear rates as the concentration increases. In the non-Gaussian chain stretching model, the critical shear rate is estimated assuming that the elastically active chains reach full extension at the onset of shear thickening. The mean square distance of the polymer chain is then on the order of molecular weight (N), and the critical shear rate is approximately $N^{1/2}/\tau$ (*25*). For the same polymer chain length, the critical shear rate will decrease with concentration if the relaxation time of network is an increasing function of concentration. As reported above, the relaxation time of network for this system increases with concentration. The two experimental results are consistent with the non-Gaussian chain stretching model.

Shear thickening was also observed in HASE system (*35*). The shear thickening effect is weaker than that in hydrophobically end-capped PEO, primarily due to the competing effect of topological entanglement and hydrophobic association at moderate shear rate. English et al. (*35*) have studied the solution behavior of comb-like HASE. Their study shows evidence of a shear-induced structure associated with shear thickening at low concentration by fluorescence measurement. It should be noted that this does not exclude the non-Gaussian chain stretching effect. At low polymer concentration, both non-Gaussian chain stretching and transition from intra- to inter-molecular association may play a role in shear thickening effect. At higher polymer concentrations, our rheology studies of hydrophobically end-capped PEO seem to be in favor of non-Gaussian chain stretching as the molecular origin of shear thickening. Due to the fact that hydrophobic association is weaker than ionic association, these model systems show only mild shear thickening. This has thus far prevented further investigation on other important parameters such as molecular weight and association strength. Since fluorocarbons associate more strongly than hydrocarbons in aqueous solution, fluorocarbons will be used as the end-groups to further investigate the mechanism of shear thickening.

In addition to the steady state studies, transient experiments also provide insight into the association of these systems. Most of the proposed mechanisms for shear-thickening can be divided into two categories based on their description of the transient response. Models based on non-Gaussian chain stretching predict a rapid increase in η^+ (the stress growth coefficient in a start-up experiment) to approach a steady-state viscosity while other mechanisms predict a structure change and a more gradual increase in η^+ prior to reaching the steady-state viscosity. Thus the transient stress growth experiment will allow us to differentiate between different mechanisms for shear-thickening. This will be discussed in future work.

Conclusions

Model polyurethane ionomers were synthesized with regularly spaced ionic groups along the polymeric backbone to allow a direct correlation of ionomer structure and properties. Light scattering and fluorescence have been applied to these ionomers in a variety of solvents. Single chains as well as aggregated ionomer chains are present in solution. It is found that in toluene, a low polarity solvent, the aggregates consist of physically crosslinked chains whereas in dimethylacetamide, a polar solvent, the aggregates consist of polyions and counterions held together due to electrostatic interactions. In N-methylformamide, a very highly polar solvent, at high polymeric ionic content, polyelectrolyte behavior is shown and at low polymeric ionic content, hydrophobic aggregation results in the formation of micelles consisting of swollen polyol chains at the core with an outer ionic shell. It is found that in addition to the polarity of the solvent, polymer-solvent interactions and polymer ionic content are critical factors in determining the solution behavior of ionomers.

The effect of deformation on the structure and chain stretching of hexadecyl-terminated associating polymers is determined through rheological studies. As concentration increases, the number of elastically active chains increases and junctions become closer and closer to each other. As a result, the relaxation time of the network increases with concentration. The zero-shear viscosity, a product of number of elastically active chains and network relaxation time, thus shows a strong concentration dependence, which is in quantitative agreement with Marrucci's prediction. PEO with higher molecular weight (>20k) shows only shear

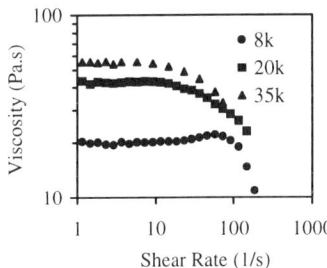

Figure 8. Effect of polymer chain length on shear thickening of hexadecyl-terminated PEO with MW of 8000.

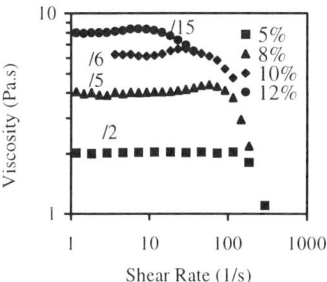

Figure 9. Effect of concentration on shear thickening of hexadecyl-terminated PEO with MW of 8000.

thinning behavior whereas PEO with low molecular weight (8000) shows mild shear thickening over a range of concentrations. The onset of shear thickening decreases with concentration as predicted by theory of Marrucci et al. Longer polymer chains or larger distances between junctions requires larger deformation (or higher shear rate) to stretch the chain into the non-Gaussian regime, in which the shear thickening effect is observable. Higher shear rates however tends to destroy the transient network resulting in shear thinning behavior. Unless the association strength is very strong, where the shear thinning effect can be postponed to much higher shear rate, the two competing effects dictate that in systems with moderate association strength such as hydrophobically end-capped PEO, only mild shear thickening occurs over a limited concentration and molecular weight range. It should be noted that non-Gaussian chain stretching effect does not exclude the "crosslinking" effect. It could well be the two effects are cooperative, especially at low polymer concentration.

Acknowledgements

The authors would like to acknowledge Dr. Chang Zheng Yang of Nanjing University for synthesizing the polyurethane ionomers studied in this work and Dr. Sam Bhargava for the data on the effect of water on the viscosity of ionomer solutions. This work was funded by the National Science Foundation through grant DMR 9531069.

References

1. Lundberg, R. D.; Makowski, H. S. *J. Polym. Sci. Polym. Phys. Ed.* **1980**, *18*, 1821.
2. Siadat, B.; Lundberg, R. D.; Lenz, R. W. *Macromolecules* **1981**, *14*, 773.
3. Gebel, G.; Loppinet, B.; Williams, C. E. *J. Phys. Chem. B* **1997**, *101*, 1884-1892.
4. Lundberg, R. D.; Phillips, R. R. *J. Polym. Sci. Polym. Phys. Ed.* **1982**, *20*, 1143.
5. Tant, M. R.; Wilkes, G. L. *JMS-Rev. Macromol. Chem. Phys.* **1988**, *C28(1)*, 1.
6. Fitzgerald, J. J.; Weiss, R. A. *J. Macromol. Sci.-Rev. Maromol. Chem. Phys.* **1988**, *C28*, 99.
7. Lantman, C. W.; MacKnight, W. J. *Ann. Rev. Mater. Sci.* **1989**, *19*, 295.
8. Lundberg, R. D. In *Structure and Properties of Ionomers*, M. Pineri, A. Eisenberg, Eds.; D. Reidel Publishing Company: Dordrecht, Holland, 1986; vol. 198, pp. 387.
9. Lantman, C. W.; MacKnight, W. J.; Sinha, S. K.; Peiffer, D. G.; Lundberg, R. D. *Macromolecules* **1987**, *20*, 1096.
10. Lantman, C. W.; MacKnight, W. J.; Sinha, S. K.; Peiffer, D. G.; Lundberg, R. D.; Wignall, G. D. *Macromolecules* **1988**, *21*, 1339.
11. Rochas, C.; Domard, A.; Rinuado, M. *Polymer* **1979**, *20*, 76.

12. Peiffer, D. G.; Lundberg, R. D. *J. Polym. Sci. Polym. Chem. Ed.* **1984**, *22*, 1757.
13. Hara, M.; Wu, J. L. *Macromolecules* **1988**, *21*, 402.
14. Hara, M.; Wu, J. L. In *Multiphase Polymers: Blends and Ionomers,* L. A. Utracki, R. A. Weiss, Eds.; American Chemical Society: Washington, D.C., 1989; vol. 395, pp. 446.
15. MacKnight, J.; Lantman, C. W.; Lundberg, R. D.; Sinha, S. K.; Peiffer, D. G. *Polym. Prepr.* **1986**, *27*, 327.
16. Hara, M. "Polyelectrolytes in Nonaqueous Solution", In *Polyelectrolytes: Science and Technology,* M. Hara, Eds.; Marcel Dekker, Inc.: New York, 1993; pp. 193.
17. Nomula, S.; Cooper, S. L. *Macromolecules* **1997**, *30*, 1355-1362.
18. Jenkins, R. D., Ph.D. Thesis, Lehigh University (1990).
19. Witten, T. A.; Cohen, M. H. *Macromolecules* **1985**, *18*, 1915.
20. Baljon, A.; Witten, T. A. *Macromolecule* **1992**, *25*, 2969-2976.
21. Vrahopoulou, E. P.; J., M. A. *J. Rheol.* **1987**, *31*, 371.
22. Ballard, M. J.; Buscall, R. *Polymer* **1988**, *29*, 1287.
23. Tanaka, F.; Edwards, S. F. *Macromolecules* **1992**, *25*, 1516.
24. Wang, S. Q. *Macromolecules* **1992**, *25*, 7003.
25. Marrucci, G.; Bhargava, S.; Cooper, S. L. *Macromolecules* **1993**, *26*, 6483.
26. Groot, R. D.; W.G.M., A. *J. Chem. Phys.* **1994**, *100*, 1657-1664.
27. Ahn, K. H.; Osaki, K. *J. Non-Newtonian Fluid Mech.* **1994**, *55*, 215-227.
28. Semenov, A. N.; Joanny, J. F.; Khokhlov, A. R. *Macromolecule* **1995**, *28*, 1066-1075.
29. Annable, T.; Buscall, R.; Ettelaie, R.; Wittlesone, D. *J. Rheol.* **1993**, *37*, 695-726.
30. Kaczmarski, J. P.; Glass, J. E. *Macromolecules* **1993**, *26*, 5149-5156.
31. Kaczmarski, J. P.; Glass, J. E. *Langmuir* **1994**, *10*, 3035-42.
32. Maus, C.; Fayt, R.; Jerome, R.; Teyssie, P. *Polymer* **1995**, *36*, 2083-2088.
33. May, R.; Kaczmarski, J. P.; Glass, J. E. *Macromolecules* **1996**, *29*, 4745-4753.
34. Tarng, M. R.; Kaczarski, J. P.; Lundberg, D. J.; Glass, J. E. In *Hydrophilic Polymers,* American Chemical Society: Washington, D.C., 1996; vol. 248, pp. 305-342.
35. English, R. J.; Gulati, H. S.; Jenkins, R. D.; Khan, S. A. *J. Rheol.* **1997**, *41*, 427-444.
36. Nomula, S.; Cooper, S. L. *Macromolecules (in preparation)*
37. Mandel, M. In *Polyelectrolytes: Science and Technology,* M. Hara, Eds.; Marcel Dekker, Inc.: New York, 1993; pp. 1.
38. Smits, R. G.; Kuil, M. E.; Mandel, M. *Macromolecules* **1994**, *27*, 5599-5608.
39. Norwood, D. P.; Benmouna, M.; Reed, W. F. *Macromolecules* **1996**, *29*, 4293-4304.
40. Nomula, S.; Cooper, S. L. *J. Coll. Int. Sci.* **1998**, *205*, 331-339.
41. Bakeev, K. N.; MacKnight, W. J. *Macromolecules* **1991**, *24*, 4578-4582.
42. Li, M.; Jiang, M.; Wu, C. *J. Polym. Sci. B: Polym. Phys.* **1997**, *35*, 1593-1599.
43. Li, M.; Jiang, M.; Zhang, Y.; Fang, Q. *Macromolecules* **1997**, *30*, 470-478.
44. Aldebert, P.; Gebel, G.; Loppinet, B.; Nakamura, N. *Polymer* **1995**, *36*, 431-434.

45. Gebel, G.; Loppinet, B. In *Ionomers: Characterization, Theory and Applications*, S. Schlick, Eds.; CRC Press: New York, 1996; pp. 83.
46. Nomula, S.; Cooper, S. L. *Polym. Prepr. (Am. Chem. Soc., Div. Polym. Chem.)* **1998**, *39*, 258.
47. Witten, T. A. *J. Phys. France* **1988**, *49*, 1055.
48. Tanaka, F.; Edwards, S. F. *J. Non-Newt. Fluid Mech.* **1992**, *43*, 289.
49. Tanaka, F.; Edwards, S. F. *J. Non-Newt. Fluid Mech.* **1992**, *43*, 247.
50. Tanaka, F.; Edwards, S. F. *J. Non-Newt. Fluid Mech.* **1992**, *43*, 273.
51. Gonzalez, A. E. *Polym. Prepr.* **1982**, *23*, 58.
52. Gonzalez, A. E. *Polymer* **1983**, *24*, 77.
53. Rubinstein, M.; Leibler, L.; Colby, R. H. *Polym. Prepr.* **1991**, *32*, 443.
54. Leibler, L.; Rubinstein, M.; Colby, R. H. *Macromolecules* **1991**, *24*, 4701.
55. Bhargava, S.; Cooper, S. L. *Macromolecules* **1998**, *31*, 508-514.
56. Bhargava, S.; Mackay, M.; Cooper, S. L. *Macromolecules (in preparation)*
57. Bhargava, S.; Cooper, S. L. *J. Poly. Sci. Polym. Phys. (in preparation)*
58. Wetzel, W. H.; Chen, M.; Glass, J. E. "Associative Thickeners: An Overview with an Emphasis on Synthetic Procedures", In *Hydrophilic Polymers*, American Chemical Society: Washington, D.C., 1996; vol. 248, pp. 163-179.
59. Yekta, A.; Duhamel, J.; Adiwidjaja, H.; Brochard, P.; Winnik, M. A. *Langmuir* **1993**, 881-883.
60. Yekta, A.; Xu, B.; Duhamel, J.; Adiwidjaja, H.; Winnik, M. A. *Macromolecules* **1995**, *28*, 956-966.
61. Xu, B.; Yekta, A.; Li, L.; Masoumi, Z.; Winnik, M. A. *Colloids and Surface A: Physicochemical and Engineering Aspects* **1996**, *112*, 239-250.
62. Maechling-Strasser, C.; Clouet, F.; Francois, J. *Polymer* **1992**, *33*, 1021.
63. Alami, E.; Almgren, M.; Brown, W.; Francois, J. *Macromolecules* **1996**, *29*, 2229-2243.
64. Green, M. S.; Tobolsky, A. T. *J. Chem. Phys.* **1946**, *14*, 80-92.
65. Annable, T.; Buscall, R.; Ettelaie, R.; Whittlestone, D. *J. Rheol.* **1993**, *37*, 695-726.
66. Jenkins, R. D.; DeLong, L. M.; Bassett, D. R. In *Hydrophilic Polymers*, J. E. Glass, Eds.; American Chemical Society: Washington, D.C., 1996; vol. 248, pp. 425-448.

Chapter 9

Determination of Aggregation Numbers in Aqueous Solutions of Hydrophobically Modified Polymers by Fluorescent Probe Techniques

Olga Vorobyova and Mitchell A. Winnik[1]

Department of Chemical Engineering, University of Toronto,
Toronto, Ontario, Canada M5S 3H6

In this chapter we examine fluorescence quenching as a methodology for determining the size of the micelle-like aggregates formed in aqueous solution by associative polymers that modify the solution rheology. The parameter of interest is N_R, the mean number of hydrophobic substituents per micelle. We cite several examples from the literature that indicate the successful application of this method to associating polymer solutions. Other examples point to shortcomings of the method and indicate that the determination of N_R values for associative polymers systems is more complex than the determination of aggregation numbers (N_{agg}) for surfactant micelles. Finally we describe in detail the use of pyrene and 1-ethylpyrene as probes to determine N_R values for three different polymer systems, a telechelic copolymer of polyethylene oxide end-capped at both ends with short alkyl chains (C_{16} and C_{18}), hydrophobically modified hydroxyethyl cellulose HMHEC, and hydrophobically modified alkali swellable emulsion polymer (HASE).

[1]Corresponding author (e-mail: mwinnik@chem.utoronto.ca)

Introduction

Associative thickeners (AT's) are members of a class of associating water-soluble polymers that contain hydrophobic substituents (*1*). When these polymers are added to water in small amounts, the viscosity increases and the solution 'thickens.' The feature of these polymers that makes them interesting is their viscosity profile at different shear rates, a property which is very sensitive to the chemical structure of the polymer and the placement of the hydrophobic substituents (*2*). Experiments carried out over the past several years have indicated that these polymers in water associate through local phase separation of the hydrophobic groups into micelle-like structures. The polymer backbone forms bridges between micelles, and in this way the polymer self-assembles to form a transient network (*3, 4*).

In order to relate rheological properties to structure for these polymers, one of the most important morphological features one needs to characterize is the mean number of hydrophobic substituents N_R that associate to form individual micelle-like aggregates. N_R is an aggregation number analogous to that (N_{agg}) used to describe the number of surfactant molecules per micelle for simple surfactants in water. For many surfactant micelles, N_{agg} has values of 60 to 100 molecules per micelle. For many traditional surfactant systems, these aggregation numbers are relatively easy to determine by scattering methods involving light, X-rays, or neutrons, or by fluorescence quenching measurements. For the micelle-like structures formed by associating polymers, the values of hydrophobe aggregation numbers are more difficult to determine. As a consequence they are known with much less confidence.

Traditional surfactant micelles in water form through a closed association process. Closed association implies not only that micelles form at surfactant concentrations above the critical micelle concentration (CMC), but that at elevated surfactant concentration, it is the concentration of micelles and not their aggregation number which increases. In open association, the micelle size increases with increasing surfactant concentration. The question of open or closed association for hydrophobe aggregation for AT's in water is not well resolved. For some telechelic polymers (polyethylene oxide (PEO) backbone, hydrophobic end groups) there is evidence for closed association over a range of concentrations. The literature on this topic is not very clear. In the first reports, during the early 1990's, different groups using different techniques reached different conclusions about the nature of the polymer association mechanism.

This situation is changing as research groups examine better characterized model polymers and develop a deeper understanding of their methods as applied specifically to associating polymers. Nevertheless, even in the best characterized systems, the results on hydrophobe aggregation numbers N_R are never as clear as the corresponding aggregation numbers N_{agg} that one can determine for surfactant micelles. In this paper, we review a number of examples from the literature and describe in detail several examples from our laboratory to indicate the type of problems one can encounter when using fluorescence quenching methods to determine N_R values of associating polymer in aqueous solution.

Micelle Structure

In the earliest picture of a spherical micelle proposed by Hartley (5) in 1936, the liquid hydrocarbon core of the micelle is surrounded by the hydrophilic surface layer of the polar head groups. Later models took into account the extent of water penetration into the micelle, so that the core of the micelle is now considered as consisting of two parts: the inner core that is essentially water-free, and the outer core or the palisade layer, where the alkyl segments close to the surface have some contact with water (6).

The size and shape of traditional surfactant micelles in water is determined by a balance of two major factors: packing considerations for the hydrophobic tail and the space requirements of the polar head group (7). For sodium dodecyl sulfate (SDS) micelles in water, the C_{12} chains pack into spheres containing ca. 60 molecules, with a very narrow size distribution, surrounded by a shell of head groups consisting of hydrated -OSO_3^- ions. At much higher surfactant concentration or in the presence of salt, the head groups are less hydrated and occupy less space. The system responds by reorganizing into cylinders.

Many non-ionic micelles are formed from surfactants having a water-soluble oligomer or polymer (such as PEO) as the head group. If this polymer is long enough, the space-filling requirements of the head group are influenced by the mean-squared radius of gyration (R_G^2) of the water-soluble polymer. When the area per chain at the surface of a micelle is smaller than R_G^2, the polymer chains become elongated in the direction normal to the surface of the micelle core. This elastic deformation entropy opposes the growth in micelle size, and leads to the formation of smaller "star-like" micelles. Non-ionic micelles often have aggregation numbers which depend sensitively on temperature. For example, Triton X-100 (a nonylphenol ethoxylate with a mean degree of ethoxylation of 9.6) has an aggregation number of 70 at 10° C, 105 at 25° C, and 127 at 34° C (8).

When hydrophobic substituents are attached to water-soluble polymers, the nature of the hydrophobic domains that they form will be profoundly influenced by the nature of the steric requirements of the polymer in the vicinity of the aggregate. Telechelic polymers with hydrophobic groups attached to the ends of a PEO or PEO-like polymer are structurally analogous to traditional non-ionic surfactants. The evidence available for these systems points to the formation of star-like spherical micelle-like aggregates. For other polymers, particularly with hydrophobic groups attached directly to the polymer backbone, the situation could be much more complicated.

Determination of N_{agg} and N_R by Fluorescence Quenching

The fluorescence quenching methodology for determining micelle aggregation numbers was developed the late 1970's for characterizing micelles formed from low molecular weight surfactants (9, 10, 11). In this approach one introduces a small amount of a fluorophore dye (D) and a larger amount of quencher (Q) into a solution containing micelles. The dye and quencher must be solubilized randomly but completely by the micelles so that a negligible fraction remains in the water phase. The principle of the method is that the dye fluoresces normally in a micelle which contains no quencher, but in micelles containing a dye and one or more quenchers, quenching competes with fluorescence. Data analysis begins with the assumption of a Poisson distribution of Q among micelles. With the assumption that neither the quencher nor the excited probe D* escape the micelle during the lifetime of D*, the expression for time resolved fluorescence quenching (TRFQ) is given by eq. (1).

$$I_D(t) = I_D(0) \exp[-t/\tau_D - n(1 - \exp(-k_q t))] \quad (1)$$

Here $I_D(t)$ is the fluorescence intensity of the dye at time t. $I_D(0)$ is its initial intensity; τ_D is the unquenched dye lifetime, and k_q is the pseudo-first order rate constant for quenching in the micelle. In experiments in which one uses excimer formation as the quenching process, $I_D(t)$ refers to the "monomer" fluorescence decay profile. In this case, the dye is commonly excited pyrene, Py*, and ground state Py serves as the quencher. Here one has to be careful to ensure that excimer formation is essentially irreversible on the timescale of the Py* lifetime.

The mean number of quenchers per micelle, n, can be related to the aggregation number through the micelle concentration: n = [Q] / [micelle] = [Q] / (C − CMC)* N_{agg}, where [Q] is the bulk molar quencher concentration; C is the total concentration of surfactant; and CMC is the critical micelle concentration. For traditional surfactants, one determines the aggregation number N_{agg}. For associative polymers, the parameter n is related to the aggregation number of hydrophobic substituents N_R by the equation

$$n = [Q]/[\text{Micelle}] = [Q] N_R / (C_{pol} q_R) \quad (2)$$

where q_R is the alkyl chain content of the polymer (in moles of alkyl groups per gram of polymer) and C_{pol} is the polymer concentration in g/L. In this equation, the critical micelle concentration (or critical association concentration, CAC) is assumed to be negligible compared to C_{pol}. The CAC is not always small, particularly in telechelic polymers with a long water-soluble backbone and relatively weak hydrophobic end groups. The expected behavior of the parameters in the Poisson quenching model is shown in Figure 1.

Aggregation numbers N_{agg} for surfactants, or N_R values for associative polymers are calculated from the micelle concentration:

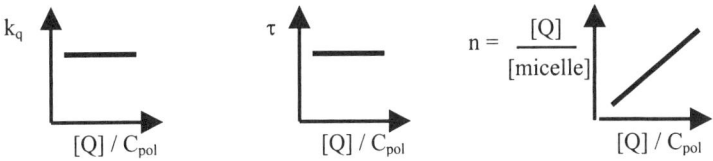

Figure 1. Expected behavior for the parameters in the Poisson quenching model, equation 1: τ is the lifetime of the unquenched probe; k_q is the pseudo-first order rate constant for quenching within the micelle; n is the mean number of quenchers per micelle

$$N_{agg} = (C - CMC) / [\text{Micelle}] \quad \text{and} \quad N_R = C_{pol} \, q_R / [\text{Micelle}] \qquad (3)$$

Novel developments extend this technique to more complicated systems by considering polydispersity effects, non-spherical micelle geometries, and dynamic aspects of the micelles (exit and entry of D* and Q during the lifetime of D*). These advances have recently been reviewed (*12*).The situation where there is a distribution of micelles of different sizes is particularly relevant for associative polymers. For the specific case of a system of micelles of different sizes, where one can assume that the fluorophore and quencher remain within their micelles during the dye excited lifetime, and where the Poisson distribution and k_q for intramicellar quenching are suitably modified by the change in micelle size *(12a)*, individual fluorescence decay experiments will give good fits to eq. 1, but the values of n and k_q will change with variation in the concentration of the quencher. An interesting system that appears to satisfy these requirements involves the micelles formed by SDS in the presence of 1-pentanol and sodium chloride investigated by J. Lang *(13)*.

In some cases, an estimate of the aggregation number may be obtained from steady-state fluorescence quenching (SSFQ) measurements through the expression

$$\ln (I_0 / I) = [Q] / [\text{Micelle}] \qquad (4)$$

where I and I_0 are the fluorescence intensities of the dye with and without quencher present, respectively. Despite the seeming simplicity of the steady-state approach, the assumptions behind equation (4) are much more restrictive than for the time-resolved method. The most significant additional assumption is that the quenching of the excited dye molecules must be so fast that micelles which contain a quencher make no contribution to the observed fluorescence. Recent experiments by Zana and coworkers (14) emphasize that while some steady state measurements provide correct values of N_{agg} for surfactants, most give values that are too small. The only way to establish that the N_{agg} values obtained by SSFQ are correct is restudy the system by TRFQ.

In the sections that follow, we will see that some values for N_R, particularly for AT's in the presence of excess surfactant, have been determined only by SSFQ experiments. In other instances, where TRFQ measurements have been carried out, the data fit eq. (1) but the systems exhibit complexities beyond those found for simple surfactant micelles (15). The message we wish to convey is that these methods for determination of aggregation numbers are by no means routine for associative polymers in water. Experiments have to be carried out over a range of probe, quencher, and polymer concentrations for the results to be accepted with confidence.

Telechelic Polymers

Association Mechanism

Of the various associative polymers that have been studied, we have the deepest understanding of the telechelic associating polymers. These are linear water-soluble polymers, normally based upon PEO, with hydrophobic end groups. Hydrophobic Ethoxylated URethane (HEUR) polymers are prepared by a chain extension reaction of an oligomeric PEO with a diisocyanate, followed by end-capping with an aliphatic alcohol. They are normally characterized by a relatively broad molecular weight distribution ($M_w/M_n \approx 2$) and not all of the polymer molecules contain two hydrophobic end groups. More recent experiments have been carried out on model polymers in which an aliphatic hydrocarbon group is attached directly (via an ether or urethane linkage) to the ends of a PEO. These polymers commonly have a very narrow molecular weight distribution ($M_w/M_n < 1.1$) and close to 2.0 end-groups per chain. Such polymers have been studied by a variety of techniques including rheometry, small angle scattering (SANS and SAXS) measurements (16, 17), calorimetry, electron paramagnetic resonance (EPR) spectroscopy (18), and by fluorescence (19, 20, 21).

A characteristic feature of some telechelic AT's is that they form flower micelles at low concentrations but above the CMC, and then undergo further association at higher concentrations to form an extended network of micelles that contain looping

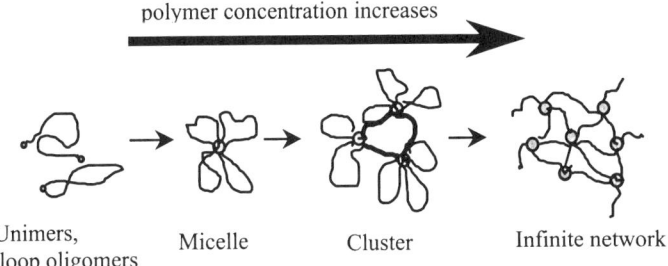

Unimers, loop oligomers   Micelle   Cluster   Infinite network

chains and chains that bridge neighboring micelles. From this point of view, one expects flower micelle formation to follow a closed association process, whereas the secondary association into finite aggregates or into a network should be an open association process. The formation of micelles consisting of looped chains is opposed by the unfavorable entropy of chain cyclization. As a consequence, flower micelles can be observed only when the groups at the chain ends are sufficiently hydrophobic compared to the length of the intervening chain backbone, to promote association well before the onset of network formation. Only for those telechelic polymers characterized by a low CMC can one find a range of polymer concentrations where discrete flower micelles can be observed by dynamic light scattering or by pulsed gradient spin-echo NMR.

A recent theoretical paper predicted that telechelic associating polymers would undergo a special transition in their associative structure (22). These polymers were predicted to form flower micelles with a narrow size distribution at low polymer concentrations. Because these flower micelles should have weak attractive interactions with no repulsive interactions beyond the hard sphere separation, they were predicted to phase separate from dilute solution into a polymer-rich phase of weakly bridged micelles. Flower micelles have been inferred from light scattering and by pulsed gradient NMR in two telechelic AT systems in water (23), and in organic solvents for various triblock copolymers with insoluble end blocks (24, 25). We now also appreciate that if the substituents are not sufficiently hydrophobic, so that the onset of association (the critical aggregation concentration, CAC) is too high, there may be no concentration range preceding the formation of extended networks where discrete flower micelles can exist (26).

End-group Aggregation Numbers N_R

In 1995 we reported N_R values for a set of HEUR polymers with $C_{16}H_{33}O-$ end groups (20). These model polymers were prepared by reacting oligomeric PEO chains with isophorone diisocyanate to elongate the chain and to permit attachment of the end groups. In this system $C_{16}H_{33}O-$ groups are attached to the chain ends via isophorone diurethane (IPDU) moieties, which serve as part of the total hydrophobic end group. The polymers had a relatively broad molecular weight distribution, with $M_w/M_n \approx 2$, and ca. 30% of the chains were substituted at only one end. We employed pyrene as a fluorescence probe, and carried out detailed fluorescence decay measurements to determine N_R values for this system. In these experiments, quenching occurs through pyrene excimer formation: excited pyrene serves as the fluorophore, and ground state pyrene serves as the quencher. Because we were concerned with the proper interpretation of the data, experiments were carried out over a range of pyrene concentrations for each polymer concentration and at three different polymer concentrations. At the lowest of the polymer concentrations, the solution was characterized primarily by flower micelles, whereas at the highest concentration, the onset of network formation was indicated by the 100-fold increase in the solution viscosity. For this particular sample ($M_n = 34,000$), we found a value of $N_R = 18 \pm 1$ independent of the polymer concentration. From this result we concluded that the primary association to form the flower micelles was a closed association process, and that the core size of the micelle-like aggregate was conserved during the secondary association into the network. Since the highest concentration examined was only about 2 wt% polymer, this conclusion applied specifically to solutions containing up to this amount of polymer.

These experiments were extended to other HEUR polymers differing in chain length (M_n 34,000 to 51,000), but all containing $C_{16}H_{33}O-$ as the hydrophobic end group. For these polymers the N_R values ranged from 18 to 28, but for any one polymer, the value of N_R did not change over a limited range of concentrations that spanned flower micelles to well-formed networks. This result leads one to conclude that the system follows a closed association model for micelle core formation over this concentration range.

Persson *et al.*(*18b*) used a method based upon EPR spectroscopy to determine the N_R value for a PEO polymer of M = 9,300 with $C_{12}H_{25}$ groups attached to the chain ends by an ether linkage. This polymer has a narrow molecular weight distribution ($M_w/M_n \approx 1.1$) and nearly complete end-group substitution. Like the fluorescence quenching technique, the EPR approach assumes a Poisson distribution of hydrophobic spin probes among micelles. In this way the authors determined that $N_R = 31 \pm 6$ at a polymer concentration of 2.5 wt%. In an earlier paper *(18a)*, they report $N_R = 28$ determined by fluorescence quenching and an onset of aggregation (the cac) of 1.5×10^{-5} M. By comparison, the non-ionic surfactant $C_{12}H_{25}$-EO_{23} has a critical micelle concentration (cmc) of 1.8×10^{-4} M and $N_{agg} = 40$. The longer EO chain length of the associating polymer leads to a lower end group aggregation number. Similar polymers were examined by SAXS and SANS by the group of J. François in France. These polymers form an ordered phase at high concentrations (> 20 wt%) (*17*). Under these circumstances, values of N_R could be estimated from the scattering data. In the context of the model they used to fit the data, the values of N_R appeared to increase with increasing polymer concentration.

Some information is available on aggregation numbers in flower micelles from diffusion coefficient measurements. For example, if dynamic light scattering (DLS) or pulsed-gradient spin-echo NMR measurements indicate a narrow distribution of diffusing species in solution at concentrations above the CMC, the size of these species can be inferred by comparison with intrinsic viscosity data. Assuming hard sphere behavior for these objects, one can combine the intrinsic viscosity with the diffusion coefficient determined to calculate the mean number of polymer molecules per flower micelle. In this way an N_R value of 30 was estimated for a PEO polymer of M = 35,000 with $C_8F_{17}O$- end groups, each coupled to the chain end via an IPDU group (27).

In each of the various experiments described above, the end-group aggregation number determined was on the order of 20 to 30. These numbers are significantly smaller than those commonly found for non-ionic micelles formed from surfactants with shorter EO_x chains. Nevertheless, these numbers are significantly larger than the values of 3 to 6 inferred indirectly from rheology measurements through calculations that did not take full account of looping chains (28).

In a recent publication, Alami et al. (29) reported fluorescence quenching measurements to determine N_R values for a $C_{12}H_{25}$ end-capped PEO of M = 20,000 using two different quenchers. Using pyrene as a probe and dimethylbenzophenone as a quencher, they found aggregation number of 28 ± 3 for polymer concentrations ranging from 2 to 7 wt%. In the same work, quenching by pyrene excimer formation lead to a very different aggregation number (16) for a 6 wt% polymer solution. To explain this difference, the authors expressed their concern about the validity of the Poisson distribution of fluorophores and quenchers in the micelle-like structures formed by their polymer. In our view, one of the most important aspects of this work is that it emphasizes the difficulties in applying fluorescence quenching methods to associative polymer solutions. When applied to surfactant micelles, these techniques are so routine that one often examines only one micelle concentration and only one or two quencher concentrations. In the case of AT's, some of the convenient assumptions used to analyze the data break down. Proper attention to control experiments is necessary to justify the conclusions drawn.

Recent Results on Urethane end-capped PEO

We have recently described measurements to determine N_R values on two PEO samples of M_n = 35,000 (M_w/M_n = 1.1), end-capped through reaction with hexadecyl isocyanate or with octadecyl isocyanate. We refer to the former as HDU and the latter as ODU.

$$R-NH-\overset{O}{\underset{\|}{C}}-(OCH_2CH_2)_{n-1}OCH_2CH_2O-\overset{O}{\underset{\|}{C}}-NH-R$$

R = $C_{16}H_{33}$ HDU
R = $C_{18}H_{37}$ ODU   n ≈ 800

ODU. At concentrations less than 2 wt% in aqueous solution, ODU phase separates (*30*), to give a low viscosity upper phase consisting of a dilute solution of the flower-like micelles, and a lower gel-like phase in the form of a network of interconnecting micelles. The decay profiles for pyrene solubilized in the upper phase could be fitted to eq. (1), but the fitting parameters do not behave as expected when the pyrene concentration is varied. As we will report elsewhere in more detail (*31*) the problem is that a significant fraction of the pyrene remains in the water phase where it emits with its unquenched lifetime τ_w. The data become well behaved if one adds a term to eq. (1) to account for the fraction of the pyrene emission (1 - R) from pyrene in water:

$$I(t) = I(0) [R \exp[-t/\tau_m - n(1 - \exp(-k_q t))] + (1-R) \exp(-t/\tau_w)] \qquad (5)$$

Excellent fits to the data are obtained. The k_q values are constant and scatter about 7 μs^{-1}. The R values are very close to 0.9, indicating that 10 % of pyrene emission is from dyes in the water phase, a very reasonable estimate. At high pyrene concentrations, the n values increase linearly with pyrene concentration, and we obtain an aggregation number $N_R = 23 \pm 1$.

HDU. Aggregation numbers HDU solutions were determined via excimer formation, using two fluorescent probes, pyrene (Py) and 1-ethylpyrene (EtPy). EtPy has a lower intrinsic solubility in water (ca. 1 x 10^{-7} M at 22° C) than pyrene (7 x 10^{-7} M) (*32*). Partitioning experiments establish conditions under which the probes are essentially completely dissolved in the micelle phase (*33*).

The TRFQ data for this system are well behaved and show good reproducibility. For pyrene, the mean lifetime was 227 ± 4 ns, and for ethylpyrene, 148 ± 2 ns. The data for k_q show some scatter, but no trend with a change in probe (quencher) concentration. The mean value for Py is 8.7 ± 0.6 μs^{-1} and for ethylpyrene, 10.4 ± 1.1 μs^{-1}. These values are in the range reported by others on similar systems with pyrene as a probe (e.g., 3 μs^{-1} in ref. 20 for C_{16}-capped PEO's with M_n 26,800 - 51,000 and 20 μs^{-1} in ref. 29 for C_{12}-capped PEO, M_n = 20,300).

The parameter of greatest interest is n, the mean number of quencher molecules per micelle. The value n should vary linearly with quencher (pyrene or ethylpyrene) concentration and be inversely proportional to the polymer concentration. Figure 2 presents plots of the probe-to-polymer ratio for pyrene and ethylpyrene. Both plots are linear, implying constant aggregation numbers, but there are three peculiarities in the data. The first unexpected result is the non-zero intercepts for both probes. This aspect was discussed in some detail in our recent publication (*33*), and can be explained in terms of a tiny amount of quencher present as an impurity in the polymer itself. Under these circumstances, the slope still gives meaningful values of N_R. A second unusual result is that the slopes of the plots are different. The pyrene data exhibit a steeper line, corresponding to a higher aggregation number ($N_R = 21 \pm 1$), compared to the ethylpyrene data ($N_R = 16 \pm 1$). The difference between these numbers cannot be attributed to random fluctuations only, and appears to represent a characteristic of the system.

We can rationalize this result by relaxing the assumption of micelle size monodispersity. If there is a distribution of micelles of efferent sizes, in which the EtPy preferentially partitions into the smaller structures, the smaller N_R values and larger k_q values can be explained. The idea of micelle polydispersity in AT's has precedence. For example, Alami *et al.* (29) suggested the possibility of oligomeric aggregates of various sizes of micelles in solutions of a poly(ethylene oxide) end-capped with $C_{12}H_{25}$ ether groups.

The third unusual result is found at much higher polymer concentrations, C_{pol} > 10 wt%. At that these high concentrations, the N_R values appear to increase substantially. This is a result that needs to be further investigated.

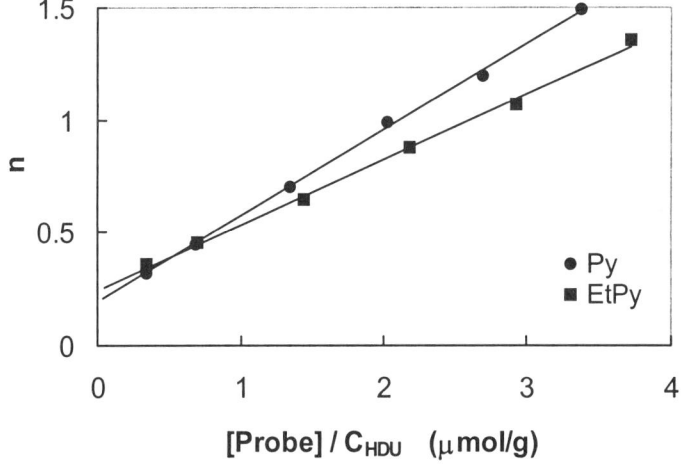

Figure 2. The fitting parameter n plotted versus probe-to-polymer ratio in 0.8 wt% HDU solutions for pyrene and ethylpyrene. From the slopes of the lines we calculate $N_R = 21$ from the pyrene data and $N_R = 16$ from the ethylpyrene data.

Hydrophobically modified Hydroxyethylcellulose

A schematic structure of a hydrophobically modified hydroxyethylcellulose (HMHEC) is given in the drawing below. These materials are prepared from HEC itself, which in turn is synthesized from cellulose, a process involving treatment with base followed by reaction with ethylene oxide. Glass *(34)* has shown that for a mean

[chemical structure diagram showing cellulose units with CH₂O(C₂H₄O)$_z$-C$_{16}$H$_{33}$, H(OC₂H₄)$_y$-O, O(C₂H₄O)$_x$H, O(C₂H₄O)$_x$H, CH₂O(C₂H₄O)$_z$H, H(C₂H₄O)$_y$-O, CH₂O(C₂H₄O)$_z$H substituents]

substituion (MS) of 2 EO units per glucose, nearly a fifth of the glucose residues in HEC remain unreacted. Normally a second ethoxylation is carried out under somewhat different conditions, and it occurs primarily at the 2- and 6- positions. Using a stochastic model, Glass priedicted that oligo(ethylene oxide) chains form part of the structure. For MS = 3.5, EO$_x$ units with x as long as 5 or 6 appear in small amounts at the 6-position. Less is known about the location of the hydrophobic groups in HMHEC. There is a fine balance between solubility in water and the extent of hydrophobe substitution.

All of the experiments which give information on aggregation numbers for hydrophobically modified cellulose derivatives such as hydroxyethyl cellulose (HMHEC) were carried out in solution in the presence an excess of ionic surfactant. Dualeh and Steiner *(35)* studied an HMHEC polymer containing 1.33 wt% $C_{12}H_{25}$ alkyl groups. This polymer is insoluble in water but rendered soluble in the presence of sodium dodecyl sulfate (SDS). The authors used SSFQ measurements to estimate the aggregation numbers of the mixed surfactant-polymer micelles, with tris(2,2-bipyridine)ruthenium (II) chloride as a fluorophore and 9-methylanthracene as the quencher. Their results suggest that at fixed C_{pol}, the total aggregation number of the micelle-like aggregates increased from 57 to 84 with increasing SDS concentration. They deduced that the number of SDS molecules per aggregate increases linearly with [SDS] from about 50 to 80. At the same time, the number of alkyl groups decreased from 8 to 5 per mixed micelle. More recently, Thuresson *et al.* *(36)* examined the binding of SDS to ethyl(hydroxyethyl) cellulose modified with 1.7 mol% of nonylphenol. They used a variety of techniques to study the system, including NMR diffusion measurements and TRFQ (pyrene as a probe, dodecyl pyridinium chloride as a quencher). The surfactant aggregation numbers determined by TRFQ were found to increase with increasing concentration of SDS. Another paper by Nilsson *et al.* *(37)* reports mixed aggregation numbers for HMHEC containing 1.7 mol% of C_{16} alkyl chains in the presence of excess SDS. The authors studied the effect of salt and cationic surfactant dodecyltrimethyl ammonium chloride (DoTAC), which are known to increase the N_{agg} of SDS micelles, on the size of the mixed micelles. They found at constant total surfactant concentration that increasing amounts of salt or the mole fraction of DoTAC leads to an increase in both the surfactant and hydrophobe aggregation numbers.

Nishikawa *et al.* *(38)* attempted to determine aggregation numbers directly on an HMHEC polymer with a molecular weight $M_W \approx 300,000$ corresponding to about 930 glucose units, with a mean degree of substitution of 3.6 ethylene oxide groups per glucose unit and on average one $C_{16}H_{33}$ group per 143 glucose. They showed that pyrene was solubilized to only a very small extent, but the pyrene molecules that were solubilized were located in a strongly hydrophobic domain ($I_1/I_3 = 1.25$). Direct determination of N_R values proved impossible because of the large fraction of pyrene (or ethylpyrene) remaining in the water phase. Panmai et al. (39) describe a very clever set of experiments on this same HMHEC derivative. The authors studied mixtures of the polymer with excess cationic surfactant under conditions where N_{agg} values could be obtained. Because they were obtained by SSFQ, the absolute N_{agg} values may be in error. The authors extrapolated the total aggregation numbers back to the CAC where HMHEC-surfactant mixed aggregates first form. The extrapolated value was on the order of an average of 2 or 3 hydrophobes per aggregate.

Aggregates containing only 2 or 3 $C_{16}H_{33}$ groups are unlikely to solubilize Py in a way that would lead to $I_1/I_3 = 1.25$. As a consequence, we imagine that HMHEC is characterized by a distribution of hydrophobes. Those attached close to the glucose backbone may not associate at all except in the presence of added surfactant, whereas those hydrophobes at the ends of EO_x oligomers would be more likely to form larger aggregates which could serve as the sites of pyrene solubilization.

Associative Polyelectrolytes

One class of commercially important associative thickeners involves a polyelectrolyte with hydrophobic pendant groups. The polymer backbone commonly consists of an ethyl acrylate-methacrylic acid copolymer with a composition such that the polymer is insoluble in water at low pH where the MAA groups are protonated, but very soluble in water at high pH where the extent of neutralization is large ($\alpha \rightarrow 1$). These materials are prepared by emulsion polymerization at low pH to yield a latex dispersion which swells and dissolves upon neutralization with base. They are commonly referred to as HASE thickeners, Hydrophobically modified Alkali-Swellable Emulsions.

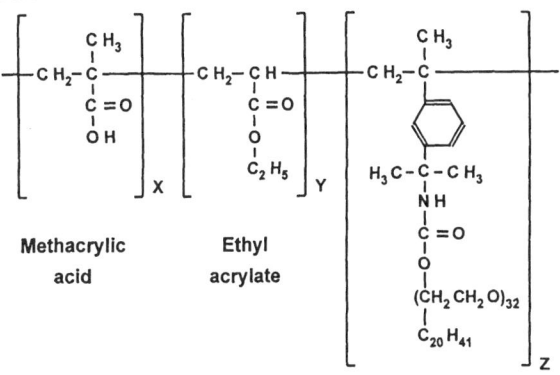

Our knowledge of the structure of these materials in solution is still very limited, but progress is being made through the study of a series of relatively well-characterized model polymers prepared by R. Jenkins (*40*) at Union Carbide. The structure of one such HASE polymer is shown above. It is a linear copolymer of ethyl acrylate (50 mol%), methacrylic acid (49.1 mol%), and a macromonomer containing a $C_{20}H_{41}$ group separated from the backbone by 32 ethylene oxide units (0.90 mol%). This polymer was prepared as a dispersion of latex particles, but dissolves as the pH of the solution is raised above 6.5. Full neutralization of the -COOH groups occurs near pH 9, accompanied by a large increase in the solution viscosity. In the most general terms, one can say that the increase in viscosity occurs through a combination of three factors: by the dissolution of the polymer to form discrete molecules in solution, by the change in shape of these molecules from a coil to a highly extended conformations due to the repulsion of negative charges at high degrees of neutralization, and by the interaction of the hydrophobes to form a network of *inter*molecular associations. Viscosification is likely to occur through entanglements as well as the formation of networks with hydrophobic domains as the junctions. Since the hydrophobes themselves ($C_{20}H_{41}$ groups) have very low solubility in water and are attached to the polymer backbone via EO_{32} spacer chains, one anticipates that some properties of the micelle-like aggregates formed will resemble those formed by linear end-capped PEO polymers.

The one complicating feature of this type of polymer structure is that the EA groups in the polymer backbone are themselves somewhat hydrophobic. The unprotonated EA-MAA copolymer is sufficiently hydrophobic that it will dissolve pyrene. At low degrees of HASE polymer neutralization ($\alpha < 0.4$), pyrene and ethylpyrene in water will partition strongly into the dispersed polymer phase.

Horiuchi *et al.* (*41*) carried out TRFQ experiments intended to determine the hydrophobe aggregation numbers in solutions of the HASE polymer, using both pyrene and ethylpyrene as fluorescent probes. Experiments were carried out at complete neutralization ($\alpha = 1$) for a variety of polymer concentrations, and for a range of probe concentrations at each polymer concentration.

These decay profiles were fitted to eq. (1). All statistical criteria (χ^2, weighted residuals and their autocorrelation) indicated excellent fits to the decay traces. Over the whole range of probe concentrations, the values of lifetime are constant: for pyrene, $\tau = 200$ ns, and for ethylpyrene, $\tau = 148$ ns. This type of behavior is expected for a well-behaved system. The other two fitting parameters, n and k_q, exhibit unusual behavior as the probe concentration is varied. These values are plotted in Figure 3. According to equation (1), one expects k_q to be constant and n to vary linearly with quencher concentration [Py] or [EtPy] (*cf.* Figure 1). Because the results were unexpected, the authors carried out a careful statistical study of the fitting results, including parameter-correlation analysis and Monte Carlo simulations to establish reliable error bounds of the parameters. This analysis indicates the parameter fits are statistically meaningful.

As a consequence, the problem is with the model itself. Features of the system differ in significant ways from the assumptions used in deriving equation (1). Some insights are available from consideration of the behavior of the fitting parameters for the two probes. In Figure 3 one can see that both k_q and n decrease with decreasing Py or EtPy concentration, with large values of n found at low fluorophore concentrations. A curious feature of the data is that this behavior does not depend on the polymer concentration. For pyrene as a probe, the results are reproducible at $\alpha = 1$ for three different polymer concentrations: 3.2, 4.8, and 6.4 g/L.

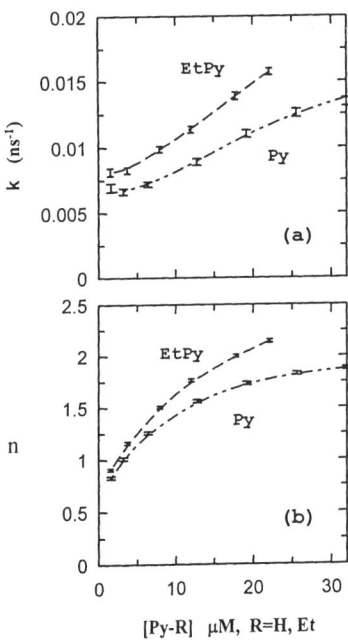

Figure 3. Variation of fitting parameters with probe concentration for pyrene and ethylpyrene in aqueous solutions of the HASE polymer described in the text (C_{pol} = 8.0 g/L; $\alpha = 1$): (a) the quenching rate constant k_q; (b) mean number of quenchers per micelle n.

From the data in Figure 3b the Horiuchi *et al.* (*41*) estimate the apparent aggregation number for the micelles formed by the hydrophobes of the HASE polymer. The values obtained are plotted in Figure 4 as a function of the probe concentration. The apparent aggregation numbers decrease drastically with increasing probe concentration. The actual aggregation numbers are probably not very meaningful, except as they point to different aspects of the system sampled by the probes at different probe concentration. One can imagine that there may be multiple binding regions in the polymer with different affinities for pyrene or ethylpyrene. As more of the probe is added to the system, the fluorescence signal becomes increasingly weighted in favor of those sites with smaller affinity for the probe. If that were the case, some of the assumptions made in the derivation of equation (1) would no longer be valid.

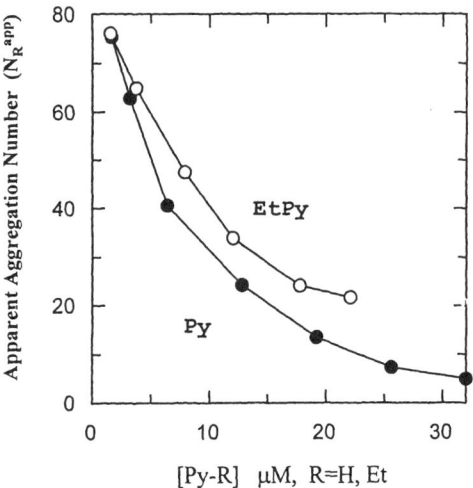

Figure 4. Apparent aggregation numbers calculated from the data in Figure 3b

Alternatively, one may be able to interpret these data in terms of a distribution of different micelles. As in the case of SDS micelles in the presence of 2-pentanol and sodium chloride described by Lang *(13)*, Horuiuchi *et al. (41)* report data in which individual decay curves fit nicely to equation 1, but N_R and k_q values vary with quencher concentration. In the steady-state fluorescence spectra for these solutions, they found a linear dependence of I_E/I_M on probe concentration for both pyrene and ethyl pyrene, but extrapolation to zero probe concentration gave a small but positive intercept. One interpretation of this result is that the system contains a distribution of hydrophobic domains, so that at low probe concentration, there is still a significant probability of finding two probes in the same micelle-like domain.

Simulations by Almgren and coworkers *(42)* of fluorescence quenching experiments in micelles with a broad size distribution indicate that mean N_{agg} values obtained with equation 1 are considerably underestimated. In this context, it is possible that the C_{20} hydrophobic groups associate to form micelle-like structures with 60 to 80 groups per aggregate. From this point of view, the initial probes added to the HASE solution partition into these large micelles formed by the C_{20} hydrophobic groups. There are other hydrophobic domains in the system, particularly regions of the polymer backbone rich in ethyl acrylate groups. As a more or less qualitative model, one can imagine a significant distribution of hydrophobic domains, ranging from large micelles formed by C_{20} hydrophobic groups to a variety of mixed micelles consisting of different mixtures of $C_{20}H_{41}$, ethyl acrylate and protonated

MAA groups. Ethylpyrene, the more hydrophobic of the two probes, preferentially partitions into the more hydrophobic of these domains. Thus for comparable concentrations of the two probes one finds larger n and k_q values for ethylpyrene than for pyrene.

Summary

TRFQ experiments are a powerful method for determining aggregation numbers for surfactant micelles. They are in principle equally powerful for the characterization of the micelle-like aggregates formed in associative thickeners, i.e., those associative polymers that act as rheology modifiers in aqueous solution. The fluorescence quenching method appears to work particularly well in characterizing the "mixed micelles" formed in solutions of AT's containing excess surfactant. The mixed aggregates that have been studied successfully are composed of many more surfactant molecules than polymer-bound hydrophobes, and contain 50 or more surfactant-plus-hydrophobe groups per aggregate. When applied to associating polymers in water in the absence of added surfactant, the technique provides reliable information for modest concentrations of linear HEUR and PEO polymers with strongly hydrophobic end groups. Here one finds hydrophobe association numbers of N_R = 20 to 30 end groups per micelle-like aggregate. These aggregates are significantly smaller than common micelles, which often have N_{agg} values of 60 to 100. On the other hand they are substantially larger than the single digit numbers on the same systems that have on occasion been inferred incorrectly from rheology measurements alone.

In some of these telechelic systems, the k_q values are smaller than those found in many traditional surfactant micelles, implying that the micellar core is less fluid than traditional micelles. Under these circumstances, fitting fluorescence decay data to equation (1) or (2) is more delicate than one might anticipate. One has to take the trouble to establish that the fitting parameters are stable to changes in probe and quencher concentration.

The first studies directed toward determining N_R values for a fully ionized HASE polymer turned out to be unexpectedly complicated. The data in individual experiments seemed well-behaved, but variations in the fitting parameters as the probe concentrations were varied pointed to deficiencies in the model. The limited evidence now available suggests that there is a distribution of sites in the polymer solution. Probes at low concentration bind preferentially to the most hydrophobic sites, and increasing the alkyl group substitution of the probe enhances its selectivity for these sites. It is likely that ethyl acrylate units along the polymer backbone participate in the formation of some of these sites. Whether hydrophobic interactions involving the EA monomer units contribute to the viscosity of the polymer solutions is not known.

In conclusion, methods based upon fluorescence quenching are capable of providing hydrophobe aggregation numbers that are essential for characterizing the structure of associative polymers in aqueous solution. Nevertheless, the high viscosity of these solutions, the relatively low concentration of micelle-like entities present at modest polymer concentration, and complicating features related to the chemical details of the polymer structure, make these experiments more difficult to interpret than similar experiments on traditional surfactant micelles.

Acknowledgments

The authors thank NSERC Canada and the Ontario-Singapore Collaborative Research Program for their support of this research.

References

1. (a) *Water-Soluble Polymers: Beauty with Performance;* Glass, J.E., Ed.; Advances in Chemistry 213; American Chemical Society: Washington, D.C., 1986. (b) *Polymers in Aqueous Media: Performance through Association*; Glass, J.E., Ed.; Advances in Chemistry 223; American Chemical Society: Washington, D.C., 1989. (c) *Hydrophilic Polymers*; Glass, J.E., Ed.; Advances in Chemistry 248; American Chemical Society: Washington, D.C., 1996. (d) *Macromolecular Complexes in Chemistry in Biology*; Dubin, P.; Bock, J.; Davis R., Schulz, D.N.; Thies, C., Eds.; Springer-Verlag: Berlin, 1994.
2. *Industrial Water Soluble Polymers*; Finch, C.A., Ed.; The Royal Society of Chemistry: Cambridge, UK, 1996.
3. Winnik, M.A.; Yekta A. *Curr. Op. Coll. Int. Sci.* **1997**, 2, 424-436.
4. Rubinstein, M.; Dobrynin, A.V. *Trends Pol. Sci.* **1997**, 5, 181-186.
5. Hartley, C.S. *Aqueous Solutions of Paraffin Chain Salts*; Herman and Cie: Paris, 1936.
6. Hiemenz, P.C.; Rajagopalan, R. *Principles of Colloid and Surface Chemistry*, 3rd ed.; Marcel Dekker: New York, 1997.
7. Israelachvili, J.N. *Intermolecular and Surface Forces*, 2nd ed.; Academic Press: New York, 1991.
8. Brown, W.; Rymden, R.; van Stam, J.; Almgren, M.; Svensk, G. *J. Phys. Chem.* **1989**, 93, 2512-2519.
9. (a)Yekta, A.; Aikawa, M.; Turro, N.J. *Chem. Phys. Lett.* **1979**, 63, 543 - 548. (b) Tachiya, M *Chem. Phys. Lett.* **1975**, 33, 289 - 292. (c) Infelta, P.P.; Grätzel, M.; Thomas, J.K. *J. Phys. Chem.* **1974**, 78, 190-195.
10. Zana, R.; in Zana R., Ed.; *Surfactant Solutions: New Methods of Investigation*; Surfactant Science Series, Vol. 22; Marcel Dekker: New York, 1987, 242-294.
11. Kalayanasundaram, K. *Photochemistry in Microheterogeneous Systems;* Academic Press: London, 1987.
12. (a) Almgren, M. *Adv. Coll .Int. Sci.* **1992**, 41, 9 – 32. (b) Almgren, M.; in Grätzel, M.; Kalayanasundaram, K., Eds.; *Kinetics and Catalysis in Microheterogeneous Systems*, Surfactant Science Series, Vol. 38; Marcel Dekker: New York, 1991, 63-113. (c) Zana, R.; Lang, J. *Colloids and Surfaces* **1990**, 48, 153-171. (d) Gehlen, M.H.; De Schryver, F.C. *Chem. Rev.* **1993**, 93, 199-221.

13. Lang, J. *J. Phys. Chem.,* **1990**, 94, 3734-3739.
14. Alargova, R.G.; Kochijashky, I.I.; Sierra, M.L.; Zana, R. *Langmiur* **1998**, 14, 5412-5418.
15. Even in the case of traditional surfactant micelles, greater than normal care is sometimes necessary. A recent example is the case of Triton X-100 micelles, where the light scattering and fluorescence quenching results on N_{agg} were seemingly in contradiction until more thorough fluorescence experiments were carried out. (a) Malliaris, A.; Le Moigne, J.; Sturm, J.; Zana ,R. *J. Phys. Chem.* **1985**, 89, 2709-2713. (b) Streletzky, K.; Phillies, G.D.J. *Langmuir,* **1995**, 11, 42-47.
16. Alami, E.; Rawiso, M.; Isel, F.; Beinert, G.; Binana-Limbele, W.; François, J. *Hydrophilic Polymers,* Advances in Chemistry Series 248, American Chemical Society: Washington, DC, 1995, pp. 344-362.
17. François, J.; Maitre, S.; Rawiso, M.; Sarazin, D.; Beinert, G.; Isel, F. *Coll. Surf. A.* **1996**, 112, 251-265.
18. (a) Perrson, K.; Wang G., Olofsson, J. *J. Chem. Soc. Faraday Trans.* **1994**, 90, 3555. (b) Persson, K.; Bales, B.L. *J. Chem. Soc. Faraday Trans.* **1995**, 91(17), 2863 – 2870.
19. Yekta, A.; Duhamel, J.; Adiwidjaja, H.; Brochard, P.; Winnik, M.A. *Langmuir* **1993**, 9, 881-883.
20. Yekta, A.; Xu B.; Duhamel, J.; Adiwidjaja, H.; Winnik, M.A. *Macromolecules* **1995**, 28, 956-966.
21. Yekta, A.; Duhamel, J.; Brochard, P.; Adiwidjaja, H.; Winnik, M.A. *Macromolecules* **1993**, 26, 1829-1836.
22. Semenov, A.N.; Joanny, J.-F.; Khokhlov, A.R. *Macromolecules* **1995**, 28, 1066-1075.
23. Rao, B.; Uemura, Y.; Dyke, L.; Macdonald, P.M. *Macromolecules* **1995**, 28, 531-538.
24. Chu, B. *Langmuir,* **1995**, 11, 414-421.
25. Raspaud, E.; Lairez, B.; Adam, M.; Carton, J.-P. *Macromolecules* **1996**, 29, 1269-1277.
26. Raspaud, E.; Lairez, B.; Adam, M.; Carton, J.-P. *Macromolecules* **1994**, 27, 2956-2964.
27. Xu, B.; Li, L.; Yekta, A.; Masoumi, Z.; Kanagalingam, S.; Winnik, M.A.; Zhang, K.; Macdonald, P.M. *Langmuir* **1997**, 13, 2447-2456.
28. Annable, T.; Buscall, R.; Ettelaie, R.; Whittlestone, D. *J. Rheol.* **1993**, 37, 695-726.
29. Alami, E.; Almgren, M.; Brown, W.; François, J. *Macromolecules* **1996**, 29, 2229-2243.
30. Pham, Q.T.; Russel, W.B.; Lau, W. *J. Rheol.* **1998**, 42, 159-176.
31. Vorobyova, O.; Winnik, M.A.; Lau, W., manuscript in preparation.
32. the vibrational fine structure in the fluorescence spectra of 1-substituted pyrene derivatives is normally insensitive to the polarity of the pyrene environment.
33. Vorobyova, O.; Yekta, A.; Winnik, M.A. Lau, W. *Macromolecules,* **1998**, 31, 8998-9007.

34. Glass, J.E.; Buettner, A.M.; Lowther, R.G.; Young, S.S.; Cosby, L.A. *Carbohydrate Res.,* **1980**, 84, 245-263
35. Dualeh, A.J.; Steiner, C.A. *Macromolecules* **1991**, 24, 112-116.
36. Thuresson, K.; Söderman, O.; Hansson, P.; Wang, G. *J. Phys. Chem.* **1996**, 100, 4909-4918.
37. Nillson, S.; Thuresson, K.; Hansson, P.; Lindman, B. *J. Phys. Chem.* **1998**, 102, 7099-7105.
38. Nishikawa, K.; Yekta, A.; Pham, H.H.; Winnik, M.A. *Langmuir,* **1998**, 14, 7119-7129.
39. Panmai, S.; Prud'homme, R.K.; Peiffer, D.G.; Jockusch, S.; Turro, N.J. *Langmuir,* **1998**, in press.
40. Jenkins, R.D.; DeLong, M.L.; Bassett, D.R. In *Hydrophilic Polymers*; Glass, J.E., Ed.; Advances in Chemistry 248; American Chemical Society: Washington, D.C., 1996; 425-447.
41. Horiuhi, K.; Rharbi, Y.; Spiro, J.; Yekta, A.; Winnik, M.A.; Jenkins, R.D.; Bassett, D.R. *Langmuir,* **1999**, 15, 1644-1650.
42. Almgren, M.; Alsins, J.; Mukhtar, E.; Van Stam, J. *J. Phys. Chem.* **1988**, 92, 4479

Chapter 10

Behavior of Branched-Terminal, Hydrophobe-Modified, Ethoxylated Urethanes

Peter T. Elliott, Linlin Xing, Wylie H. Wetzel, and J. Edward Glass

**Polymers and Coatings Department,
North Dakota State University, Fargo, ND 58105**

Hydrophobically-modified, Ethoxylated URethanes of narrow molecular weight (uni-HEURs) were synthesized with hydrophobic groups of different structure to determine their effect on solution properties. A linear hydrophobe ($C_{12}H_{25}$) and a branched hydrophobe (b-($C_{16}H_{34}$)), of comparable hydrophobicity, were coupled to the ends of a 29,500 M_n polyoxyethylene (POE) via a real telechelic process using 4,4' methylenebis(cyclohexyl isocyanate), H_{12}MDI. The aggregation numbers of these model uni-HEUR solutions increase with polymer concentration and plateau, at a similar value, at high polymer concentrations. However, at higher concentrations the solution viscosities of the uni-HEUR thickeners with the branched hydrophobes are higher than those with linear hydrophobes of the same effective carbon length, but lower than linear hydrophobes with an equivalent number of carbon units per linear chain. It is the number of hydrophobes in the aggregate and not the number of chain ends that is important in building solution viscosities. This is reflected by the longer relaxation times of the branched terminal hydrophobe uniHEUR in oscillatory rheology studies. When H_{12}MDI was used to couple larger hydrophobes to POE the solution viscosity increased dramatically and soft gels were observed; therefore, the influence of branching was examined in greater detail using hexamethylene diisocyanate (HDI) to couple larger branched hydrophobes. Multiple branched hydrophobes in groups of six, varying in size from $C_{10}H_{21}$ to $C_{16}H_{33}$, were prepared. These hydrophobe groupings were used to prepare terminal position and comb-architecture uniHEURs and are discussed.

Introduction

To understand the results observed in these studies, the observations noted in prior investigations of conventional branched surfactants are reviewed below. In one of the first studies dealing with branched surfactants (1) in relation to petroleum recovery, Wade and coworkers observed that the surfactant structure among ethoxylated alkyl phenols had a significant effect on the solubilization parameter and on the interfacial tension of microemulsions. If the hydrophilic and lipophilic moieties of the surfactant were increased simultaneously (e.g., as the number of ethoxylates was increased, the surfactant's hydrocarbon tail length also was increased to maintain a constant Hydrophilic-Lipophilic Balance [HLB]), a group of surfactants was obtained that exhibited optimum petroleum microemulsion behavior with respect to given salinity and cosurfactant levels (2). In the latter study, hydrophobe branching was used in concert with the correct ethoxylation level.

In studies reported (3) from Exxon's laboratories, hydrocarbon chain branching resulted in a higher CMC, due to steric hindrance to aggregation in the bulk phase. The critical micelle concentration (CMC) of branched sodium dodecyl sulfate (SBDS) was observed to be ~ 2 times higher than with the linear analog (SDS). The CMC of SBDS, however, was 1 order of magnitude lower than that of sodium octyl sulfate; indicating that a decrease in chain length due to hydrocarbon chain branching is compensated by increased hydrophobicity.

In additional studies of C_{16} sulfates, two different types of branching were studied (Figure 1) (4).

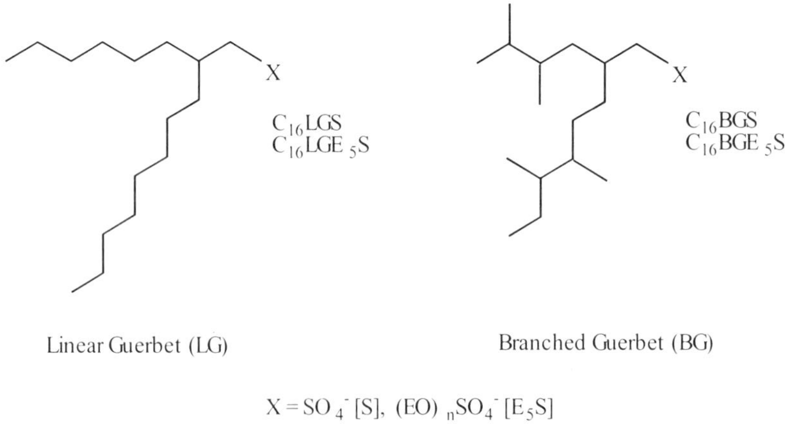

$C_{16}LGS$
$C_{16}LGE_5S$

$C_{16}BGS$
$C_{16}BGE_5S$

Linear Guerbet (LG)   Branched Guerbet (BG)

$X = SO_4^-$ [S], $(EO)_nSO_4^-$ [E_5S]

Figure 1. Surfactant Structures used in prior Branching Studies.

The first series consisted of surfactants in which an alkyl substituent is attached to the carbon atom beta to the head group (referred to as "linear" Guerbet surfactants); then numerous methyl groups are placed on the resulting two "linear" segments of the Guerbet surfactants (referred to as "branched" Guerbet surfactants). The head group (hydroxyl functionality) was then modified with sulfates and monodispersed ethoxysulfates, producing two different types of Guerbet surfactants. Both types of branched surfactants exhibited higher CMCs.

Our prior studies of Hydrophobically-modified, Ethoxylated URethanes (HEURs) have included: examination of different geometries with short oxyethylene spacing *(5,6)*, linear polymers with variable terminal hydrophobe sizes, separated by larger ethoxylate spacings *(7)*, and the influence of broad molecular weight distributions with different effective terminal hydrophobe sizes along with underivatized oxyethylene components *(8,9)*. This contribution addresses the influence of branching of the terminal hydrophobes on the fluorescence and rheology of uni-HEUR solutions.

Experimental Section

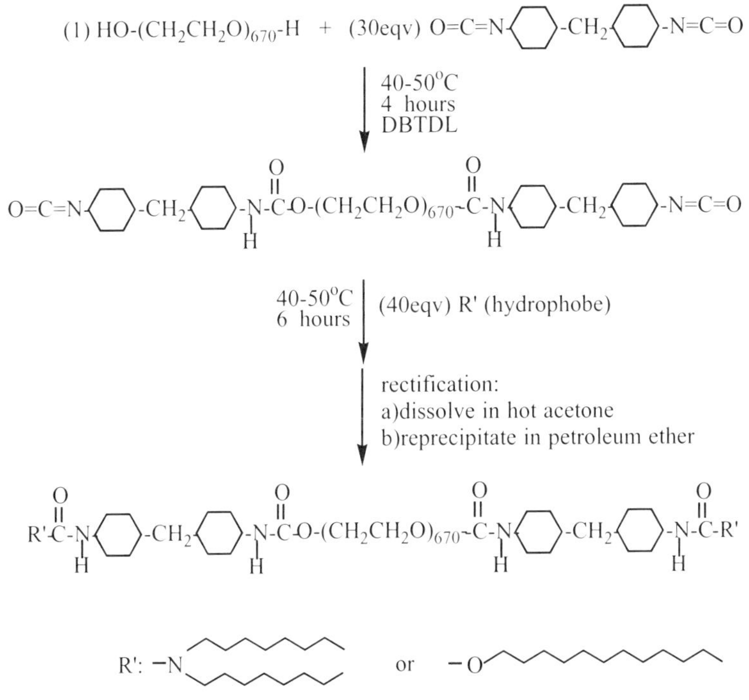

(1) HO-(CH$_2$CH$_2$O)$_{670}$-H + (30eqv) O=C=N-⟨ ⟩-CH$_2$-⟨ ⟩-N=C=O

↓ 40-50°C
4 hours
DBTDL

O=C=N-⟨ ⟩-CH$_2$-⟨ ⟩-N-C-O-(CH$_2$CH$_2$O)$_{670}$-C-N-⟨ ⟩-CH$_2$-⟨ ⟩-N=C=O
                         | || || |
                         H   O   O H

↓ 40-50°C   (40eqv) R' (hydrophobe)
6 hours

↓ rectification:
  a) dissolve in hot acetone
  b) reprecipitate in petroleum ether

R'-C-N-⟨ ⟩-CH$_2$-⟨ ⟩-N-C-O-(CH$_2$CH$_2$O)$_{670}$-C-N-⟨ ⟩-CH$_2$-⟨ ⟩-N-C-R'
 || | | || || | | ||
  O H   H   O   O H   H   O

R': —N⟨(alkyl branches)⟩   or   —O⟨(alkyl chain)⟩

Scheme 1. General synthetic reaction for uni-HEUR associative thickeners with variable terminal hydrophobe sizes.

The synthesis of the linear uniHEURs (Scheme 1 above) followed the procedure described first in an ACS PMSE preprint *(5)* and later in greater detail in a 1994 Langmuir article *(6)*. The synthesis of the dibranched uni-HEURs involved the reaction of disubstituted amines instead of alcohols with the diisocyanates and without the addition of dibutyltin dilaurate catalysis that would have retarded the amine reaction. The characterization procedures described earlier were followed, and the degree of hydrophobe modification was 95% or higher in all materials. Hydrophobe modification was determined using ^{1}H-NMR with 1,4-difluorobenzene as the standard. Molecular weight distributions of model-HEURs were measured through size exclusion chromatography (SEC) in tetrahydrofuran with poly(ethylene oxide) as standards.

Fluorescence Studies

Solutions for critical micelle (or aggregate) concentration measurements were prepared according to the following procedure. The surfactant or thickener solution was weighed in a 20-mL vial and diluted to a 1 g total weight with distilled water. This solution was further diluted to 10 g with a filtered pyrene saturated solution so that the concentration of pyrene was kept constant and less than 10^{-6} M.

Solutions for quenching experiments were prepared as follows. A surfactant or thickener stock solution was prepared by first dissolving the powder in pyrene-saturated water and then putting it on the roller in the dark for 24 hours. A portion of these solutions was saturated with benzophenone, and the mixtures were put on the roller in the dark for 24 hours. After centrifugation or filtration, aliquots of the benzophenone-saturated solutions were added to the surfactant or thickener/pyrene solutions to cover a benzophenone concentration range of 0 to $2*10^{-4}$ M.

UV absorption spectra were recorded with a Hewlett-Packard 8450A diode array spectrometer. Steady-state fluorescence spectra were recorded on a SPEX 2T2 Fluorolog fluorometer using a xenon lamp as a light source and equipped with a DM3000F data system. The UV absorption of benzophenone at 255 nm was measured to calculate the concentration of the quencher in each sample, dilutions being performed for solutions of absorbance greater than 1.3.

Fluorescence emission spectra (λ = 350-450 nm) were recorded at room temperature for the probe (pyrene). Excitation wavelength was set at 334 nm and the bandwidths set to 3 nm for excitation and 1.5 nm for emission. The intensity ratios of the first vibronic peak (λ = 372 nm) and the third vibronic peak (λ = 383 nm) in the pyrene emission spectra, I_1/I_3, were used to determine the critical aggregation concentration of thickeners. This approach is discussed in several chapters in this text. In the determination of mean aggregation numbers, N, the ratios of fluorescence intensities in the absence and presence of quencher (benzophenone) were calculated as the ratios of the integrated spectral areas (in wavelength units, from 350 to 450 nm).

Rheological Measurements

Steady state, low shear viscosities (2 s^{-1}) were measures on a Brookfield cone and plate viscometer. Shear rate profiles and oscillatory measurements were recorded on a Carri-Med CSL100 controlled stress rheometer fit with a cone and plate (4 cm and 2° cone angle) geometry. Torque and frequency sweeps were run prior to oscillatory measurements to ensure that the experiments were run in the linear viscoelastic region (amplitude~ 10^{-4} rad).

Results and Discussion

Synthesis

A large excess of diisocyanate is used to produce narrow molecular weight distributions by what is a true telechelic process (Scheme 1), not a half breed variation of a step-growth synthesis. Previous studies have shown that HEURs with narrow molecular weight distributions produce solutions with higher viscosities at the same concentration of thickener *(9)*. This is related to the low molecular weight segments preferring intra-hydrophobic associations in a step-growth synthesis mixture of products, and to the fact that a portion of the step-growth product is not substituted and therefore will not contribute to network formation.

The use of H$_{12}$MDI in this scheme adds to the difficulty in obtaining a clean HEUR for the diisocyanate is a mixture of 3- and 4- isomeric isocyanate derivatives, and the alicyclic substituents also produce a mixture of equatorial-equatorial and equatorial-axial isomers. However, H$_{12}$MDI contributes significantly to the overall size of the terminal hydrophobe that is added to the telechelic isocyanate and therefore to its viscosity. A more extensive rectification is required compared to a direct addition of a monoisocyanate, but the prior telechelic approach allows for variation in the effective terminal hydrophobe size, and a method of producing large branched terminal hydrophobe HEURs with relatively narrow molecular weights.

Effect of Hydrophobe Structure on Low Shear Viscosity

From a surfactant chemist view, one CH$_2$ group, introduced as a branch off a linear hydrophobe, contributes approximately half to an increase in hydrophobicity to that noted when it is included in the linear chain *(10)*. An ideal comparison would therefore be between a linear C$_{12}$H$_{25}$ and a di-C$_8$H$_{17}$ (i.e., branched-C$_{16}$H$_{34}$). The influence of the terminal hydrophobe structure on the low shear rate (2 s^{-1}) aqueous solution viscosity of 29,500 M$_n$ H$_{12}$MDI-based uniHEURs as a function of thickener concentration is illustrated in Figures 2. As noted in the original synthesis of uni-HEURs, even with large oxyethylene (EO = 670) spacer lengths, it is necessary to add a small amount of surfactant (0.002M SDS) to obtain solubility in water when the terminal urethane linkages are fully substituted *(7)*.

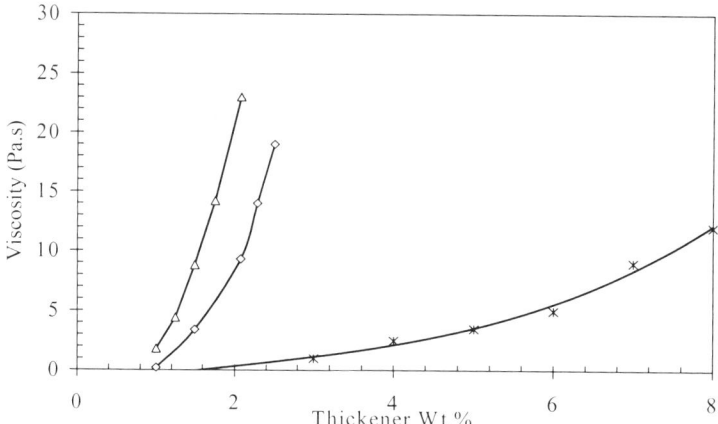

Figure 2. Low shear rate viscosity ($2s^{-1}$) of $H_{12}MDI$-based, 29500 M_n uniHEUR thickeners with different hydrophobe structures as a function of concentration. (△) b-($C_{16}H_{34}$); (◊) l-$C_{12}H_{25}$; (✕) b-($C_{12}H_{26}$); all with .002M SDS.

Based on prior fluorescence studies of $H_{12}MDI$ uni-HEURs, hydrophobic associations (7) occur at ca. 0.005 to 0.05 wt.% with the $C_{12}H_{25}$- and C_6H_{13}- terminal groups, but there are insufficient associations to promote networking and viscosity build until concentrations of ca. 1.0 and 2.0 wt.% l-$C_{12}H_{25}$- and l-C_8H_{17}-$H_{12}MDI$-EO_{468} HEURs have been added. The Y-branched b-($C_{12}H_{26}$)-$H_{12}MDI$-EO_{670} HEUR in this study is less effective than the equal carbon number l-$C_{12}H_{25}$ in facilitating viscosity build (Figure 2). The b-($C_{12}H_{26}$) HEUR begins to increase in viscosity only at a concentration greater than 3 wt.%. This is similar to prior studies of a linear C_8H_{17}-$H_{12}MDI$-EO_{468} HEUR. Consistent with those studies is the observation that a higher concentration of b-($C_{16}H_{34}$)-$H_{12}MDI$-EO_{670}- HEUR, used in this study, is required to facilitate the viscosity increase compared to the linear $C_{16}H_{33}$ (not reported). It is also shown in Figure 2 that at equivalent hydrophobicities, according to the rule of half contribution, the branched hydrophobe (b-$C_{16}H_{34}$) is more viscosifying than a comparable linear hydrophobe (l-$C_{12}H_{25}$).

The Effect of Chain Length and Hydrophobe Structures on the CAC

The I_1/I_3 intensity ratio for pyrene fluorescence as a function of model HEUR, b-$C_{16}H_{34}$, concentration is shown in Figure 3. In contrast to the sharp decrease in I_1/I_3 at the CMC of normal surfactants, such as SDS and Tergitol 15-S-9, the I_1/I_3 transition for the model HEUR solutions occurs over a broader concentration range (from 10^{-7} to $2*10^{-5}$ M). Past studies on step-growth HEURs have suggested this is related to broad molecular weight distributions. However, our studies, as well as those of Francoise, et. al. *(11)*, on model narrow molecular weight distribution HEURs also observe the broad I_1/I_3 transition. Considering the broadness of the association, we took the critical aggregation concentration (CAC) of the model HEURs at the inflection point of the I_1/I_3 downsweep. The branched $C_{16}H_{34}$- HEUR has a similar CAC ($2*10^{-5}$ M), within experimental error, to the linear $C_{12}H_{25}$- one. The aggregation behaviors of the model branched HEURs are different from the branched surfactants that were studied in Exxon's laboratories *(3)*. For branched surfactants, it is reported that hydrocarbon chain branching resulted in a higher CMC due to steric hindrance to aggregation in the bulk phase. The steric effect from the long oxyethylene (EO) spacer length in the micelle packing causes a gradual association in model HEUR solutions and the hydrophobe branching does not therefore exhibit a significant hindrance in the formation of micelles.

As was already mentioned and summarized in one of Rosen's reviews *(10)*, according to the general rule of half contribution, a comparison between the model HEUR with the linear $C_{12}H_{25}$ and the one with the branched $C_{16}H_{34}$ (i.e., 8+4) at the same EO spacer length is appropriate for assessing the influence of branching.

The branched $C_{16}H_{34}$-HEUR provides a less hydrophobic domain than the linear $C_{12}H_{25}$-HEUR, reflected by a higher I_1/I_3 value in the plateau region. At concentrations above 10^{-4} M this behavior can be explained if the branching of the hydrophobe results in a more porous micellar aggregate with water penetration of the branched hydrophobic domains. Of course, a more porous structure would lead to greater distribution of the pyrene in the aqueous phase as discussed in the preceding chapter. This would also explain the difference.

The Dependence of Mean Aggregation Number of HEUR Solutions on Polymer Concentration

When the concentrations of the model HEUR aqueous solutions are increased, it is observed that the aggregation numbers for both the linear and branched hydrophobes increase and finally approach a plateau at a higher HEUR concentration. The different hydrophobe structures (linear vs. branched) may affect the aggregation numbers at the higher concentrations (Figure 4). It seems that the HEUR with the linear hydrophobe end (l-$C_{12}H_{25}$) has a higher aggregation number than the HEUR with the branched $C_{16}H_{34}$- hydrophobe at the same EO spacer length. However, the difference is small and could arise from the greater steric hindrance of a branched hydrophobe or from a greater distribution of pyrene in water with the branched $C_{16}H_{34}$- thickener.

Although there is a small, or no, difference in the aggregation number between the linear and branched hydrophobe end groups, we believe that the internal structures of the aggregates for branched hydrophobes are different from linear ones. The two C_8H_{17} hydrophobes of the branched $C_{16}H_{34}$- are capable of forming micelles in aqueous solution, and with the H_{12}MDI connecting unit it would be aided in inter-associations in the hydrophobic domain. The number of chain ends for the *l*-$C_{12}H_{25}$

and b-$C_{16}H_{34}$- are essentially the same, but this provides twice the number of hydrophobes in the

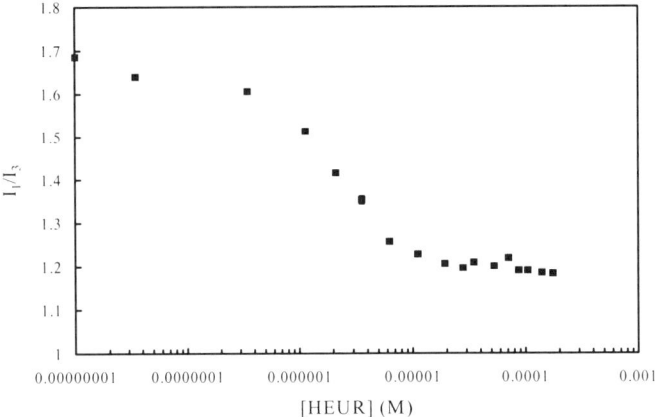

Figure 3. The influence of the hydrophobe on the I_1/I_3 of aqueous pyrene solutions as a function of 29,500 M_n, H_{12}MDI-based HEUR concentration. (■) branched $C_{16}H_{34}$-.

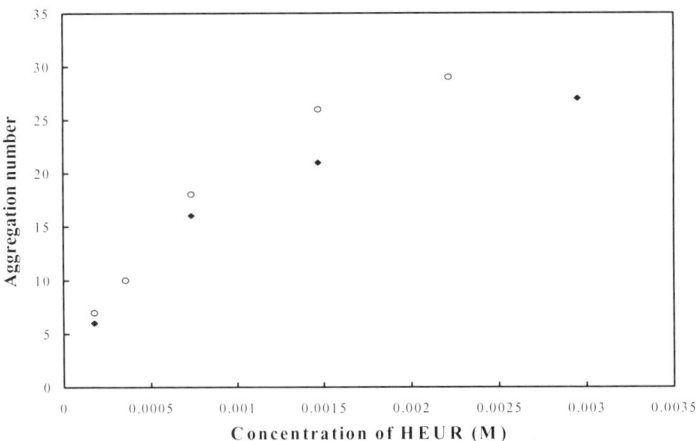

Figure 4. Aggregation number as a function of concentration of 8,600 M_n, H_{12}MDI based HEURs with various hydrophobes. (O) linear $C_{12}H_{25}$; (♦) b-$C_{16}H_{34}$.

micellar aggregate (Figure 5). This would give rise to a greater number of associations, and a longer residence time in the aggregates for the b-$C_{16}H_{34}$-HEUR, as well as, a greater viscosity (Figure 2). Oscillatory rheometry studies (Figure 6) discussed below support such a mechanism.

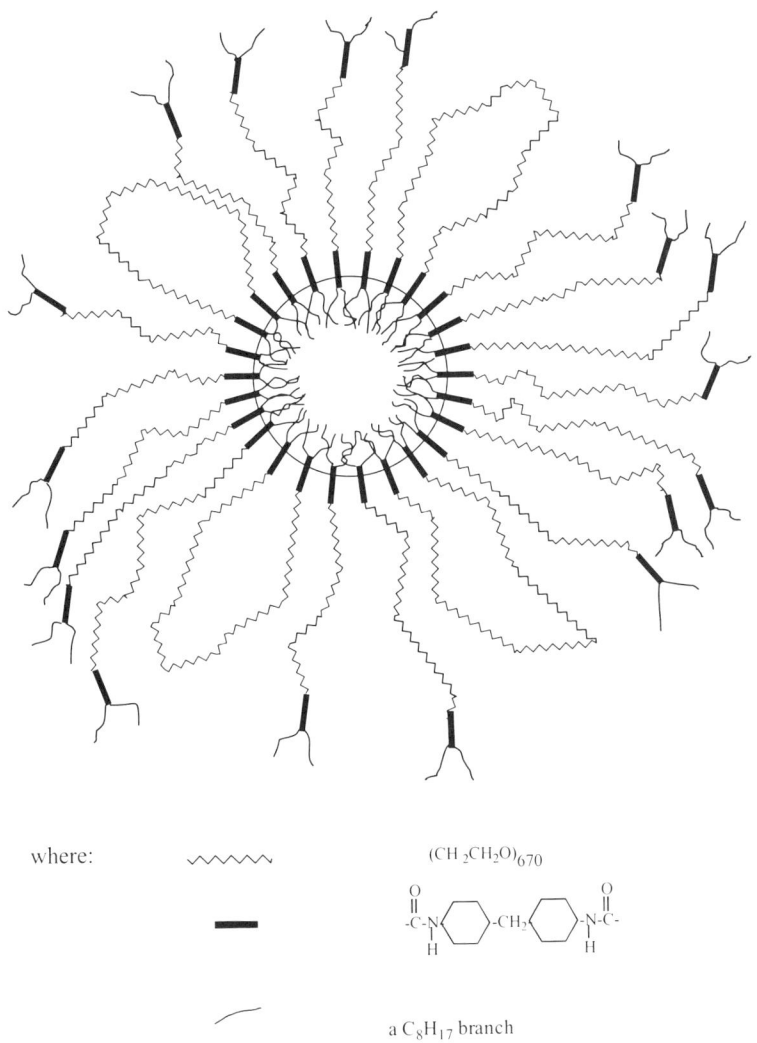

Figure 5. Representation of a branched b-$C_{16}H_{34}$-H_{12}MDI EO_{670}- uniHEUR micelle. (note: this figure is not drawn to scale).

Oscillatory Measurements

The use of oscillatory rheometry provides additional insight into what is contributing to the viscosity of polymer systems. A fluid that possesses a significant elastic component (reflected by the storage modulus, G') typically forms a network either by associations (hydrophobic in this case) or by chain entanglement. In oscillatory shear measurements, a sinusoidally varying shear deformation is imposed on a fluid, and the amplitude of the resulting shear stress or strain and the phase angle between the imposed force and resulting stress is measured. From this measurement, the storage moduli, G', and the loss moduli, G", can be determined (elastic and viscous components).

The oscillatory response at concentrations approximate to the beginning of the viscosity increase for the linear $C_{12}H_{25}$-H_{12}MDI-HEURs with large EO (670) spacings is illustrated in Figure 6A. The dominant response is the loss modulus, G", or the viscous component. The branched b-($C_{16}H_{34}$)- analog (Figure 6B), even at a slightly lower concentration to approximate viscosities, gives a much greater elastic response, indicating that two C_8H_{17}- hydrophobes per chain end creates a greater network strength than one $C_{12}H_{25}$- hydrophobe per chain end.

When larger hydrophobes, a linear $C_{16}H_{33}$- and branched $C_{20}H_{42}$, are attached to POE with H_{12}MDI, aqueous solutions containing 2 mM SDS produce viscous gel like solutions. In order to extend the branching study to include these larger hydrophobes, a smaller diisocyanate, hexamethylene diisocyanate (HDI), which has a lower vapor pressure and is therefore more toxic than H_{12}MDI, was used. Neither of these products (i.e. l-$C_{16}H_{33}$-HDI or b-$C_{20}H_{42}$-HDI HEURs) was an efficient thickener (Figure 7). There is more than the seven carbon difference between HDI and H_{12}MDI involved in this variation in viscosity, but to date, we have not delineated what this difference in magnitude is related to.

Terminal Multi-Branched HEURs

A set of multiple branch hydrophobes were synthesized to attach to the telechelic HDI derivatized POE (Scheme 2A). The multiple branch hydrophobe oligomers were synthesized by methoxide-initiated oligomerization of 1,2 epoxydodecane; 1,2 epoxytetradecane; 1,2 epoxyhexadecane; and 1,2 epoxyoctadecane to produce oligomers with an average of six hydrophobes per grouping (Table 1). The multiple branch hydrophobes were then attached to 29,500 M_n POE with HDI as the diisocyanate using the uniHEUR process (Scheme 2B).

Table 1. Number of Hydrophobes in a Group

Hydrophobe	from OH#	from NMR	average (OH# and NMR)	Mn
($C_{10}H_{21}$)	5	5.7	5.4	5.6
($C_{12}H_{25}$)	4.8	8.9	6.8	5.8
($C_{14}H_{29}$)	5.7	8	6.8	6.0
($C_{16}H_{33}$)	4.4	10	7.2	5.2

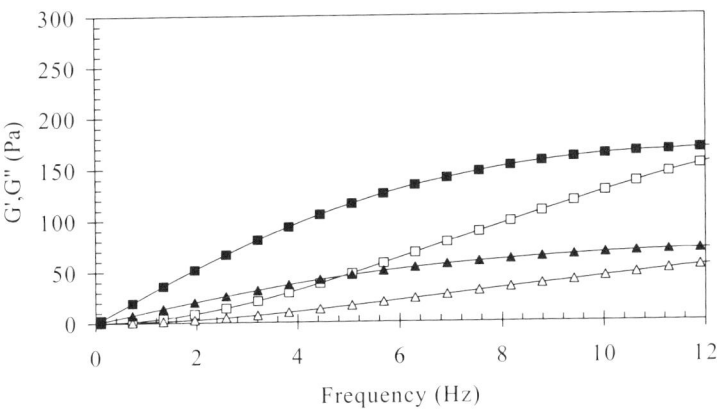

A. (□) 2.0 wt.% thickener; (△) 1.5 wt.% thickener, linear, l- $C_{12}H_{25}$, hydrophobe

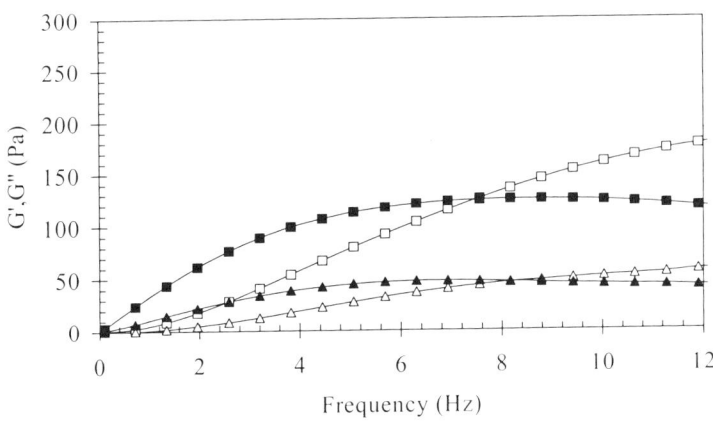

B. (□) 1.5 wt.% thickener; (△) 1.25 wt.% thickener, branched, b-$C_{16}H_{34}$, hydrophobe

Figures 6 A & B. Storage modulus (G') and loss modulus (G") dependence on frequency (Hz) for 29,500 M_n H_{12}MDI-based uniHEURs with different size hydrophobes. Open symbols: G'; Closed symbols: G".

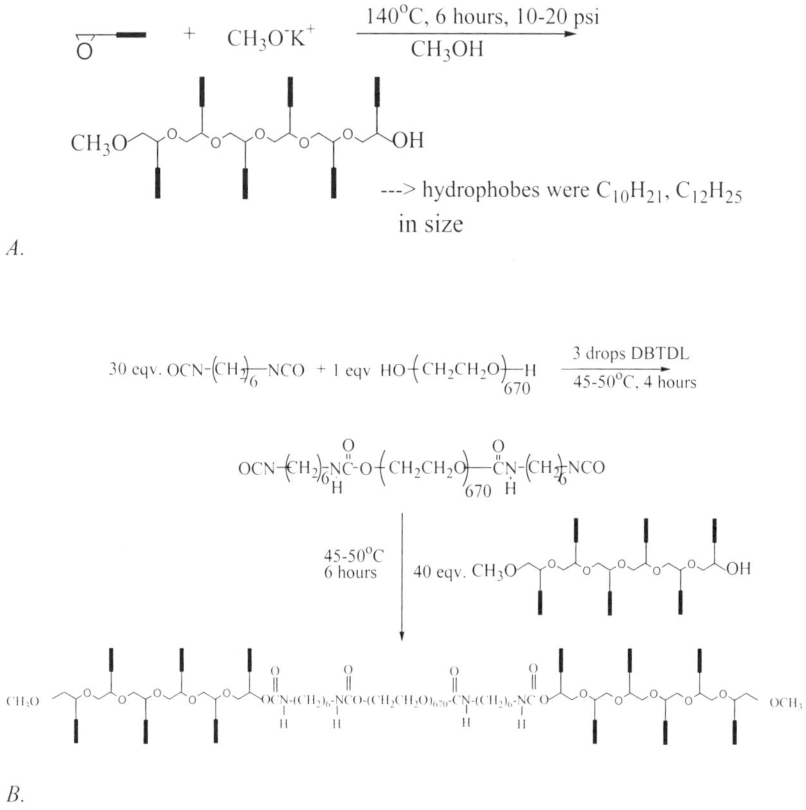

Scheme 2. (A) Synthesis of multiple branch hydrophobe; (B) Addition of multiple branch hydrophobes to 29.500 M_n POE with HDI.

The multi-branched, $(C_{10}H_{21})_6$, hydrophobe samples did not viscosify as much as was expected (Figure 7). One possible explanation are the ether linkages between the hydrophobes and using the reasoning in the conventional branched surfactants, perhaps both this and the multiple branching make the multiple branch hydrophobes relatively hydrophilic. A second point to note is that all four HEURs were run in aqueous solutions with 10mM SDS, above the anionic surfactant's CMC and above the concentration where many HEURs exhibit a viscosity maximum. These need to be reevaluated for efficiency at a lower SDS concentration (4 mM). Our studies are continuing in this area.

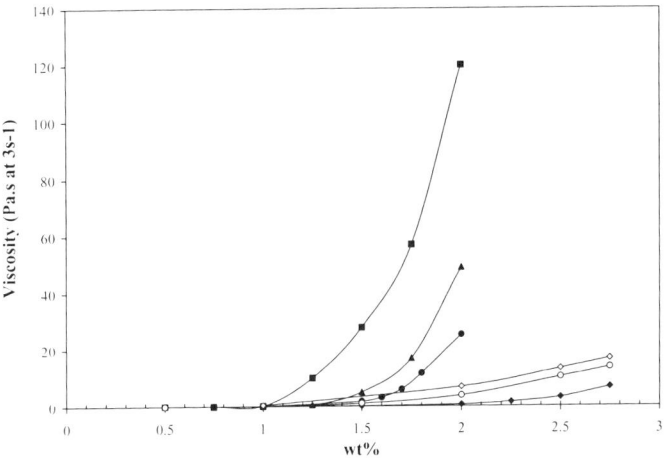

Figure 7. Influence of hydrophobe structure on the low shear rate ($2\ s^{-1}$) viscosity of aqueous solutions of 29,500 M_n HDI-based uniHEURs as a function of thickener concentration. (■) $[C_{16}H_{33}]_6$; (▲) $[C_{14}H_{29}]_6$; (●) $[C_{12}H_{25}]_6$; (◇) b-$C_{20}H_{42}$; (○)l-$C_{16}H_{33}$; (◆) $[C_{10}H_{21}]_6$. (open symbols are data from other work[12])

Multiple-Branching in Comb Architecture HEURs

Hydrophobe groupings also were incorporated in a comb geometry. The first series synthesized is illustrated in Scheme 3.

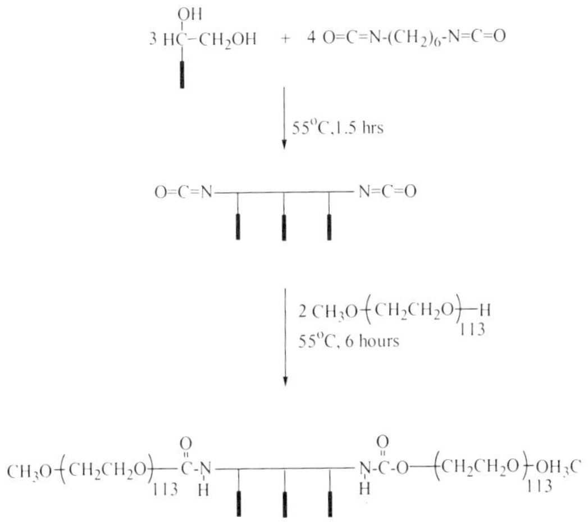

Scheme 3. Synthesis of model internal comb hydrophobe thickener.

We found it more difficult to oligomerize epoxides with hydroxyl initiation than with methoxide anions, so hydrophobe groupings were prepared by reacting diols with HDI, and then reacting the terminal isocyanates with methoxy-initiated POE. This approach resulted in low molecular weight models that did not viscosify water. A second attempt involved chain extension via a step-growth reaction of the 3-branched hydrophobe with POE (Scheme 4) using a slight excess of POE.

The reaction product was then split, and part was end capped by reaction of the terminal hydroxyls with octadecyl monoisocyanate. Aqueous solution viscosities were not achieved without terminal position substitution (Figure 8).

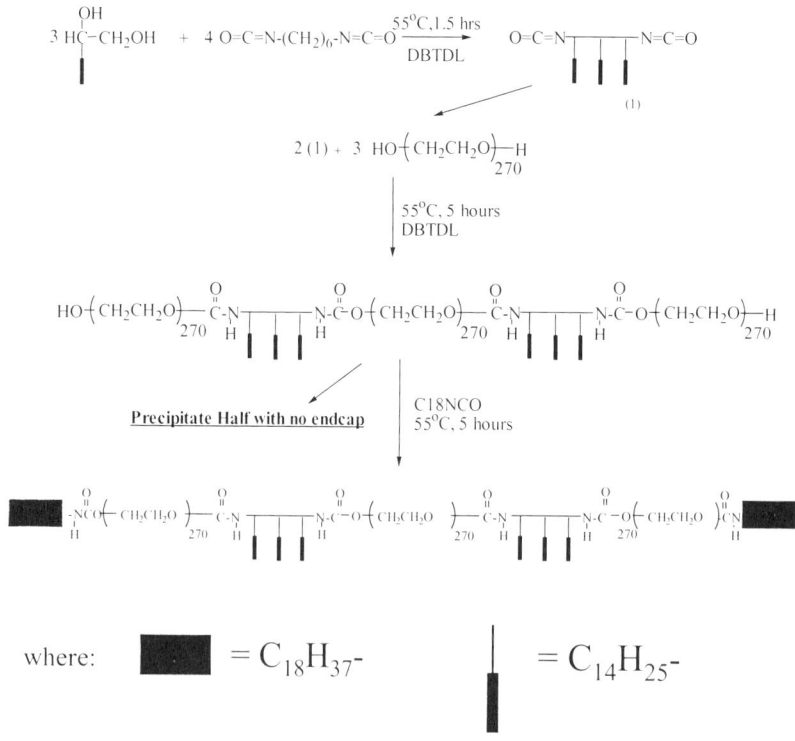

Scheme 4. Synthesis of step-growth internal comb hydrophobe HEUR thickener

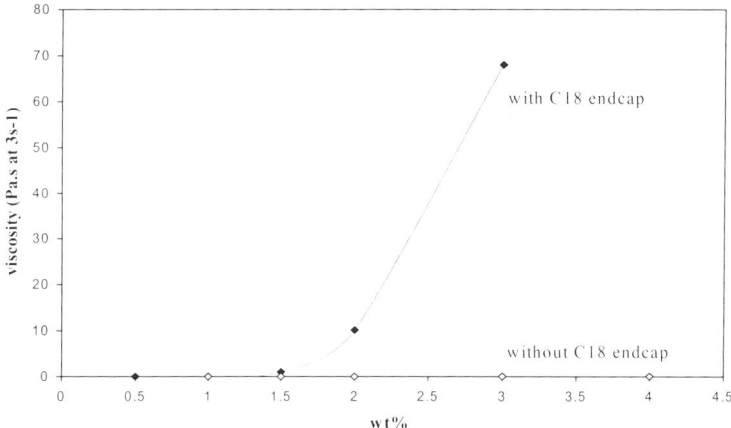

Figure 8. Low shear viscosity (Pa.s at $2s^{-1}$) dependence on concentration (wt.%) of step-growth internal comb HEUR thickeners: (◆) with $C_{18}H_{37}$ endcap; (◇) without $C_{18}H_{37}$ endcap (see Scheme 4 above).

Conclusions

Comparisons between narrow molecular weight, linear (l-$C_{12}H_{25}$-) and branched (b-$C_{16}H_{34}$-) 29,500 M_n, H_{12}MDI-based uni-HEUR associative thickeners revealed a dependence of the solution viscosity on hydrophobe structure. Branched hydrophobe containing uni-HEURs displayed higher viscosities than linear hydrophobes with equivalent hydrophobicities (ie. l-$C_{12}H_{25}$- vs. b-$C_{16}H_{34}$-). Fluorescence studies indicate that branched and linear model uni-HEURs have a similar critical aggregation concentration (CAC=$2*10^{-5}$ M) and steady-state fluorescence quenching studies demonstrate that the model uni-HEURs, at the concentration of $1.77*10^{-4}$ M, have the same aggregation number (ie. number of chain ends), N= 5-7. As the uni-HEUR concentration is increased, the aggregation number increases and then plateaus at higher concentrations. The difference in the plateau value for the l-$C_{12}H_{25}$- and b-$C_{16}H_{34}$- hydrophobes, with the same EO spacer length, is small and could arise from the greater steric hindrance of a branched hydrophobe or from a greater distribution of pyrene in water with the branched thickener. To explain the greater viscosity increase for the branched hydrophobe uni-HEURs, it is proposed that the greater number of hydrophobes on essentially the same number of chain ends leads to a greater extent of association. This is reflected by the longer relaxation times of the branched terminal hydrophobe uni-HEUR in oscillatory rheology studies.

Studies of larger linear and branched hydrophobes (l-$C_{16}H_{33}$ and b-$C_{20}H_{42}$) were conducted with HDI as the diisocyanate coupler to circumvent the gel-like solutions produced with these hydrophobes when $H_{12}MDI$ was used as the diisocyanate. Neither of these thickeners was very viscosifying. Multiple branched hydrophobes were synthesized and placed either at the terminal position of 29,500 Mn POE with HDI as the diisocyanate coupler or placed internally as comb structures along the backbone of the HEUR thickener. The terminal position, multiple branched hydrophobes were viscosifying; however the multi-branch, $(C_{10}H_{21})_6$, in comparison with the branched, b-$C_{20}H_{42}$-, HEUR (with HDI as the coupler) displayed viscosities lower than expected and this could be related to surfactant sensitivity (multi-branch at 10mM SDS) or an increase in hydrophilicity with the increased branching and ether units in the hydrophobe backbone. When placed as comb structures along the backbone of the HEUR thickeners, viscosity was only achieved when terminal hydrophobes were placed on the HEUR. Our studies are continuing in this area to better define the trends observed.

Literature Cited

1. Graciaa, A.; Fortney, L. N.; Schechter, R.S.; Wade, W.H.; Yiv, S., *Soc. Pet. Eng. SPE*, **1980**, 5815.
2. Abe, Masahiko,; Schechter, David; Schechter, R.S.; Wade, W.H.; Weerasooriya, Upali; Yiv, Seang, J., *Colloid Interf. Sci.*, **1986**, *114(2)*, 342.
3. Varadaraj, R.; Bock, J.; Zushma, S.; Brons N., *Langmuir*, **1992**, *8*, 14.
4. Varadaraj, R.; Bock, J.; Zushma. S.; and Thomas, R., *J. Phys. Chem.* **1991**, *95*, 1671.
5. Lundberg, D. J.; Glass, J. E.; Eley, R. R., *Proc. ACS Div. Polym. Materials: Sci. & Engin.*, **1989**, *61*, 533.
6. Lundberg, D. J.; Brown, R. G.; Glass, J. E.; Eley, R. R., *Langmuir* **1994**, *10(9)*, 3027.
7. Kaczmarski, J. P.; Glass, J. E.; *Macromolecules* **1993**, *26*, 5149.
8. Kaczmarski, J. P.; Glass, J. E; *Langmuir* **1994**, *10(9)*, 3035.
9. May, R.; Kaczmarski, J. P.; Glass, J. E., *Macromolecules,* **1996**, *29(13)*, 4745.
10. Rosen, M.J., *Surfactants and Interfacial Phenomenon*, 2nd *Edition*, John Wiley & Sons, New York, New York, **1989**.
11. E. Alami, M. Rawiso, F. Isel, G. Beinert, W.Q. Binana-Limbele, and J. Francois, Glass, J.E., Ed.; Advances in Chemistry 248; American Chemical Society: Washington, D.C., 1996, Chapter 18.
12. Chris Anderson, M.S. Thesis, North Dakota State University, 1999.

Chapter 11

Synthesis and Characterization of Well Defined End-Functionalized Hydrocarbon and Perfluorocarbon Derivatives of Polyethyleneoxide and Poly(N,N-dimethylacrylamide)

Thieo E. Hogen-Esch, Huashi Zhang, and David Xie

Loker Hydrocarbon Research Institute,
University of Southern California, Los Angeles, CA 90089

Abstract

Water-soluble poly(dimethylacrylamide)(PDMA) derivatives with hydrophobic end groups were synthesized by reaction of one or two-ended living PDMA anions with octadecanoyl- and perfluorooctanoyl chlorides or with an alphaphenyl acrylate monomer containing an octadecyl group attached via an oligooxyethylene spacer to the acrylate functionality. Similar perfluorocarbon (C_6F_{13} and C_8F_{17}) end-functionalized PEO's were synthesized by reaction of the PEO terminal OH group(s) with R_F-functionalized isocyanates. Size exclusion chromatography and proton and ^{19}F NMR studies indicate that the end-functionalizations were close to quantitative and proceeded with few side reactions.

PEO derivatives of molecular weights of approximately 5 and 10K and end-functionalized with C_6F_{13} groups in aqueous solutions show strong association of the perfluorocarbon end groups into dimers as indicated by ^{19}F NMR. Dimerization is favored by lower PEG molecular weight higher temperatures and addition of NaCl. Hydrophobic association is confirmed by a positive change in the entropy of association, ΔS, of 21 cal mole^{-1} K^{-1}. Kcal/mole. In sharp contrast to the the C_6F_{13} derivatives, the C_8F_{17} functionalized PEO's appear to associate into micelles.

Viscometry of the mono and bis end-functionalized telechelic PDMA derivatives indicate that the association is consistent with hydrophobe dimerization. The presence of hydrophilic spacers between the octadecyl and the acrylate end group of the end-functionalized PDMA was shown to greatly enhance the hydrophobic association. The dependence of the solution viscosity of the perfluorooctanoyl end-functionalized polymers on the concentration of added perfluorooctanoate surfactant indicates the occurrence of inter-polymer micellar bridging.

Introduction

The synthesis and characterization of hydrophobically modified (HM) water-soluble polymers has been an area of considerable interest and activity in recent years.[1-4] Such polymers have been prepared by chemical modification of water-soluble polymers such as cellulose derivatives with pendent hydrophobic groups or by radical copolymerization of hydrophilic vinyl monomers such as acrylamide with hydrophobic comonomers in the presence of surfactants that are needed in order to disperse the comonomers. In aqueous media, strong associations occur between the hydrophobic groups of the polymers and the added surfactants resulting in the formation of "mixed micelles" giving transient networks thereby enhancing the solution viscosity. In these polymers, the precise sequences of the pendent hydrophobic groups on the chain are unknown or are inferred either from the copolymerization kinetics or from a correlation of their synthesis and their chararterization by rheological and other measurements.

Another type of HM polymers, the so called HEUR segmented copolymers are prepared by polycondensation of a low molecular weight PEO with a bis-isocyanate.[5-9] In this case the hydrophobic groups are present in the chain backbone and their relative placements are better defined. In this case also the presence of surfactants in the synthesis of the polymers is avoided and thus the role of of surfactants in the formation of micelles may be more readily evaluated.

Experimental Section

Synthesis

PEO derivatives were synthesized as reported earlier.[10-11] Octadecyldeca(ethyleneoxy)-2-phenylacrylate(ODPA) was synthesized from the reaction of 2-phenylacryloyl chloride with octadecyl-deca-(ethyleneoxy) alcohol. Structural characterization of the intermediates and products was carried out by proton and carbon-13 NMR and by elemental analysis. PDMA derivatives were synthesized by anionic polymerization[12] and end-functionalized by reaction of the living PDMA's with the hydrocarbon or fluorocarbon acid chlorides or by reaction with ODPA.[13,14]

Characterization

Apparent molecular weights and molecular weight distributions were determined by size exclusion chromatography (SEC) on a Waters HPLC system equipped with ultra-µ-styragel columns. Narrow distribution PEG standards and polystyrene standards ("Polysciences") were used for the PEO and PDMA derivatives respectively. The ^{1}H and ^{13}C spectra were obtained in CDCl$_3$ at 30 °C using a Bruker AM-360 spectrometer operating at 360 and 90 MHz, respectively. The chemical shifts are referenced to tetramethylsilane (TMS). Fluorine-19 NMR measurements were carried out as reported earlier. The reduced viscosity in deionized water of end-functionalized polymers and homopolymers was determined by capillary viscometry using an Ubbelhode viscometer (Cannon no.100).

Results

Synthesis of PEO derivatives

The synthesis of the fluorocarbon functionalized PEG derivatives FP605M, FP805M, FP610M and FP810M is reported elsewhere [10] (**Table 1**) using reported methods.[15,16] The first digit gives the carbon number of the R$_F$ group, the second two digits give the molecular weights in thousands and the letters M and T stand for mono- and bis end-functionaliztion respectively. The extent of incorporation of the perfluorocarbon groups was determined by ^{19}F NMR spectroscopy using sodium triflate (CF$_3$SO$_3$Na) having a resonance at -80.8 ppm as internal standard.

Synthesis of PDMA derivatives

The mono- and difunctionalized polymers were prepared at -78 °C in THF by anionic polymerizations initiated with triphenylmethylcesium (TPMCs) or with naphthylcesium (NapCs), respectively (**Scheme 1**).[12] In each case a fraction of the living polymer solution was protonated with MeOH to give the matching PDMA homopolymer for the purpose of characterization by SEC and viscometry. The rest of the living polymer was reacted with an excess (5-10 times) of perfluorooctanoyl chloride, octadecanoyl chloride or ODPA by addition of the living polymer to the THF solution of the electrophile. Since the electrophile is in large excess at all times

Table 1: Synthesis and Characterization of Perfluorocarbon End Functionalized Polyethyleneglycols

Sample[e]	Mp[a]	MWD[a]	Mole% Rf[b]	k_H[c]	$K_{ass} \times 10^3$	n[d]	$CMC \times 10^6$ [d]
PEG5K	5,250	1.12	-	.37	-		
PEG10K	10,500	1.13	-	.35	-		
FP605M	5,840	1.10	92	4.7	18	2	
FP610M	11,200	1.10	90	4.6	.80	2	
FP805M	6,020	1.09	98	4.0	-	>10	1.5
FP810M	11,500	1.10	95	4.0	-	>10	7.0
FP810T	11,300	1.10	98	25	-	-	

a. Peak MW determined by SEC using PEG standards. b. Degree of end-functionalization determined by ^{19}F NMR in methanol using $CF_3 SO_3Na$ as internal standard. c. Huggins constant determined by capillary viscometry. d. Association number, n, and CMC determined by ^{19}F NMR (see text). e. PEG5K and PEG10K are homopolymers precursors for one-end functionalized PEG's: FP605M/FP805M and FP610M/FP810M respectively; T indicates two-ended (telechelic) PEG. f. Association constants in the presence of 0.4 and 0.6M NaCl were $4.1 \cdot 10^4$ and $1.0 \cdot 10^5$ repectively.

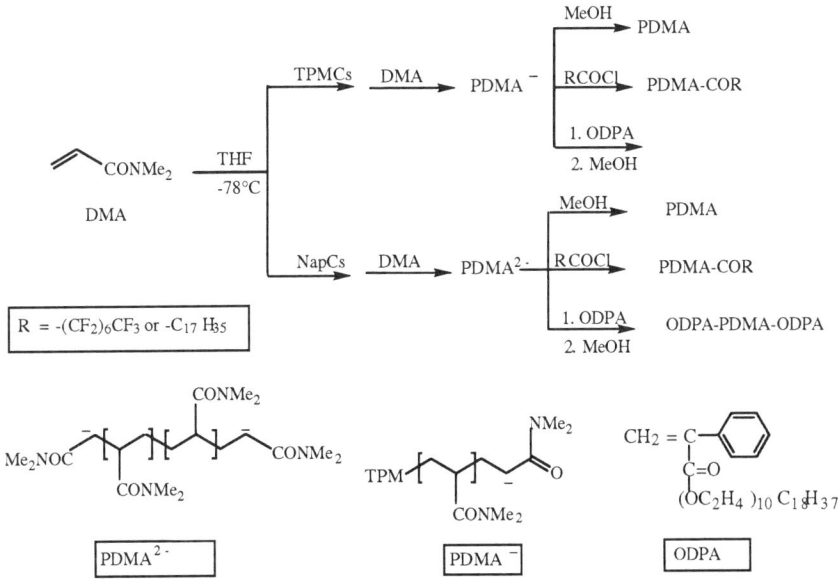

Scheme 1 End Functionalization of living PDMA with fluorocarbon or hydrocarbon groups.

the possible subsequent reactions of the polymer anion with the perfluorooctanoyl end functionalized PDMA are avoided.

Tables 2 and 3 give the molecular weights and molecular weight distributions (MWD) for the perfluorocarbon- and hydrocarbon end-functionalized PDMA polymers and their homopolymer precursors. The number average molecular weights (Mn), obtained by SEC and based on polystyrene standards, were found to be in good agreement with the calculated values. SEC analysis of the end-functionalized polymers also showed the expected shift toward higher molecular weights approximately by the mass of the end group. The molecular weight distributions of the end functionalized and precursor polymers were narrow in most cases (MWD≤1.13).

The comparison of the molecular weights calculated from the ^{19}F NMR spectrum with that determined by SEC indicates an almost quantitative reaction of perfluorooctanoyl chloride with the living PDMA anions (**Table 2**).

The end functionalization of living PDMA with a low ceiling temperature monomer ODPA is illustrated in **Scheme 1**. The living mono- and dianion PDMA precursors prepared using TPMCs and NapCs as initiators were reacted at 0 °C for 30 minutes with two equivalents of ODPA. The immediate color change from colorless to orange indicated a rapid addition of the PDMA carbanion onto the double bond of phenylacrylate. After termination with MeOH the resulting polymer was precipitated in hexane and was purified as described above.

The degree of R_F functionalization was found to be .91 and 1.85 for mono- and bis-functionalized polymer respectively (**Table 2**). Near quantitative bis end-functionalization using octadecanoyl chloride was also demonstrated by ^1H by integration of the methyl end group of the hydrocarbon vs. the dimethylamino group of the polymer chain and by ^{13}C NMR (**Table 3**). The molecular weights of the ODPA functionalized calculated by ^1H NMR from the relative ratios of the $N(CH_3)_2$ and the CH_2CH_2O protons in the mono end-functionalized low MW PDMA's were in a good agreement with the calculated values and with that determined by SEC indicating essentially quantitative end functionalizations (**Table 3**). This good agreement indicates that end-functionalization is essentially quantitative and that only one unit of the α-phenylacrylate was added to the chain end anion. These results are in agreement with the findings of Hopff et al. who reported the ceiling temperature of methyl phenylacrylate (MPhA) to be about -10 °C in THF at a concentration of one molar.[17] and were consistent with results by Quirk and Renwho reported mono end-functionalization of PMMA when reacted with 1-2 equivalents of MPhA via group transfer polymerizations in THF at 0°C.[18]

Association Studies of PEO derivatives

Fluorine-19 NMR

The fluorine-19 spectra of all of the fluorine atoms of the perfluorocarbon end groups show pronounced changes with concentration. Above $1.7 * 10^{-3}$ M, only a single broad CF_3 absorption of the one-ended FP605M(Table 1) in aqueous solutions is

Table 2: Synthesis and characterization of perfluorooctanoyl end-functionalized PDMAs.

No	Init.[a]	M_n^b (Calc.)	M_n^c (SEC)	MWD^c	Ex^f	% End Group[d] Functionalization	k_H^e
1.	TPMCs	15,600	13,900	1.07	H^+	100	-
2a.	TPMCs	48,100	48,600	1.08	H^+	100	.55
2b.	TPMCs	48,500	47,100	1.08	R_F	91	.70
3a.	NaphCs	46,800	45,000	1.06	H^+	100	-
3b.	NaphCs	47,200	46,700	1.08	R_F	185	1.0

a. Triphenylmethyl-or naphthylcesium initiators giving one- and two-end functionalized PDMA's respectively. b. Calculated from monomer/initiator ratio. c. Polystyrene standards. d. Determined by ^{19}F NMR using sodium triflate as internal reference. e. Huggins constant determined by capillary viscometry. f. Termination by $CH_3OH(H^+)$, or by $C_7H_{15}CO(R_F)$ ODPA see text.

Table 3: Synthesis and characterization of octadecanoyl or ODPA end- functionalized PDMA s.

No Init.[a]	M_n^b (Calc.)	M_n^c (SEC)	MWD[c]	Ex[f]	% End Group[d] Functionalization	k_H^e
4a. TPMCs	2,000	1,800	1.13	H^+	100	-
4b. TPMCs	2,300	2,100	1.18	R_H	96	-
5a. NaphCs	48,200	46,900	1.10	H^+	100	0.5
5b. NaphCs	48,700	47,600	1.11	R_H	-	1.7
6a. NaphCs	48,400	46,200	1.13	H^+	100	-
6b. NaphCs	49,700	48,000	1.14	ODPA	-	5.6
7a. NaphCs	218,600	174,600	1.55	H^+	100	0.5
7b. NaphCs	219,000	176,000	1.72	ODPA	172	1.0

a. Triphenylmethyl-or naphthylcesium initiators giving one- and two-end functionalized PDMA's respectively. b. Calculated from monomer/initiator ratio. c. Polystyrene standards. d. Determined by ^{19}F NMR using sodium triflate as internal reference. e. Huggins constant determined by capillary viscometry. f. Termination by $CH_3OH(H^+)$, by octadecanoyl chloride (R_H) or by ODPA, see text.

visible at about -82.1 ppm (**Figure. 1**). At a lower concentration ($1.7*10^{-3}$M) the presence of a second species is revealed by the broadening and the presence of a distinct shoulder corresponding to a lower MW species. Upon further dilution a separate downfield absorption (at -80.2 ppm) appears, the relative magnitude of which increases at lower concentration. At about $5*10^{-5}$ M, the downfield absorption is about 5 times larger than that of the upfield absorption. These changes are fully reversible and reproducible and appear to indicate changes in the degree of association upon dilution.

Clearly the absorptions at -80.2 and -82.1 ppm may be assigned respectively as the dissociated and associated forms of FP605M. For a simple association of n R_F functionalized PEG's (P) into a micellar aggregate (P_n) we can write:

$$n\,P \xrightleftharpoons{K} P_n \tag{1}$$

$$K = [P_n]/[P]^n \tag{2}$$

$$[P_n] = [P_a]/n = K\,[P]^n \tag{3}$$

$$\log[P_a] = n \log[P] + \log(K*n) \tag{4}$$

where [P_n] and [P] represent the concentrations of the aggregated and non-aggregated PEG's, respectively, and K is the equilibrium constant (eqns 1-4). The concentration of P_n is evaluated from the magnitude of the upfield resonance, [P_a] (eqn. 3). Thus, a plot of log [P_a] versus log [P] is expected to give a straight line with a slope equal to the association number (n). The data of **Figure 2** give a slope of close to 2 consistent with a monomeric–dimeric equilibrium in the 10^{-5}–10^{-3} region. The corresponding value of K is equal to $1.8 \pm 0.5 * 10^4$ M^{-1} (**Table 1**).

The ^{19}F NMR spectra for FP610M gives the same characteristic -82.1 and -80.2 ppm absorptions as observed for the lower MW derivative (FP605M) and a similar concentration dependence of the relative magnitudes of the two absorptions was observed Based on the same analysis as used for FP605M, again a value for the association number of about 2 was obtained. However the association constant of about 800 M^{-1} is more than 20 times lower than for FP605M (**Table 1**) and this is consistent with the expected greater excluded volume of the 10K PEO chain upon formation of this dimer. Similar effects have been observed for hydrocarbon PEG surfactants. Thus for surfactants of the structure: $C_9H_{19}C_6H_4(OCH_2CH_2)_nOH$ the CMC of the surfactant with n=50 was $2.8*10^{-4}$M whereas that of the surfactant containing PEG of a higher molecular weight (n=100) was about 4 times higher.[19]

At higher concentrations of FP605M the downfield resonances broaden and move upfield (**Figure 1**). This is also observed for FP610M. We interpret this interesting behavior as due to an exchange between the monomeric and dimeric species on the NMR time scale. Since the formation of the dimer(P_2) from the monomer (P) is bimolecular, the lifetime of the monomeric species, being inversely proportional to the concentration of P decreases with increasing monomer- and thus with the total polymer concentration. However, the dissociation of the dimer into monomer is unimolecular so that the half-life of the dimer is concentration

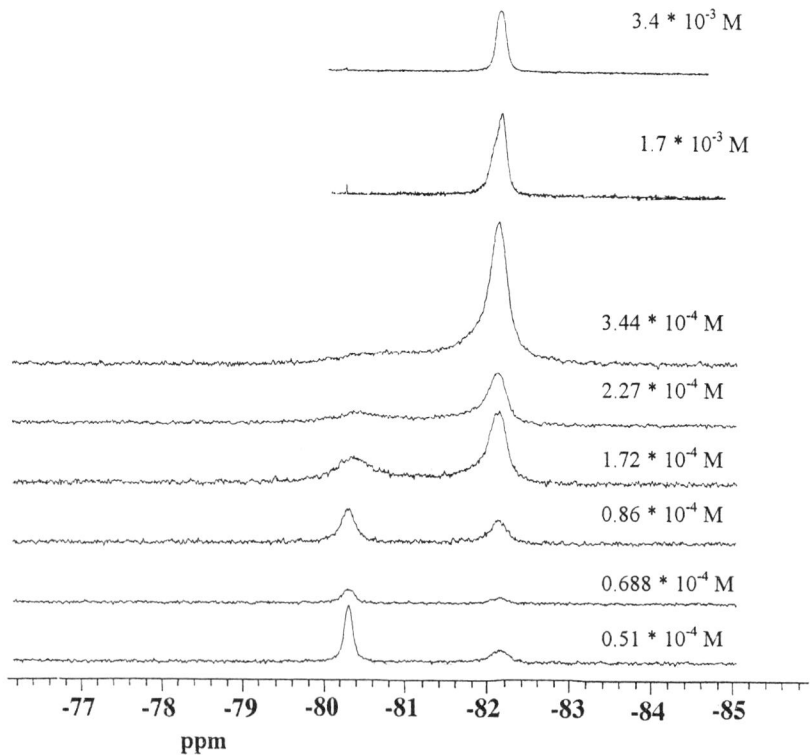

Figure 1. Concentration dependence of the ^{19}F NMR resonances of the CF_3 group in aqueous solutions of FP605M. Reproduced with permission from reference 10.

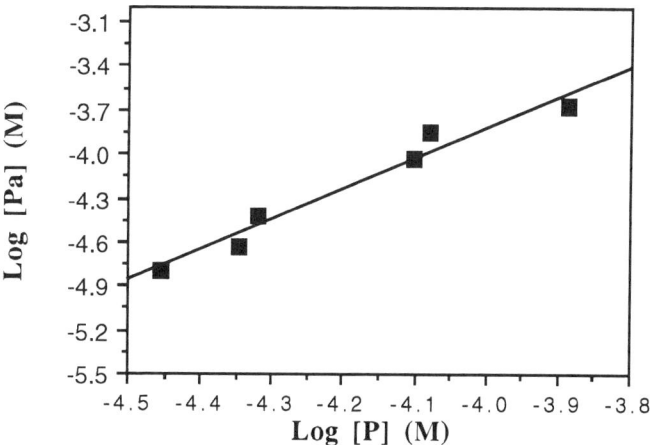

Figure 2. Log-lot plot of the intensities of up- and downfield CF_3 resonances of PF605M as a function of concentration. Reproduced with permission from reference 10.

independent. Thus at higher polymer concentrations the monomeric species is expected to shift upfield whereas the chemical shift of the dimer is unchanged.[20]

In sharp contrast with the C_6F_{13} derivatives, the CF_3 resonances of FP810M are less broad. The same is the case for the other resonances of this polymer and for FP805M. This would be consistent with a slower monomer-aggregate exchange on the NMR time scale and a slower rate of micellar dissociation due to the stronger interaction between the longer perfluorocarbons. The fluorine-19 NMR spectra of the telechelic FP810T show similar concentration dependent changes in the low and high field resonances of the CF_3 end groups as seen with the one ended derivatives.

In contrast to the C_6F_{13} derivatives, the concentration of the dissociated forms of FP810M corresponding to the downfield resonance was observed to be *independent* of the total polymer concentration whereas the concentration of the associated form increased with total polymer concentration. This indicates that FP810M behaves as a typical surfactant. The NMR data gave a CMC value of $7 * 10^{-6}$ M which equals the concentration of the dissociated species.[21]

Because of NMR sensitivity limitations the ^{19}F spectra of FP805M below 10^{-5} M could not be obtained. Only at a total polymer concentration of about $1*10^{-5}$ M, does the resonance of the dissociated form become visible (about 10%). Assuming that, as for the case for FP810M, the magnitude of the downfield resonance is concentration independent, the CMC of FP805M thus is estimated as about $1 * 10^{-6}$ M (**Table 1**). [22]

The magnitude of the upfield CF_3 absorption of FP605M compared to the downfield absorption increases with increasing NaCl concentrations. Thus at a concentration of FP605M of about 10^{-4}M the ratio of the two species is about one but this ratio increases to three and five at salt concentrations of .4M and .6M repectively. These changes in association are consistent with salt effects expected for hydrophobic association and in agreement with our previous results in other hydrophobically associating fluorocarbon modified water soluble polymer systems. The corresponding association constants using an association number of 2.0 are listed in **Table 1**.

At a concentration of $1.3 * 10^{-4}$ M. of FP605M and at 4.8 °C the two CF_3 resonances of the the monomeric and dimeric form are of about the same intensity. Upon heating, the magnitude of upfield resonance increases at the expense of the downfield resonance, consistent with an entropy driven aqueous hydrophobic association process.[23,24] A plot of the log of the dimerization constant K (n=2) defined in equations (1)-(4) against 1/T between 5°C and 27°C gives a ΔH value for the change in enthalpy of the formation of the dimer of + 6.9 Kcal mole^{-1} K^{-1} and a corresponding ΔS value of + 42 cal mole^{-1} K^{-1} corresponding to values of +3.45 Kcal. and +21 cal mole^{-1} K^{-1} per polymeric surfactant. The large positive value of ΔS is comparable to that of fluorocarbon non-ionic surfactants (30 to 33 cal mole^{-1} K^{-1}), and confirms the hydrophobic nature of the entropy driven association process.

Viscosity Studies

As expected the end-functionalized PEGs (FP605M and FP805M) have higher reduced viscosities than the unmodified precursors, indicating the existence of hydrophobic association (**Figure 3**). Between 0.75 and 2.0 g/dL, (1.3-3.4 * 10^{-3} M) FP605M shows a reduced viscosity that is similar to that of the corresponding unmodified PEG (MW=10,000) consistent with the predominantly *dimeric* association of this polymer as indicated by NMR. Thus the NMR spectra indicate that dimerization is largely complete at about $3.0*10^{-3}$M.) that is three times lower than the crtical concentration that from the intrinsic viscosity would be expected to be on the order of 6g/dl or about $1.2*10^{-2}$ M.

At or below 1g/dL the reduced viscosity of FP805M is similar to that of FP605M. Above 2.5 g/dL the reduced viscosity of FP805M, having a larger C_8F_{17} end group, surprisingly, is *smaller* compared to that of the C_6F_{13} one-ended derivative. This difference is consistent with micellization of FP805M (see NMR studies). [25] However, above 2.5 g/dL the reduced viscosities of FP805M are much larger than the estimated value of its dimer, the PEG of a molecular weight of 10,000. This appears to be inconsistent with only micellization and may be due to intermolecular interactions of the R_F groups with the partially hydrophobic PEO.

Similar reduced viscosity vs. concentration profiles are seen for the FP610M and FP810M polymers (**Fig. 3b**). Thus both exhibit higher reduced viscosities compared with the unmodified PEG precursor. Assuming intrinsic viscosities that are identical to that of the unmodified (10K) precursor, the apparent Huggins constants of all four PEO derviatives are about four to five times larger than for the PEO precursors.

The Huggins constants for FP605M and FP610M below a concentration of about 4g/dL are about 4.5 more than an order of magnitude larger than that of the PEG precursors. Similar large increases in k_H are seen for the FP805M and FP810M polymers. This is not consistent with simple dimerization-micellization expected on the basis of the NMR data (see Discussion).

The reduced viscosity vs. concentration profile of the telechelic C_8F_{17} end-functionalized PEO(FP810T), as expected, shows a much stronger increase of viscosity with concentration (**Figure 4**). At 2.5g/dL the reduced viscosity is at least one order of magnitude larger than the comparable one-ended FP810M. The much larger (25 times), apparent Huggins constant of FP810T compared with that of the isobaric PEG precursor between 0.2 and 1.0 dL/g, is also consistent with strong intermolecular association.[22] It is worth noting that this polymer below a concentration of 0.40g/dL shows reduced viscosities that are lower than that of the unmodified PEO consistent with intramolecular association of the R_F end groups.

PDMA derivatives

The much higher MW of the PDMA makes fluorine NMR measurements difficult so that in this case viscosity measurements were used only. Preliminary capillary viscometry measurements have indicated that, at least for the PDMA derivatives, the shear dependence is quite small especially at the lower concentrations.[26]

Figure 5 shows the reduced viscosity vs. polymer concentration for a one- and a two fluorocarbon end-functionalized polymer with the same number-average molecular weight of about 47,000 (**Table 2**, #2b and #3b respectively) and their

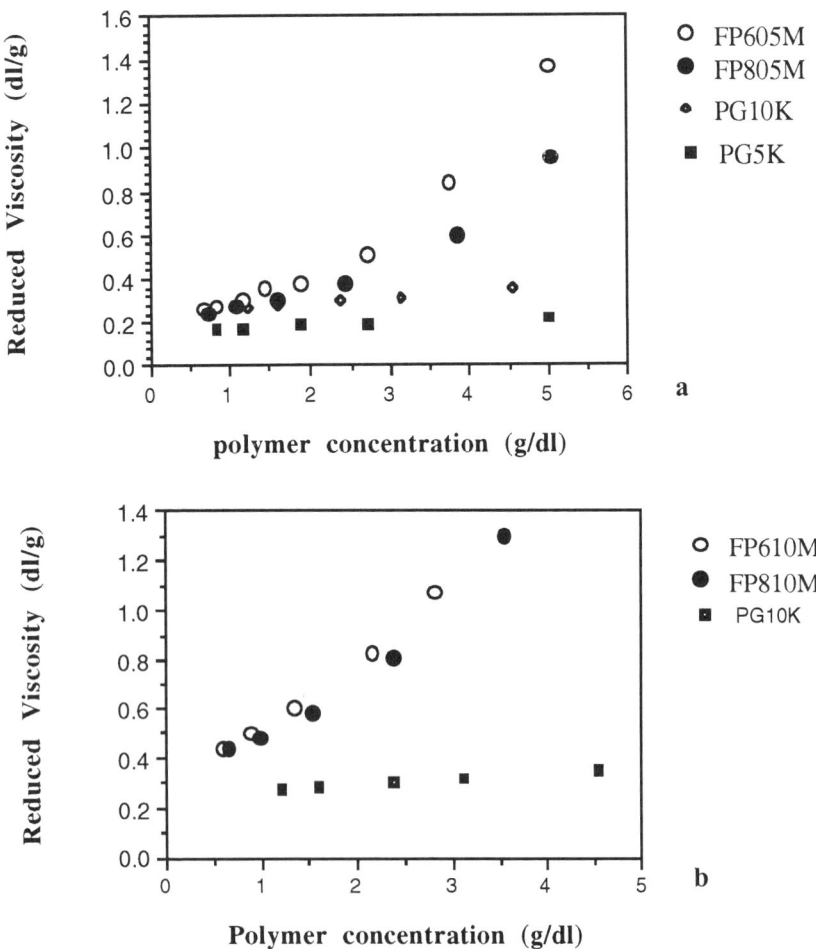

Figure 3. a. Reduced viscosity vs. concentration profiles of PEG5K (■), PEG10K (◆), FP605M(O) and FP805M (●). b. Reduced viscosity vs. concentration profiles of FP610M, (O) FP810M (●) and PG10K (■).

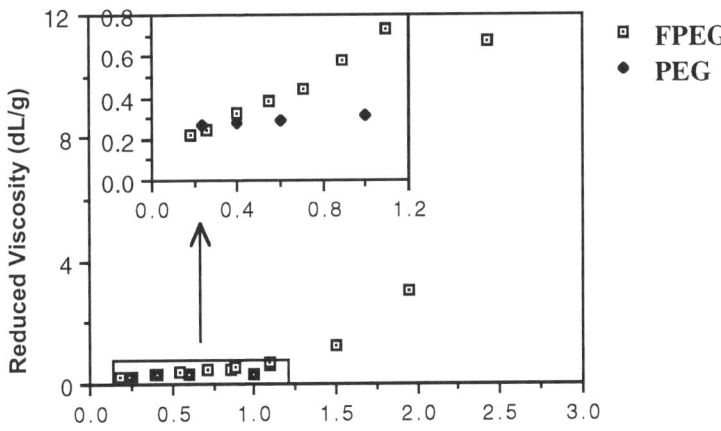

Figure 4. The reduced viscosity vs. concentration profile of an aqueous solution of FP810T (□); PEG10K (◆). Reproduced with permission from reference 11.

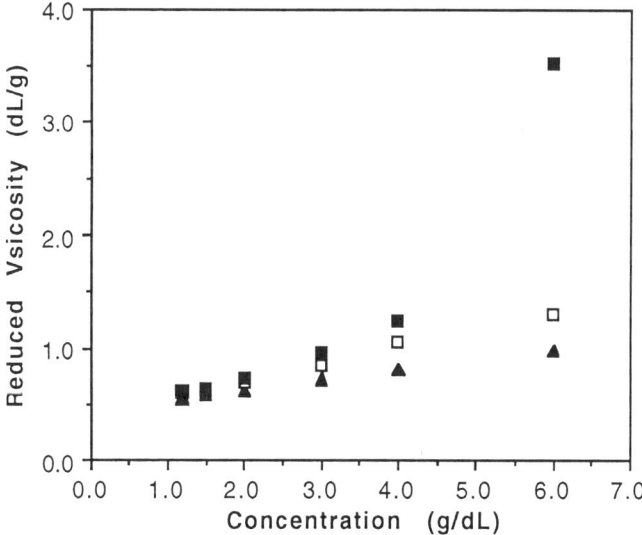

Figure 5. Reduced viscosity as a function of concentration of PDMA homopolymer precursor (▲) Table 2, #2a) and its matching end-capped perfluorooctanoyl end-functionalized polymer (□) Table 2, #2b) and the telechelic R_F end-capped (■), PDMA (Table 2, #3b) in water at 25.0 °C.

homopolymer precursor with the same degree of polymerization (**Table 2**, #2a). Below approximately 2g/dL linear plots are observed for the PDMA homopolymer and for the end-functionalized PDMA's. The intercepts are essentially identical, consistent with the absence of significant intermolecular association below a concentration of 1.0g/dL.

The onset of weak association for the one-ended perfluorooctanoyl -modified PDMA above a concentration of about 2g/dL is visible and the apparent Huggins constant of this sample (k_H = 0.7) is somewhat higher than that of the precursor PDMA(k_H = 0.5).The telechelic bis-functionalized PDMA shows a strongly curved plot and much higher reduced viscosities, especially above 4 g/dL (See Discussion).

Figure 6 shows the reduced viscosity-concentration profiles of a PDMA homopolymer with a number-average molecular weight that is nearly identical to R_F derivatives of the previous section and bisfunctionalized with an octadecanoyl end groups or with an ODPA unit containing a PEO spacer. The telechelic octadecanoyl end-capped PDMA (**Table 3**, #5b) and the PDMA homopolymer of about the same molecular weight (**Table 2**,#5a) appear to have roughly the same intercept consistent with the absence of association at very low concentrations but the viscosity increase with concentration is greater consistent with stronger asociation as also shown by the approximately three times larger Huggins constant (k_H=1.7 vs k_H=0.5).

The PDMA of about the same molecular weight but bis-functionalized with ODPA (**Table 3**, # 6b), shows a much higher reduced viscosities than the octadecanoyl funtionalized sample even below 0.5 g/dl and much larger apparent Huggins constant (K_H = 6.2) consistent with the much stronger interpolymer association (see below). Similar enhancements of viscosity due to the presence of hydrophilic spacers have been demonstrated earlier.[27-29] In methanol the reduced viscosity plots of the ODPA bisfunctionalized PDMA and its unmodified precursor are virtually the same consistent with the hydrophobic nature of the association.[13]

Figure 7 shows the reduced viscosity plots of ODPA bis-functionalized PDMA's and their homopolymer precursors with number average MW's of about 175,000. (**Table 3**, #7a and #7b respectively). The reduced viscosity plots of the isobaric samples #6b and #5a are shown for comparison. The increase in reduced viscosities of sample #6b compared to its isobaric homopolymer(# 5a) is much greater than that between the higher MW PDMA isobaric pair (#7a and 7b) and this is reflected in the much higher apparent Huggins constant of sample #6b (k_H= 5.6) compared to sample #7b (k_H= 1.0). The larger k_H value of the lower MW associating PDMA appears to reflect the stronger association of this polymer brought about by the greater concentration of hydrophobic end groups at comparable polymer concentrations and possibly by the weaker excluded volume effects in the lower MW pair(see below). The Huggins constant of the PDMA homopolymer is not very MW dependent as the lower MW PDMA within experimental error has the same k_H as the higher MW PDMA (Tables 2 and 3).

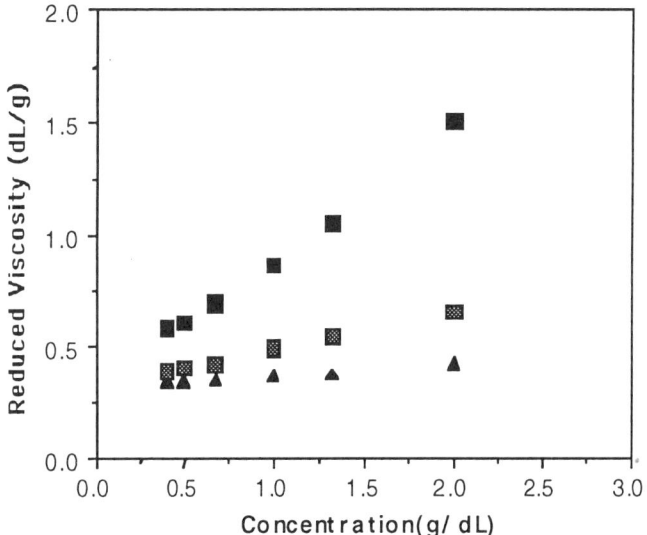

Figure 6. Reduced viscosity in water at 25 °C as a function of the concentrations of PDMA precursor (▲) (Table 3, #5a) and its two-ended octadecyl-funtionalized PDMA: (□) (Table 3, #5b); ODPA bisfunctionalized PDMA (■) (Table 3, #6b).

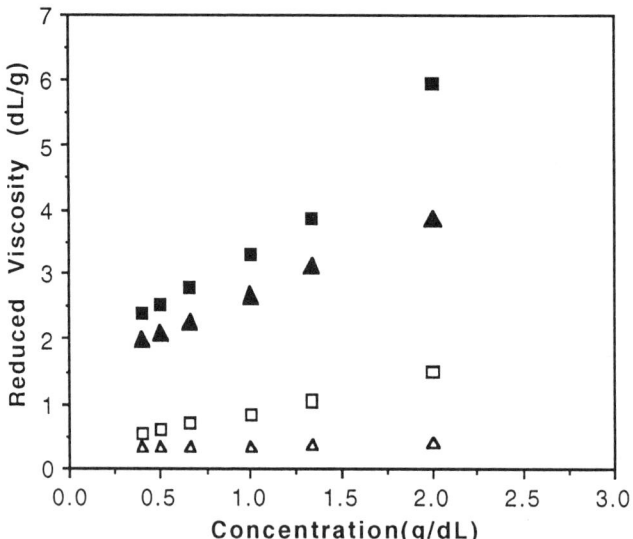

Figure 7. Reduced viscosity as a function of concentration of ODPA bis-functionalized PDMAs and their homopolymer precursors: (▲) PDMA homopolymer, Mn=174,600 (Table 3, #7a); PDMA ODPA bis-functionalized (■) Mn=176,000 (Table 3, #7b) (□) ODPA bis-functionalized, Mn=48,000 (Table 3, #6b); (△) its matching homopolymer, Mn=46,200, (Table 3, #6a).

Surfactant Effects

The dependence of the reduced viscosity of a 2.0 weight % solution of a telechelic C_7F_{15} end-functionalized PDMA (**Table 2**, #3b) on the concentration of ammonium perfluorooctanoate (APFO) is shown in **Figure 8**. Upon addition of APFO to the polymer solution, the reduced viscosity increases, passes through a maximum, and then decreases to values close to the original ones. The maximum occurs at a surfactant concentration that is about 15 times that of the end groups (See Discussion).

Discussion

PEG derivatives

It would appear that the dimerization model at low concentrations demonstrated for the above one ended FP605M and FP610M does not apply to the C_8F_{17} end-functionalized PEO's at the higher concentrations where the high reduced viscosities do not seem consistent with simple dimeric or micellar association of perfluorocarbon groups suggested by the NMR data. Thus the apparent Huggins constants of the *one-ended* PEG derivatives is on the order of about 4-5 and that of the telechelic PEG derivative FP810T is evn greater at about 25 (Table 1). Such Huggins constants are far larger than predicted based on simple dimerization and/or simple micellization. Although it is known that perfluorocarbon and hydrocarbon groups are not completely compatible,[30] the large apparent hydrophobicity of the perfluoroalkyl groups may lead to hydrophobic bonding of the perfluorocarbon groups to the hydrophobic portion of the PEO's (**Scheme 2**). Such a process could explain the much greater than expected viscosity increases at higher polymer concentrations. It is also possible that association occurs of the initially formed dimers into extended micelle-like structures with a large hydrodynamic volume but this would seem less likely.[22] The present data do not seem to allow a resolution of this problem.

With the telechelic PEO derivatives at very low concentrations there is the possibility of intramolecular association and this would tend to reduce the hydrodynamic volume of the telechelic PDMA's giving smaller intercepts (**Scheme 3**). This was observed also for well defined mono- and bis -functionalized PEG's carrying hydrophobic end groups.[31]

The influence of perfluorocarbon size on the hydrophobic association of one-ended fluorocarbon modified PEG derivatives as measured by ^{19}F NMR is larger than expected. Other rheology studies of similar telechelic C_6F_{13} and C_8F_{17} modified PEG (MW = 35,000) derivatives indicate that the C_6F_{13} derivative has a much faster relaxation time (1.4 vs 66 ms)[32] than that of the C_8F_{17} derivative in agreement with the results obtained from the above ^{19}F NMR studies (see above). The relaxation time corresponding to the exit of the hydrophobic group from the micelle is the slowest

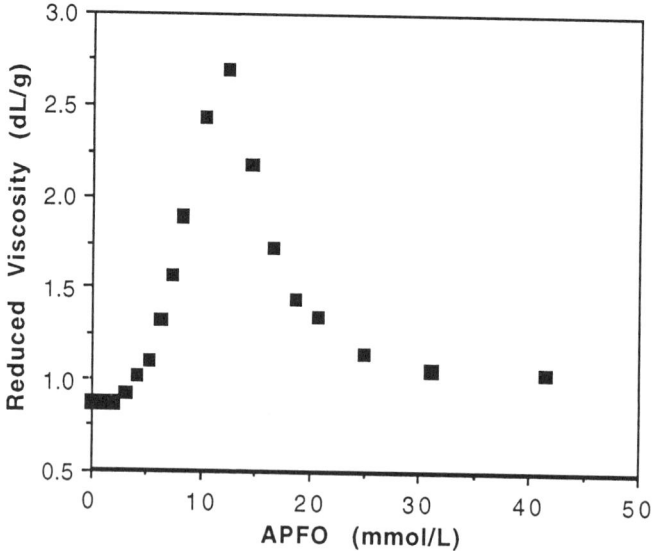

Figure 8. Reduced viscosity of a 2 wt % two-ended perfluorooctanoyl functionalized PDMA (Table 2, #3a) as a function of APFO concentration at 25. 0°C.

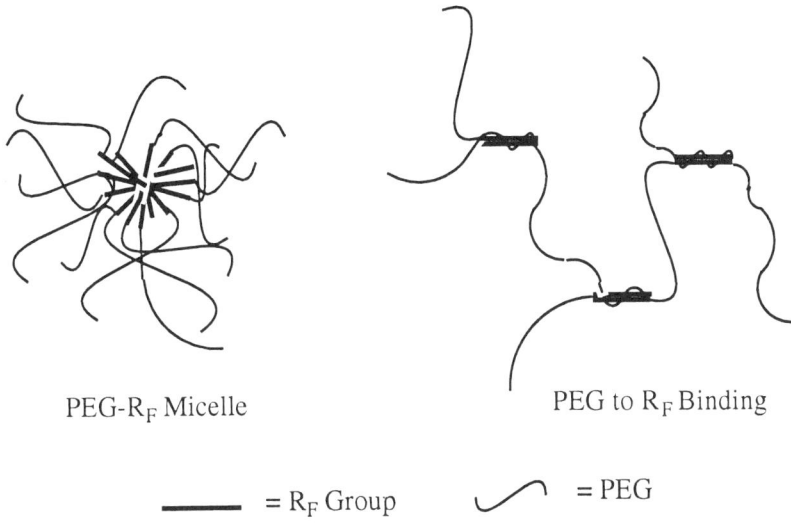

Scheme 2. Modes of Association of R_F End-functionalized PEG's

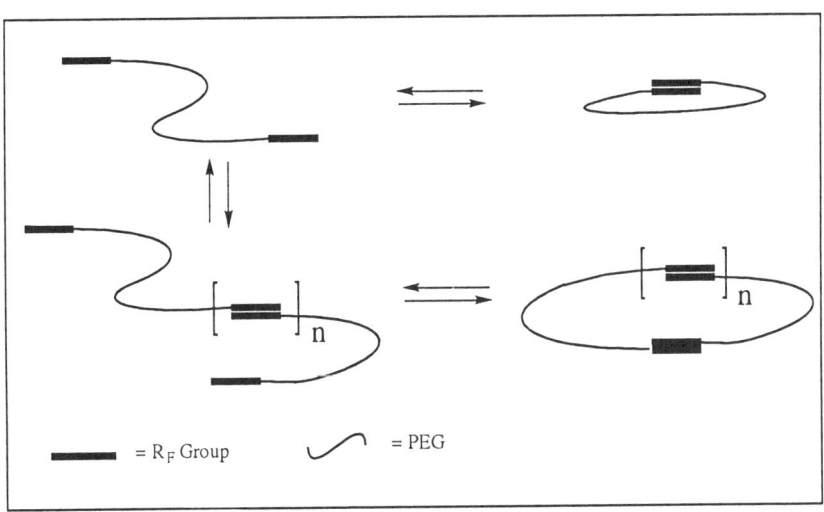

Scheme -3. Association modes of hydrophobically modified telechelic polymers.

PDMA Polymers

The reduced viscosities of the isobaric telechelic R_F and R_H PDMA derivatives 3b and 4b are similar as shown in Figures 5 and 6 and as indicated by the similar k_H values (Tables 2 and 3). This similarity indicates again the much larger apparent hydrophobicities of the fluorocarbon versus the hydrocarbon groups of the same carbon numbers.[28,29]

The Huggins constant of the one ended C_8F_{17} PDMA derivative(k_H =0.7) is much less than that of the corresponding one ended PEG's that are at least five to six times larger. This may be due, at least in part, to the larger molar masses of the PDMA chains giving rise to a larger excluded volume effect and perhaps to as yet poorly understood differences in hydrophobicity of PEG and PDMA.

Because of the pronounced curvature of the reduced viscosity plots of some of the hydrophobe-modified PDMA's shown in **Figures 5-7** it would seem possible to ascribe the curvature of this plot to an increase in Huggins constant. However the changes, if any, of the Huggins constant with the MW of PDMA are not likely to be large [34] as also indicated in Tables 2 and 3. Following the expression (5) for the reduced viscosity :

$$\eta_{sp}/c = [\eta] + k_H [\eta]^2 c \qquad (5)$$

$$(\eta_{sp}/c)_m = [\eta]_1 w_1 + [\eta]_2 w_2 + k_H ([\eta]_1 w_1 + [\eta]_2 w_2)^2 \cdot c \qquad (6)$$

the presence of the dimer having a larger intrinsic viscosity will now contribute to the increase of the viscosity through both terms of equation (5) particularly the term containing $[\eta]^2$. For the concentration dependent equilibrium mixture of a polymer, P_1 and its dimer P_2 having the same Huggins constant the reduced viscosity is given by (6) where w_1 and w_2 denote the weight fractions of the two polymers and where the total polymer concentration, c, equals the sum of the concentrations of the two polymer components.[35] As the weight fraction of the "dimer", w_2, increases with c, according to (6), the slope of the reduced viscosity plot increases eventually to $k_H [\eta]_2^2$ and there will be a non-linear increase of the slope in the region where the increase in w_2 is most significant.

Thus, a change in k_H with MW is not required to account for some of the reduced viscosity plots where the increases in reduced viscosity are small and the reduced viscosity vs. concentration profiles for the one-ended PDMA's and for the telechelic PDMA 's at the lower concentrations shown in **Figures 5 and 6** may be simulated rather well using such a simple dimerization model.[14] The larger increases for the two-ended R_F and R_H derivatives 2b and 5b below about 3g/dL also would appear to be consistent with hydrophobe dimerization leading to chain extension.

Spacer Effects

The reaction of the living PDMA with the octadecanoylchloride or perfluorooctanoyl chloride generates a beta-ketoamide structure that is expected to lead to strong coupling between the hydrophobic group and the PDMA backbone. This should lead to a greater loss of mobility and thus of entropy upon hydrophobic association of the PDMA hydrophobic end groups lacking flexible spacers. The much stronger association behavior of the ODPA-modified PDMA compared with the octadecanoyl- and the perfluorooctanoyl -modified PDMA's of the same MW is most likely due to the presence of the deca(oxyethylene) spacer that serves to decouple the motion of the hydrophobic end group from that of the chain backbone. A similar increased extent of association was demonstrated earlier for polyacrylamides containing pendent hydrophobic 1,1-dihydroperfluorooctyl groups separated from the chain backbone by oligoethyleneoxide [$(-CH_2CH_2O-)_n$ (n=1, 2, 3)] spacers.[28,29] Viscosity increases due to the presence of spacers were demonstrated earlier by Schulz et al.[27] for hydrophobically modified associating polyacryamides having pendent nonylphenyl groups liked to the chain backbone by PEO spacers of lengths varying from 10 to 40 ethylene oxide units.

The effect of molecular weight on the association of the telechelic PDMA's is illustrated in **Figure 7** where the reduced viscosity vs. concentration plots are shown for two ODPA functionalized telechelic PDMA's (**Table 3** #6b and #7b) that differ in MW by as factor of almost four. The larger apparent Huggins constant of the low MW telechelic PDMA #6b,(5.6 vs.1.0) is attributable to the larger concentrations of the hydrophobic groups in this polymer and possibly to smaller excluded volume effects. Such effects have been observed for HEUR's by Glass et al for telechelic octadecyl end-functionalized PEO's.[6a]

Surfactants Effects

Bell shaped curves such as that shown in **Figure 8** illustrating the effects of surfactant addition on the viscosity of hydrophobically modified polymers have been observed by several other groups upon addition of hydrocarbon surfactants to hydrocarbon-modified water-soluble polymers [5,36] or upon addition of perfluorocarbon surfactants to PDMA with pendent R_F groups [37] and have been ascribed to micellar bridging. The effect of surfactant on hydrophobic association can be illustrated by a similar model. As the concentration of surfactant increases, small mixed micelles are formed that are relatively stable compared to the associated structures without the presence of surfactants. However, at higher surfactant concentrations, the large excess of surfactant micelles relative to hydrophobic groups of the polymer increasingly leads to the formation of mixed micelles containing single polymer hydrophobic groups and crosslinking diminishes resulting in a decrease in viscosity.

For the present case, at the viscosity maximum observed for 2.0 weight percent solutions, the concentration of the APFO surfactant is on the order of $12 * 10^{-3}$ M which is about 15 times larger than the concentration of the perfluorocarbon end groups. It would appear then that the corresponding bridging micelles would contain a number of surfactants per PDMA end group that would be equal or smaller than that number.

Acknowledgments

The financial support of this work is acknowledged from the U. S. Department of Energy, Office of Basic Energy Sciences. We wish to thank Dr. J. Zoeller (Hoechst, Germany) for a gift of octadecyl-deca(ethylene glycol).

Key Words: perfluorocarbons, polyethyleneoxide, polyethyleneglycol polydimethylacrylamide, hydrophobic association, fluorine-19 NMR.

References

1. Glass, J.E., Ed. *"Hydrophilic Polymers, Performance with Environmental Acceptance"*; Advances in Chemistry Series 240; American Chemical Society: Washington, DC, **1996**.
2. Schulz, D. N.; Glass, J. E., Eds. *Polymers as Rheology Modifiers*; ACS Symposium Series 462; American Chemical Society: Washington, DC, **1991**.
3. McCormick, C. L.; Bock, J.; Schulz, D. N. In *Encyclopedia of Polymer Science and Engineering*, 2nd ed.; Mark, H. F., Bikales, N. M., Overberger, C. G., Menges, G., Eds.; Wiley-Interscience: New York, **1989**; Vol. 17, pp 730.
4. Glass, J. E., Ed. *"Polymers in Aqueous Media: Performance through Association;"* Advances in Chemistry 223: American Chemical Society: Washington, DC, **1989**.
5. a. Wetzel,W.H.; Chen, M. ; Glass, E.J. Reference 1, Chapter 10; b. Karunasena, A.; Glass, J. E. *Prog. Org. Coat.* **1989**, *17*, 301.
6. Jenkins, R. D. Ph.D. Dissertation, Lehigh University, Bethlehem, PA, **1990**.
7. *The Function of Associative Thickeners in Water-borne Paints*; Huden, M., Sjoblon, E. Bostrom, P., Eds.; XXI Fatipec Congress, Amsterdam, **1992**.
8. Hulden, M. *Colloid Surf. A* **1994**, *82*, 263.
9. Yekta, A.; Duhamel, J.; Brochard. P.; Adiwidjaja, H.; Winnik, M. A. *Macromolecules* **1993**, *26*, 1829.
10. Zhang, H. ; Pan, J.; Hogen-Esch, T.E. *Macromolecules* **1998**, *31*, 2815.
11. Zhang, H.; Hogen-Esch, T.E. ; Boschet F.; Margaillan, A. *Langmuir* **1998**,*14(18)*, 4972.

12. Xie, X. Y.; Hogen-Esch, T. E. *Macromolecules* **1996**, *29*, 1746.
13. Xie X, Ph.D. Thesis University of Southern California, Dec. 1994.
14. Xie, X.; Hogen-Esch, T. E. In preparation.
15. a) Lundberg, D.J.; Brown, R.G.; Glass, J.E.; Eley, R.R. *Langmuir*, **1994**, *10*, 3027 b) Kaczmarski, J.P.; Glass, J.E. *Langmuir*, **1994**, *10*, 3035.
16. Xu, B.; Lie, L.; Yenta, A.; Misaim, Z.; Kanagalingam, S.; Winnik, M. A.; Zhang, K.; Macdonald, P.; Menchen, S.; *Langmuir*, **1997**, *13*, 2447.
17. Hoppf, H.; Lussi, H.; Borla, L. *Makromol. Chem.* **1965**, *81*, 268.
18. Quirk, R. P.; Ren, J. *Polymer International* **1993**, *32*, 205.
19. (a)Schick M. S.; Atlas, S. M. ; Eirich F. R.; *J. Phys. Chem.* **1962**, *66*, 1326 ; (b)Hsiao, L. ; Dunning , H. V. ; Lorentz, P. B. *J. Phys. Chem.* **1956**, *60* , 657.)
20. *Applications of dynamic NMR spectroscopy to organic chemistry, chapter 1*, Michinori, Oki. Ed. VCH. Publishers, **1985**.
21. Myers, D., "Surfactant Science and Techology" VCH Publishers Inc., New York, **1988**, Chapter 3.
22. For recent papers on the association of telechelic R_F end-functionalized PEO's see: a. Menchen, S.; Johnson, B.; Winnik, M. A.; Xu, B. *Chem. Mater.* **1996**, *8*, 2205. b.Preuschen, S.; Menchen, S.; Winnik, M. A.; Heuer, A.; Spiess, H. W. *Macromolecules* **1999**, *32*, 2690.
23. Ravety, J.C.; Stebe, M.J. *Colloids Surfaces A: Physicochemical and Engineering Aspects* **1994**, *84*, 11.
24. Jiang, X. K. *Acc. Chem. Res.* **1988**, *21*, 362.
25. Zimm B. H.; Stockmayer W. H. *J. Chem. Phys.* **1949**, *17*, 1301.
26. Tomczak S.; Hogen-Esch, T. E. Unpublished results.
27. Schulz, D. N.; Kaladas, J. J.; Maurer, J. J.; Bock, J.; Pace, S. J.; Schulz, W. W. *Polymer* **1987**, *28*, 2110.
28. Hwang, F. S.; Hogen-Esch, T. E. *Macromolecules* **1995**, *28*, 3328.
29. Hwang, F. S.; Hogen-Esch, T. E. *Macromolecules* **1993**, *26*, 3160.
30. a. Kunitake, T.; Ihara,H.; Hashiguchi,Y *J.Am. Chem. Soc.***1988**, *106*, 1156. b. Asakawa,T.; Johten, K. Miyagishi, S.; Nishida, M. *Langmuir*,**1988**, *4*, 136.
31. Kaczmarski J. P.; Glass J. E. *Macromolecules* **1993**, *26*, 5149.
32. Menchen, S., Johnson, B., Winnik, M. A., Xu, B. *Chem. Mater.* **1996**, *8*, 2205.
33. Annable, T., Buscall, R., Ettelaie, R., Whittlestone, D. *J. Rheol.* **1993**, *37*, 695
34. Huggins, M.L. *J. Am. Chem. Soc.* **1942**, *64*, 2716.
35. Krigbaum., W.; Wall, F. *J. Polym. Sci.* **1950**, *4*, 505. b.Cragg, L.H. *J. of Colloid Science* **1946**, *1*, 261; c. Cragg, L.H.; Bigelow, C. *J. Polym. Sci.* **1955**, *16*, 177.
36. a. Biggs, S.; Selb, J.; Candau, F. *Langmuir* **1992**, *8*, 838. b. Volpert, E.; Selb, J.; Candau, F. *Polymer* **1998**,*39(5)*, 1025. c. Nilson, S. ; Thuresson, K.; Hansson, P.; Lindman, B. *J. Phys. Chem. B* ,**1998**, *102*, 7099.
37. Xie, X. Y.; Hogen-Esch, T. E. *Macromolecules* **1996**, *29*, 1734.

Adsorption Studies:
POE, HEUR, and HMPAM

Chapter 12

Dynamics in Adsorbed Polymer Layers

Maria M. Santore, Zengli Fu, and Ervin Mubarekyan

Department of Chemical Engineering, Lehigh University,
111 Research Drive, Bethlehem, PA 18015

The dynamic features of polymer layers are considered in the context of the adsorption process for a model polyethylene oxide (PEO) system adsorbing onto silica. The findings with PEO are then used to interpret the dynamic behavior of hydroxyethyl cellulose (HEC) on silica. Rapid adsorption is transport limited and controlled by the hydrodynamic size of the adsorbing chains. In polydisperse systems, therefore, short chains adsorb more rapidly than long ones. For homopolymers, however, long chains are preferred on the surface at equilibrium, such that at intermediate stages of adsorption, an exchange between long and short chains occurs. The exchange rate between free and adsorbed chains is a strong function of the age of the chains already on the surface and their molecular weights. For long chains of both PEO and HEC, interfacial relaxations immobilize the entire adsorbed layer; however for short PEO chains, a mobile population remains in layers that are fully relaxed. Aging typically occurs on the timescale of several hours and, therefore, has a potentially significant impact on the adsorption process itself. The influence of aging is more pronounced with polydisperse systems because the molecular weight driven exchange is hindered in mature layers, retarding their approach to equilibrium.

Introduction

When it comes to polymers adsorbed on dispersed colloidal particles for rheological modification of coatings or the stabilization of the particles themselves, two aspects of interfacial dynamics are critical: First, the timescales of chain diffusion to and sticking onto particles are relevant to successful formulation. If unstable chain configurations occur during adsorption, for instance chain bridging between particles, the dispersion stability may be permanently compromised. This can occur, for instance, when adsorption is slower than the rate of particle collisions. Second, following the initial adsorption, adsorbed chains reconfigure at constant interfacial mass and established layers respond to stimuli in the bulk solution such as temperature changes, the addition of new components, and shear. These adsorbed layer dynamics are critical to dispersion rheology and the robustness of a particular formulation.

In the past decade, studies of adsorbed polymers have shifted from an emphasis on interfacial structure[1-3] and associated network morphology[4,5] to dynamic behavior. As a result, certain general ideas concerning the dynamics of physically adsorbed polymers have developed. First, it is established that adsorption often occurs quickly and is frequently diffusion limited.[6-8] It has also been observed that adsorbed polymer layers are resistant to removal by solvent washing when the solvent in question is the same as that during the original adsorption.[9,10] That is, removal of the free polymer chains from a bulk solution does not, on reasonable experimental timescales, cause chains to desorb from a surface to take their place in the bulk solution. Desorption is very slow, even though the physical adsorption of the chains is fundamentally reversible. The reversibility of the adsorption is evidenced by the ability of polymer chains in solution, identical to those originally adsorbed on the surface (except for a radiotracer label) to trade places with chains in an adsorbed layer.[11] The timescale for this "self exchange" usually lies somewhere between that of adsorption and desorption.

The explanation of the relative timescales for these processes was put forth by deGennes.[12] First, desorption is slow relative to adsorption because in the case of physisorbed polymers about 50-100 segment-surface contacts, each contributing about a kT to the energy barrier, must be broken for a chain to leave the interface. Further perspective on slow desorption kinetics, in the context of a polymer's adsorption isotherm, was provided by Dijt et al.[13] When homopolymers adsorb, the plateau of the true adsorption isotherm extends down to extremely dilute bulk solution concentrations.[3] The surface is still saturated when the free solution contains 10^{-10} ppm or fewer chains. As a result, when one considers the fastest desorption kinetics possible, corresponding to the diffusion limit with local equilibrium between the surface and the free solution near the surface, diffusion away from the surface has an extremely small driving force. This driving force is the concentration of polymer chains near the surface [10^{-10} ppm (or less)] minus that in free solution [zero]. While this small driving force persists, the surface is nearly saturated as a result of the local equilibrium. Though the form of the diffusion-limited desorption depends on the detailed flow geometry near the surface, logarithmic decays tend to result.[13] In contrast to the slow desorption kinetics for physisorbed polymers, self exchange dynamics are relatively more rapid because the energy barrier is thought to involve only a few segment-surface contacts.[12] Free chains approach those in the adsorbed layer and trade places a few segments at a time.

While polymer adsorption is thought to be rapid and even transport-limited, quantitative reports of adsorption kinetics have appeared predominantly within the past decade because these measurements require precise, time-resolved detection and controlled hydrodynamic conditions near a surface: either a quiescent fluid, shear, or impinging jet flow. The latter two yield the most precise kinetic data because perfectly quiescent conditions are difficult to maintain. Adsorption kinetics have been reported for a variety of systems; however, some of the most quantitative interpretations of adsorption kinetics include Dijt's studies of narrow molecular weight standard polystyrene (PS),[14] and Dijt's and our work with polyethylene oxide (PEO),[14,15] and Robertson's,[16,17] Lenhoff's,[18] and Lyklema and Norde's studies[19] of protein adsorption kinetics. In these cases, the transport-limit has been established exactly and used as a basis of comparison for adsorption kinetics. For the particular cases of PEO on silica and several of the protein-surface combinations, the transport-limited kinetics from gentle shearing or impinging jet flow were manifest as a constant adsorption rate, persisting up to surface saturation.[14-18] The constant (linear) rate is consistent with the development of a pseudo steady-state concentration gradient in the solution near the surface that remains fixed while the surface remains unsaturated. Then, the adsorption rate is proportional to the bulk solution concentration and scales on the flow rate to a power which depends on the flow geometry.

A number of other cases do not exhibit linear adsorption kinetics. Three possible reasons for the deviations exist: First, linearity requires that all the adsorbing species are identical. Polydispersity leads to a hydrodynamic size range such that the rate of chain arrival at the interface depends on the molecular weights of the various populations. Further, depending on the polymer architecture and chemistry, the surface may prefer long or short chains, such that adjustments in the adsorbed populations occur after the surface is nearly saturated with mass.[20] This leads to exchange processes during the final stage of adsorption, which give a rounded kinetic shoulder. Even in the absence of polydispersity or chemical heterogeneity, nonlinear (rounded) adsorption kinetics may be observed as a result of a low surface capacity.[17] When the equilibrium adsorbed amount is sufficiently small that the time to saturate the surface is shorter than the time needed to set up the pseudo-steady state concentration gradient near the surface, the kinetics appear rounded with a maximum slope less than the mass transport limit. This can also occur when the bulk solution concentration is relatively high; however, linear kinetics may be seen for the same system at more dilute concentrations.[15] A third potential reason for rounded kinetics may be that the rate of interfacial chain arrival and the rate of interfacial reconfiguration are similar. In that case, saturation of the surface may occur slowly, as the initially adsorbed chains rearrange to make room for the late-arriving chains. With PEO adsorbing on silica, the linearity of the adsorption kinetics up to saturation implies that the adsorbed chains quickly reconfigure to accommodate new arrivals.[14,15] Dijt's observations suggest that PS kinetics are rounded as a result of slower PS motions on surfaces compared with PEO.[14] Up until recently, it was thought that PEO was a special case where the surface mobility was sufficiently high that local equilibrium was always achieved.[14,15]

The third contribution, above, for rounded adsorption kinetics motivates a closer look at the interfacial chain reconfigurations themselves. Reconfigurations within adsorbed layers occur in response to changes in bulk solution conditions such as changes in temperature;[21,22] and in many cases, reconfigurations also occur in the final stages of polymer adsorption. The latter were first studied by Pefferkorn *et al.* for polyacrylamides adsorbing onto glass spheres.[11] More recent quantitative kinetic studies include Granick's work with PS relaxation on silica in a number of different solvents.[23,24] In the case of polystyrene, replacement of preadsorbed protio-PS by deuterio-PS in quiescent conditions follows a single exponential decay while for higher molecular weights, two distinct stages of the exchange are visible. The self exchange process becomes slower as the originally adsorbed layer ages, and the approach to the final relaxed state is history dependent. In the case of displacement of PS by PMMA, stretched exponential kinetics are observed,[25,26] attributed to pinning of the PS on the surface by PMMA and removal of the system from equilibrium.

Our work addresses the relative timescales of adsorption and relaxation for PEO on silica in an aqueous environment. Until recently PEO was thought to be mobile on a silica surface or to exhibit rapid interfacial relaxations, due to its highly flexible backbone.[14,15] The current study shows that even with highly mobile polymers such as PEO, segment-surface interactions can drastically increase the timescales for chain reconfiguration. The current work then goes on to compare the behavior of narrow molecular weight standard PEO samples with the polydisperse hydroxyethyl cellulose (HEC) system. From an understanding of the relaxation times of HEC, an interpretation of the initial adsorption kinetics and the apparent equilibrium isotherm is put forth. Our approach is to examine adsorption and relaxation in gentle shearing flow, sufficient to set up a constant rate of interfacial chain arrival without significantly perturbing adsorbed configurations. These controlled hydrodynamic conditions facilitate more precise quantification of interfacial timescales, especially relevant to systems with rapid interfacial motions.

Experimental

Narrow molecular weight standard PEO was purchased from Polymer Laboratories, Inc., and, in some cases, used as is. For fluorescent tracer experiments, a coumarin tag (Molecular Probes) was attached to the hydroxyl end of each chain via isocyanate chemistry, yielding a urethane linkage. The procedures for the reaction, purification, and assessment of the tagging yield (80-90%) for coumarin and PEO have been described previously.[27] Natrosol 250GR HEC was provided by Hercules, and where appropriate, used directly. This grade of HEC has a weight average molecular weight near 300K and a polydispersity of 2-3. In studies requiring fluorescent labels, fluorescein tags were attached randomly via isothiocyanate chemistry to hydroxyls on the HEC backbone. The labeling densities for various samples ranged from 0.5 - 2.3 tags per 100K molecular weight of backbone, for different batches. The procedures for the fluorescein - HEC labeling reaction, purification, and assessment of labeling density followed our existing protocols for fluorescein.[28,29]

In the experiments described here, polymers were dissolved in dilute phosphate buffer (0.002 M KH_2PO_4 and 0.008 M Na_2HPO_4), needed to control the silica surface charge and the fraction of dissociated silanols. This is important because PEO and HEC adsorption are thought to result from hydrogen bonds between the polymer's ether linkages and non-dissociated silanols. This hypothesis is reasonable because increases in the silanol dissociation reduce polymer adsorption.[30] Additionally with HEC, hyrdroxyls from the polymer may hydrogen bond with ionized surface silanols.

Acid etched glass microscope slides (exposed to concentrated sulfuric acid *in-situ* for several hours, followed by rinsing with copious amounts of deionized water) were used as the substrate. XPS studies of these surfaces revealed a silica layer,[29] and optical characterization suggested that the silica region is approximately 13 nm thick for acid exposure times on the order of several hours.[15]

Adsorption measurements were conducted in a slit-shear flow cell made of a black teflon base into which a thin (0.5 mm) channel was machined.[28,31] A microscope slide comprised one wall of the channel, and adsorption onto it was measured either optically or via fluorescence. Optical internal reflection near-Brewster reflectivity, employing a polarized Helium-Neon laser (633 nm) provided a measure of the evolving surface excess, independent of any fluorescent labeling. The set-up for this technique and the calibration procedures were described previously.[15] Total internal reflectance fluorescence (TIRF) comprised a second means of tracking adsorption. In TIRF, an Argon ion (488 nm) laser beam is totally internally reflected within the substrate, setting up a standing surface light wave (evanescent wave) whose intensity decays normal to the interface on a length scale of ~ 100 nm. Fluorescent tags on adsorbed polymers and those on chains in solution near the surface are excited by the evanescent wave and this fluorescence is tracked in real time.[28,32] The free chain contribution to fluorescence is often small relative to that from the adsorbed chains. The combination of TIRF and reflectivity allows one to measure the kinetics of specific interfacial populations and interpret the results in terms of the overall interfacial evolution.[20,27]

Results

Adsorption Isotherms

Prior to an investigation of adsorption kinetics, one must first consider the statics, summarized by the isotherms in Figure 1. The adsorbed amounts were measured using reflectivity, and the free solution concentrations are known from the solution formulation. The bulk solution concentration is fixed since it flows continuously through the cell. For two different molecular weights of PEO, in Figure 1A, the isotherm exhibits a flat plateau extending into the dilute regime for several decades.

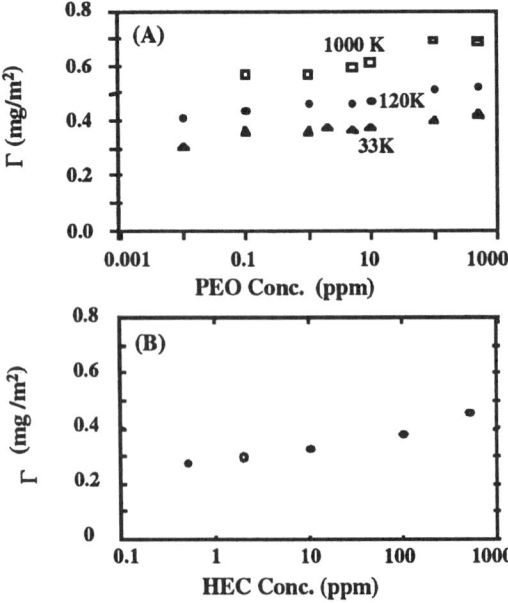

Figure 1. Adsorption isotherms for (A) PEO (Ref 15) and (B) HEC.

While higher molecular weights of PEO give slightly higher coverages, the physisorption is strong. Figure 1A is consistent with theoretical predictions and the observation of extremely slow desorption.[14,15] PEO resists removal into DI water and phosphate buffer because the plateau of the isotherm corresponds to solutions at least as dilute as 0.01 ppm, giving very little driving force for diffusion from the surface. In the case of hydroxyethyl cellulose, in Figure 1B, there is a slighter greater influence of free concentration on the adsorbed amount, a result of polydispersity. Adsorbed HEC is also resistant to desorption in buffer or DI water over a period of days.

Effects of Molecular Weight on Adsorption and CoAdsorption Kinetics

The influence of polydispersity on adsorption kinetics was addressed using binary mixtures of narrow molecular weight standard PEO, in Figure 2. Figure 2A shows the transport-limited adsorption kinetics from reflectivity for two separate runs with 33K and 120K PEO, both of which employ a concentration of 2.5 ppm and a wall shear rate of 7.2 s^{-1}. The lower molecular weight sample adsorbs more quickly, but to a lower ultimate coverage per Figure 1A. The adsorption rates correspond to transport-limited adsorption, which for the case of shearing flow, follows the Leveque solution to the convection diffusion equation.[33]

$$\frac{d\Gamma}{dt} = 0.538 \, (\gamma/L)^{1/3} \, D^{2/3} \, C \tag{1}$$

Equation 1 applies when the surface has a large adsorption capacity such that a pseudo-steady state concentration profile has sufficient time to develop in the solution near the interface. The constant rate of adsorption, dG/dt, is then proportional to the bulk solution concentration, C, the diffusion coefficient, D, to the 2/3 power, and the wall shear rate, γ, to the 1/3 power. L is the distance from the cell inlet to the point of observation. The adsorption kinetics of the single-component runs in Figure 2A quantitatively adhere to equation 1, and calculation of the diffusion coefficient from the observed adsorption rate yields values in agreement with the literature for several PEO molecular weights.[15] In Figure 2A the transport-limited rate persists from the initiation of adsorption up to surface saturation, suggesting that even at the highest coverages, the chains already adsorbed to the surface do not hinder the adsorption of the last chains needed for surface saturation. This in turn suggests that the populations of PEO adsorbed onto the surface at short times are mobile relative to the rate of chain arrival in equation 1.

Figure 2B illustrates the coadsorption kinetics for a mixture of 33K and 120K PEO chains, each with a concentration of 2.5 ppm such that the total PEO concentration is 5 ppm. Reflectivity reveals the evolution of the total surface excess while the TIRF data show the adsorption of the two populations within the mixture. (TIRF data were obtained over a series of experiments involving mixtures of labeled 33K PEO and unlabeled 120K PEO, and vice versa.) At short times, both short and long chains adsorb independently at their individual transport-limited rates and the total surface coverage is simply the sum of the two transport limited adsorption traces, per equation 1. When the total surface coverage is equal to the saturation coverage for the short chains, near 4 minutes, the surface appears saturated at short chains. Beyond

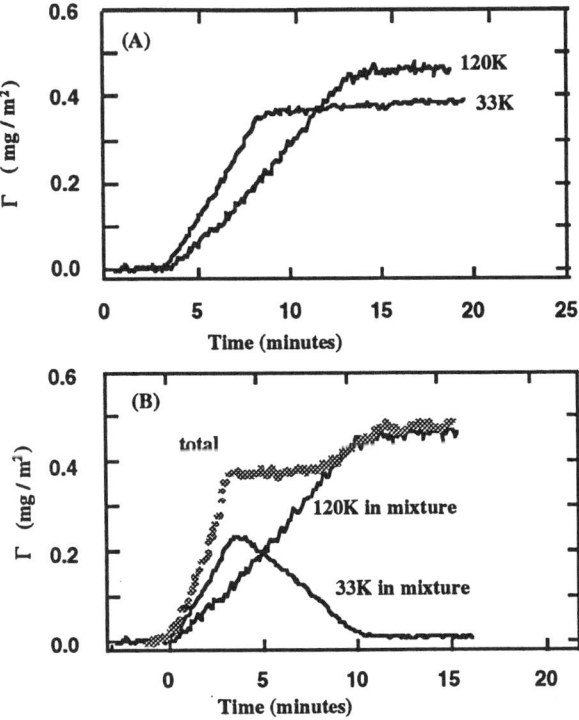

Figure 2. (A) Single component adsorption kinetics for narrow molecular weight PEO standards. (ref 20) (B) Adsorption kinetics for a mixture of long and short chains. (ref 27)

this time, the TIRF data reveal that the long chains continue to adsorb at their transport limited rate, displacing the short chains. During most of this exchange period, the total surface coverage is nearly constant. Once most of the short chains have been displaced, the long chains continue to adsorb towards equilibrium, causing a slight increase in the total surface coverage towards the end of the run.

The entire coadsorption run in Figure 2B takes place at a diffusion limited rate, as confirmed by repeated runs with different bulk solution concentrations. The diffusion limit implies local equilibrium between the adsorbed chains and those in the fluid elements nearest the surface, providing information about the competitive adsorption isotherm. First, from Figure 2 it is obvious that in competitive situations involving homopolymer chains of different lengths, long chains are preferred on the surface over short ones. The surface selectivity for long chains results from the smaller translational entropic loss (in the bulk solution) to saturate surface sites with long chains as opposed to short ones, and has been confirmed by calculations[14,34,35] and other experiments.[36-38] Figure 2B also suggests that while long chains are ultimately preferred on the surface, there exists on the binary adsorption isotherm a region where short chains and long chains coexist on the surface. This mixed surface regime occurs with little effect on the total adsorbed amount (evidenced by the constant total coverage during the exchange) and corresponds to a solution primarily containing short chains with very few long chains present at all (as evidenced by the continued transport-limited adsorption which implies the long chain concentration near the surface is nearly zero.) At higher interfacial long chain concentrations, long chains dominate the surface. These interpretations of our experimental results are consistent with Dijt's mean field calculations for binary adsorption isotherms.[14]

From an examination of the step-wise evolution of the total surface coverage for the bimodal mixture in Figure 2B, one can imagine that the kinetic adsorption trace for a polydisperse system would be rounded as a result of an infinite number of chain populations of varying lengths giving rise to an infinite number of little steps. The exact shape of the kinetics would depend on the molecular weight distribution and would give the appearance of a slow approach towards equilibrium. In comparison with the linear kinetics in Figure 2A for the narrow molecular weight standards, polydisperse samples can give the false impression of a slow kinetically hindered approach to the maximum coverage. In reality, a transport-limited exchange processes may dominate the final adsorption.

Also relevant to the adsorption of polydisperse samples is whether the exchange process following the initial adsorption proceeds at the mass-transport limited rate, as was the case in Figure 2B with PEO. If the initially adsorbing polymers adhere with sufficient strength, as will be shown to occur with HEC, then when long chains from bulk solution attempt to displace short ones already on the surface, the kinetically-hindered exchange may become protracted and equilibration may exceed experimental timescales. In such instances, isotherms may appear rounded, and not represent the true equilibrium. In experiments such as ours with polydisperse materials like HEC, measurement of the true equilibrium coverage becomes a waiting game for completion of the exchange step; however, when equilibrium is approached, the molecular weight distribution of the adsorbed chains may be dramatically different from that in the bulk solution.

Surface Relaxations

The kinetics of surface relaxations and their influence on the exchange between adsorbed and free chains was investigated via fluorescence self-exchange studies.[39] In these experiments, a layer of polymer is adsorbed, aged in flowing solvent (with minimal desorption), and then exposed to unlabeled chains that are identical to those originally adsorbed, with the exception of a fluorescent label. As the labeled chains are displaced by the unlabeled ones, the TIRF signal decreases. The procedure is also run in reverse with the unlabeled chains adsorbed first.

The exchange kinetics are interpreted in the context of a diffusion exchange model, which has been derived in detail elsewhere.[39,40] To summarize, chains diffuse from the flowing bulk solution to the fluid nearest the interface, with mass transfer coefficient $M = 0.538(\gamma/L)^{1/3}D^{2/3}$ (per equation 1). The same M applies to both labeled and unlabeled chains, since fluorescent tagging does not affect the hydrodynamic coil size.

In series with the bulk mass transport is a surface exchange step between the labeled and unlabeled chains. Replacement of a labeled chain by an unlabeled one involves a first order rate constant k while the reverse reaction has kinetic constant k'.

$$\text{PEO(free)} + \text{C-PEO (adsorbed)} \underset{k'}{\overset{k}{\rightleftharpoons}} \text{PEO(adsorbed)} + \text{C-PEO(free)} \quad (2a)$$

$$\frac{d\Gamma_{PEO}}{dt} = -\frac{d\Gamma_{C-PEO}}{dt} = k\, C_{PEO}\text{(local)}\, \Gamma_{C-PEO} - k'\, C_{C-PEO}\text{(local)}\, \Gamma_{PEO} \quad (2b)$$

The two rate constants may not be exactly equal if the surface has a slight preference for the labeled species, which has been found to be the case to a slight extent for coumarin tagged PEO. The surface selectivity, K=k/k', was found to be 0.4 for 33K PEO with coumarin tags, and 0.83 for 120K PEO with coumarin labels. This influence of the label is almost an order of magnitude less than the effect of deuteration on PS adsorption[41] studied by neutron reflectivity and IR. In the case of the fluorescently-tagged HEC adsorbing on silica, the forward and reverse rate constants were found to approach each other such that K=1 within experimental error.

With the mass transport and surface exchange steps in series so that all chains diffusing to or away from the interface undergo the self-exchange reaction, the following form results for the fluorescence decay:[39]

$$\tau = -\left(\frac{1}{\lambda} + \frac{1}{K}\right) \ln(F) + \left(1 - \frac{1}{K}\right)(1-F) \quad (3)$$

The analytical solution, equation 3, resulting from the coupled differential equations is implicit in the normalized evolving fluorescence signal, F (which is initially unity and decays to zero with complete exchange.) The dimensionless time, τ = MCt / Γ(total), is that needed to saturate the surface at the mass-transport limited rate. The kinetic parameter, λ=kΓ(tot)/M, is the ratio of the forward surface exchange rate to the rate of chain diffusion to the interface. Fitting equation 3 to the data and solving for λ facilitates a determination of k. In the case of K=1, equation 3 reduces to a single exponential decay: F = exp(-τ/Λ) where Λ = (λ+1)/λ.

Figure 3 shows self exchange experiments for 33K PEO layers, in both the forward and reverse protocols. In Figure 3, following 10 minutes for the initial adsorption, the layers are incubated for 10 minutes in flowing phosphate buffer and then challenged with the appropriate labeled or unlabeled chains. The subsequent exchange is slightly slower than the transport-limited rate, with k= 600 cm^{3}g^{-1}s^{-1} and k'=1500 cm^{3}g^{-1}s^{-1}, describing the first 80% of the exchange process. The last chains to be displaced are removed from the surface at a slightly slower rate. In Figure 3, the forward and reverse self-exchange curves are skewed slightly upwards due to the slight preference of the surface for labeled chains. This feature is taken into account explicitly in equation 3, because a finite value of K= 0.4 (measured independently) has been employed.

Figure 4 shows the influence of layer age on the displacement of coumarin-tagged PEO by unlabeled chains of the same molecular weight. In Figure 4A the molecular weight is 33K; in Figure 4B it is 120K. Layers are adsorbed from flowing PEO solution for 10 minutes, and then incubated in flowing buffer for various times. In Figures 4A and B, older layers are slower to self-exchange; however, there is an

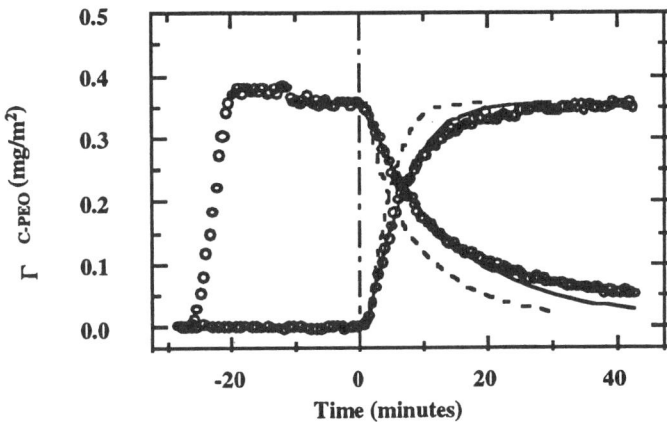

*Figure 3. Forward and reverse self-exchange for young 33K PEO layers.
(- - - -) transport limited exchange rate; (———) best fit to equation 3. (ref 39).*

influence of molecular weight. First, for the 33K PEO sample, regardless of the layer age, the initial displacement of the labeled chains by unlabeled ones is nearly transport-limited, following the kinetics in Figure 3. For the 33K PEO chains, the primary effect of layer age is to develop a tightly bound population which is more resistant to self-exchange. After about 10-20 hours, the adsorbed layer appears fully relaxed, with about 25% of its chains still able to self exchange near the transport-limited rate.

In contrast to the behavior of the 33K PEO chains, aging causes 120K PEO chains to self-exchange more slowly. With the 120K chains, there is no population which readily self exchanges. Rather as the layer ages, all the chains become more difficult to displace until finally after 10-20 hours, the chains do not self-exchange at a rate that is experimentally accessible. Another important difference, therefore, between the 33K and 120K chains, besides the rates of self exchange is that in the fully relaxed state, low molecular weight chains are still self exchangeable while moderate and higher molecular weight chains become permanently trapped on the surface. These conclusions are summarized in Table 1.

The initial stages of self exchange for low molecular weight layers, and the exchange of young layers of higher molecular weight PEO chains can be explained physically in the context of a first order surface rate constant (which implies a lack of cooperativity in the motions of the adsorbed chains). The form of the long-time self exchange kinetics is, however, still open to interpretation. While stretched exponential forms have been reported for the displacement of chains trapped in nonequilibrium conformations,[25,26] stretched exponential kinetics do not describe the data in Figure 4. A closer approximation to Figure 4 is a bi-exponential form where the more rapid decay is described by equation 3, and the slow decay is a second exponential, within the resolution of the current data. In order to argue that the 2nd decay were stretched rather than regular exponential, data would be needed for several more decades in time,

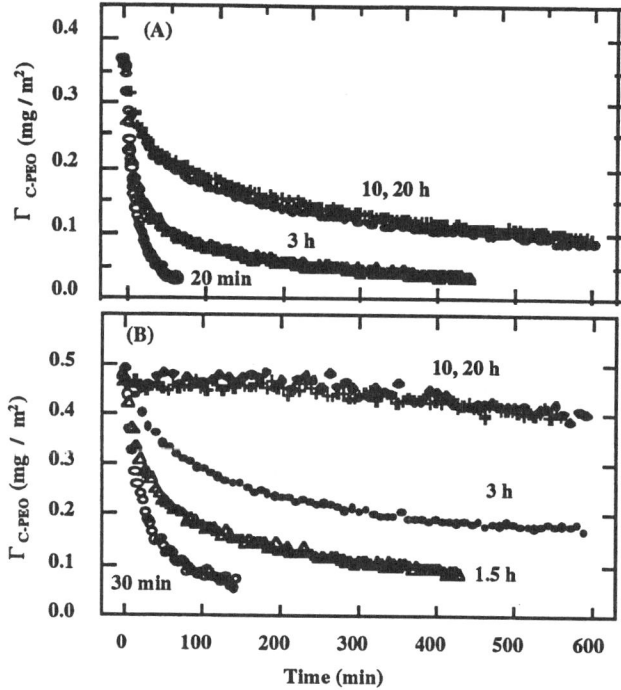

Figure 4. Self exchange kinetics after various aging periods for (A) 33K and (B) 120K PEO layers.

an impracticality for wet-adsorption experiments. The acquisition of data over a broader range of molecular weights is a focus of ongoing work which may highlight the role of lateral entanglements in surface mobility.

The results in Figures 3 and 4 and Table I for the model PEO system suggest that layer age and molecular weight are important in establishing the surface mobility of adsorbed chains. For practical reasons it is desirable to extend this concept to other systems which may not be as uniform in their molecular characteristics, for instance HEC in Figure 5. Prior to the self exchange data shown in Figure 5, fluorescein-tagged HEC layers have been adsorbed from flowing HEC solutions for 10 minutes. After different aging times in flowing buffer, unlabeled HEC is introduced to initiate the exchange and, subsequently, the signal decays as shown. The data are compared in Figure 5 with the predictions from equation 3 with K=1. In the case of HEC, the transport limited self exchange rate is much faster than any of the observed kinetics. The best fits of equation 3 to the initial exchange kinetics were used to determine k, also summarized in Table I.

Table I. Summary of Surface Exchange Kinetics

Sample Age (hours)	k (initial exchange kinetics) cm^3/(g s)	mobile fraction
PEO 33K		
0.33 h	600	0.95
3	600	0.9
10	600	0.65
20	600	0.65
PEO 120K		
0.5 h	500	0.9
1.5	310	0.8
3	50	0.6
10,20	∞	0
HEC		
0.25	3.1	0.9
1	1.6	0.85
2.5	1.3	0.75
4	1.1	0.75
7.5	0.78	0.5
16	∞	0

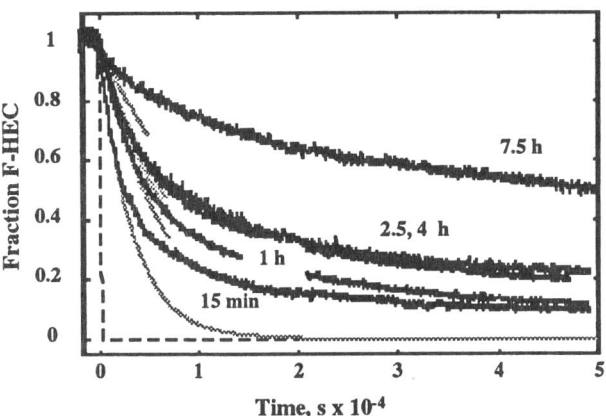

Figure 5. Self exchange kinetics for HEC layers of various ages. (- - --) transport limited exchange rate; (gray lines) best fits to equation 3. Note: Layers aged beyond 16 hours did not self exchange at all.

The relaxation behavior of HEC is qualitatively similar to that for the 120K PEO; however before interpretations are made, one must keep in mind that the self exchange kinetics for the HEC are more complex than those of PEO as a result of the HEC polydispersity. When the HEC layer is adsorbed, chains of different molecular weights adsorb initially and then a molecular weight-driven exchange process sets in, leading to an increase in the average molecular weight of the adsorbed chains. For the HEC sample, this molecular weight driven exchange is not apparent in the fluorescence signal because the HEC is randomly labeled: each unit mass of HEC contains the same number of fluorescent tags.

Because of the evolving molecular weight of the adsorbed HEC chains, in our studies it was important to end the adsorption step as quickly as was reasonable to curtail the molecular-weight exchange. Therefore adsorption times were kept to 10 minutes. Following incubation in pure solvent, the intent during the self exchange step was to challenge the adsorbed chains with a bulk solution of nearly the same molecular weight distribution. This could not, however, be accomplished rigorously, and it is likely that at the start of self exchange, the adsorbed molecular weight was on the high side relative to the solution. Further, self exchange serves to increase the surface molecular weight, leading to a tailing of the self exchange kinetics themselves. Also, to the extent that surface relaxations depend on molecular weight, as suggested by Granick,[23,24] aging of the mixed surface may be difficult to interpret.

With that said, the most striking feature of the HEC self exchange kinetics is that even for young layers, the self exchange kinetics are orders of magnitude slower than the transport-limited exchange rate. This behavior cannot be attributed to the polydispersity, because the polydispersity effects tend to lead to tailing. Figure 5 indicates that the short time HEC self-exchange is slow, even when the transport limited rate is based on a polydispersity-averaged mass-transfer coefficient, M, that was calculated from the initial HEC adsorption kinetics.

Despite the slow HEC self exchange, HEC aging is, in other ways similar to that of the high molecular weight PEO. First, the relaxation time for the HEC layer is about 15-20 hours, similar to that for PEO. This was found despite the fact that HEC is a much stiffer molecule than PEO. Also, with HEC, aging appears to immobilize all the chains in the layer, similar to the 120K PEO. Finally, like PEO, the HEC self exchange traces are neither exponential nor stretched exponential. The closest approximation is a bi-exponential decay, where the fast process is described by equation 3 and a second slower process prevails at long times. Even the binary form is not, however, an exact quantitative description of the observations.

The results in Figure 5 have implications for the interpretation of HEC adsorption isotherms, in a previous section. For HEC layers that are about 30 minutes or older, the exchange between the interface and the solution is slow compared to the transport limited rate, and grows slower with time. Polydispersity provides an additional complication: For an adsorbed layer to equilibrate, the surface must saturate with the highest molecular weight population of chains which may be present in bulk solution at extremely low levels, such that the transport-limited equilibration rate would be several hours (even for a bulk solution of total concentration near 100 ppm, because the high molecular weight chains may only contribute 1-2 ppm.) During this time, the initially adsorbed chains may age and may resist displacement on a reasonable experimental timescale. Therefore the adsorption isotherms, like those in Figure 1B may not represent the true equilibrium situation from the perspective of the adsorbed amount or the molecular weight of the adsorbed chains.

In contrast, for PEO narrow molecular weight standards, minimal self exchange is needed for the surface to achieve its equilibrium molecular weight distribution. We have also observed that the coverage levels are independent of history. Therefore, the PEO isotherms in Figure 1A represent the equilibrium coverage and surface molecular weight. Figure 4 suggests that equilibration (or relaxation) also involves chain reconfigurations, but for the case of PEO, these relaxations apparently do not affect coverage. One must therefore conclude that even though the adsorbed layers in Figure 1A may not be equilibrated microscopically, the information in Figure 1A represents the equilibrium adsorbed mass.

Summary

When homopolymers adsorb from solution onto solid surfaces, the initial surface composition is determined by the bulk solution composition and the relative diffusion rates of populations of differing molecular weights. At long times; however, the longest chains are preferred on the surface, giving rise to a molecular weight driven exchange after the initial adsorption. For bimodal distributions, if local equilibrium is achieved between the adsorbed chains and those in solution near the interface, the exchange will take place at nearly constant interfacial mass. For polydisperse systems, even with local equilibrium, the adsorption kinetics may appear rounded as a result of many competing populations of varying chain lengths.

Also going on while adsorption occurs is interfacial chain relaxation. For PEO, the impact of relaxation on the adsorbed layer mobility is highly molecular-weight dependent, though the relaxation times were similar for the two layers studied. Lower molecular weight PEO layers, in the fully relaxed state, contained significant populations that could exchange with the bulk solution. For higher molecular weight relaxed PEO layers, the exchange between the adsorbed layer and the bulk solution was minimal. Despite the polydispersity-driven complexities for interpreting the relaxation behavior of HEC chains, they appeared to age in a manner very similar to that of the higher molecular weight PEO samples.

For the model narrow molecular weight standard PEO, the primary effect of layer age was to reduce the self-exchangeability of the layer. We found no evidence for any influence of layer relaxations on the adsorbed amount, since with narrow molecular weight standards, self-exchange was not needed to adjust the average molecular weight of the adsorbed chains. In contrast with HEC, relaxations potentially influenced the adsorbed amounts because of the sample polydispersity. As the layer ages, the self-exchange process becomes hindered, making it difficult for long chains to replace short ones. This gives a more protracted shoulder on the adsorption kinetic trace, which may extend beyond experimentally accessible timescales or make it difficult to recognize the final level of coverage. Therefore with polydisperse systems, the dilute region of the experimental adsorption isotherm can be exceedingly difficult to measure.

Acknowledgments.

This work was supported by the National Science Foundation through grants CTS-9209290 and CTS-9817048.

References

1. Scheutjens, J.M.H.M.; Fleer, G.J. *J. Phys. Chem.* **1979**, *83*, 1619.
2. Scheutjens, J.M.H.M.; Fleer, G.J. *Colloids Surfaces*, **1986**, *21*, 285.
3. Fleer, G.J.; Cohen Stuart, M.A.; Scheutjens, J.M.H.M.; Cosgrove, T.; Vincent, B. *Polymers at Interfaces*; Chapman and Hall, London, 1993.
4. Farinha, J.; d'Oliveira, J.; Martinho, J.; Xu, R.; Winnik, M.A. *Langmuir*, **1998**, *14*, 2291.
5. Rager, T.; Meyer, W.H.; Wegner, G.; Winnik, M.A. *Macromolecules.*, **1997**, *30*, 4911.
6. Malmsten, M.; Tiberg, F. *Langmuir* **1993**, *9*, 1098.
7. Takahashi, A.; Kawaguchi, M.; Hirota, H.; Kato, T. *Macromolecules*, **1980**, *13*, 884.
8. Kawaguchi, M.; Hayakawa, K.; Takahashi, A. *Macromolecules*, **1983**, *16*, 631.
9. Lee, J.; Fuller, G.G. *Macromolecules*, **1984**, *17*, 375.
10. Lee, J.; Fuller, G.G. *J. Colloid Interface Sci.*, **1985**, *103*, 569.
11. Pefferkorn, E.; Carroy, A.; Varoqui, R. *J. Polym. Sci., Part B.* **1985**, *23*, 1997.
12. deGennes, P.G. *Adv. Colloid Interface Sci.* **1987**, *27*, 189.
13. Dijt, J.C.; Cohen Stuart, M.A.; Fleer, G.J. *Macromolecules*, **1992**, *25*, 5416.
14. Dijt, J.C.; Cohen Stuart, M.A.; Fleer, G.J. *Macromolecules*, **1994**, *27*, 3219.
15. Fu, Z.; Santore, M. M. *Colloid Surfaces A. Physiochem. Eng. Aspects*, **1998**, *135*, 63.
16. Cheng, Y.L.; Darst, S.A.; Robertson, C.R. *J. Colloid Interface Sci.*, **1987**, *118*, 212.
17. Lok, B.K.; Cheng, Y.L.; Robertson, C.R. *J. Colloid Interface Sci*, **1983**, 91, 104.
18. Shibata, C.T.; Lenhoff, A.M. *J. Colloid Interface Sci.*, **1992**, *148*, 485.
19. Shirahama, H.; Lyklema, J.; Norde, W. *J. Colloid Interface Sci.* **1990**, *139*, 177.
20. Fu, Z.; Santore, M.M. *Macromolecules*, **1997**, *30*, 8516.
21. Zhu, P.W.; Napper, D. H. *Phys. Rev. E.* **1998**, *57*, 3010.
22. Zhu, P.W.; Napper, D.H. *J. Phys. Chem. B.*, **1987**, *101*, 3155.
23. Frantz, P.; Granick, S. *Phys. Rev. Lett.* **1991**, *66*, 899.
24. Frantz, P.; Granick, S. *Macromolecules*, **1994**, *27,.* 2553.
25. Douglas, J.F.; Johnson, H.E.; Granick, S. *Science*, **1993**, *262*, 2010.
26. Douglas, J.F.; Frantz, P.; Johnson, H.E.; Schneider, H.M.; Granick, S. *Colloids Surfaces A. Physiochem. Eng. Aspects*, **1994**, *86*, 251.
27. Fu, Z.; Santore, M.M. *Macromolecules*, **1998**, *31*, 7014.
28. Kelly, M.S.; Santore, M. M. *Colloids and Surfaces A: Physiochem Engr. Aspects*, **1995**, *96*, 199.
29. Rebar, V.A.; Santore, M.M. *J. Colloid Interface Sci.*, **1996**, *178*, 29.
30. Eremenko, B.V.; Sergienko, Z.A. *Colloid J. USSR*, **1977**, *41*, 422.
31. Rebar, V.A.; Santore, M.M. *Macromolecules*, **1996**, *29*, 6273.
32. Rebar, V.A.; Santore, M.M. *Macromolecules*, **1996**, *29*, 6263.
33. Leveque, M. *Ann. Mines*, **1928**, *13*, 284.
34. Cohen Stuart, M.A.; Scheutjens, J.M.H.M.; Fleer, G.J. *J. Polym. Sci., Polym. Phys. Ed.*, **1980**, *18*, 559.
35. Roe, R. *Polym. Sci. Tech. Ser.* **1980**, 12B, 629.
36. Kawaguchi, M.; Maeda, K.; Kato, T.; Takahashi, A. *Macromolecules*, **1984**, *17*, 1666.
37. Cohen Stuart, M.A.; Fleer, G.J.; Bisterbosch, B.H. *J. Colloid Interface Sci.* **1982**, *90*, 310.
38. Hlady, V.; Lyklema, J.; Fleer, G.J.; *J. Colloid Interface Sci.*, **1982**, *87*, 395.
39. Fu, Z.; Santore, M.M. *Macromolecules*, **1999**, *32*, 1939.
40. Fu, Z.; Santore, M.M. *Langmuir*, **1998**, *14*, 4300.
41. Frantz, P.; Leonhardt, D.C.; Granick, S. *Macromolecules*, **1991**, *24*, 1868.

Chapter 13

Dispersions Containing PEO with C_{16} Hydrophobes: Adsorption and Rheology

Q. T. Pham[1], J. C. Thibeault[2], W. Lau[2], and W. B. Russel[1]

[1]Department of Chemical Engineering,
Princeton University, Princeton, NJ 08544
[2]Research Laboratory, Rohm and Haas Company,
Spring House, PA 19477-0904

The rheology of polymer latices containing 35 kg/mole polyethylene oxide chains with C_{16} terminal hydrophobes is related quantitatively to the composition. Correcting the solution concentration and the particle volume fraction to account for adsorption of polymer permits a clear distinction between the contributions the solution and those from the particles. The relaxation spectrum then clearly direct interactions between the particles that increase the low shear viscosity and produce a power law spectrum that controls the modulus at low frequencies.

Introduction

The rheology of latex dispersions containing associative polymers depends not only on the dynamics of the associated solution but also on polymer-particle interactions that control the structure and stability of the dispersion (1,2,3). Associative polymers adsorb onto polymer latices via hydrophobic interactions, forming dense layers that increase the hydrodynamic size of the particles while reducing the polymer concentration in solution. These processes have opposite effects on the magnitude of the stresses, but generally preserve the relaxation times as those of the associated solution. Attractions between dense layers on interacting particles, e.g. due to bridging chains, also increase the stress, while introducing slower relaxations that require diffusion of the particles. The overall dispersion rheology depends qualitatively and quantitatively on the relative importance of these processes.

Our previous work (4) demonstrates that narrow distribution polyethylene oxide (PEO) chains with C_{18} terminal hydrophobes (ODU=octadecyl unimers) adsorb on poly(methylmethacrylate) (PMMA) latices from water as dense layers of moderately stretched chains in very dilute solutions and appear to reorganize as whole or hemi-micelles at higher concentrations. In either case chains in an adsorbed layer associate hydrophobically with micelles in solution and with adsorbed layers on other particles. At finite particle volume fractions ϕ, adsorption increases both the intrinsic viscosity

[η], by an amount consistent with the measured layer thickness, and the Huggins coefficient k_h. The former indicates strong coupling with the associated solution, equivalent to a no-slip boundary condition. The value of k_h exceeds that of hard spheres, implying direct coupling between adsorbed layers on interacting particles. Thus, the observed enhancement of the dispersion viscosity arises from the high viscosity of the polymer solution, the increased hydrodynamic volume of the particles, and direct interactions between adsorbed layers. Likewise, the viscoelasticity of the dispersion resembles that of the associated solution with elasticity at high frequency governed by a single relaxation time of $O(5\text{-}15 \text{ ms})$, while the particles impart a power law distribution of longer relaxation times that dominate the elasticity at low frequencies. The latter is not sufficiently strong to affect the shear rate dependence of the viscosity.

Here we examine the effect of hydrophobe size on the behavior with the same dialyzed PMMA latices (diameter $2a$=224 nm) and PEO backbones (35 kg/mole) but with C_{16} (HDU=hexadecyl unimers) instead of C_{18} hydophobes. To facilitate direct comparison between the two hydrophobes, we perform the same experiments on adsorption and rheology and use the data as a second test for the validity of the correlations presented previously. The materials and methods are described in detail elsewhere (4,5).

Adsorption

Adsorbed amounts for HDU, ODU, and PEO generally increase monotonically with the polymer concentration c_s in the aqueous phase and approach a plateau Γ_m at high concentration (Table I) that increases with hydrophobe size, translating into a reduced area per chain for the bigger hydrophobe (6). Unlike ODU, for which a small initial plateau was observed at low c_s, HDU and PEO increase more smoothly and steeply to the apparent plateau (Figure 1).

Table I: Adsorbed amounts and layer thicknesses

Polymer	Γ_m (mg/m^2)	$1/\sigma$ (nm^2/chain)	δ (nm)	$0.35N(\sigma v l^2)^{1/3}$ (nm)
ODU	2.69	21.6	26±5	10-15
HDU	1.79	32.4	18±6	9-14
PEO	0.90	64.9	10±4	---

Unmodified PEO adsorbs on latex particles at many points along its backbone, producing loops, trains, and tails with the tails governing the hydrodynamic thickness (7). On polystyrene particles with a=120 nm, monodisperse PEO (40 kg/mol) has a plateau absorbed amount Γ_m = 0.68±0.50 mg/m^2 and a layer thickness δ = 12±2 nm

(8). Both Γ_m and δ increase monotonically with molecular weight (9). In good solvents, δ is about 1.5 to 2 times the radius of gyration of PEO in bulk solution, indicating a moderately stretched configuration for the adsorbed chain. Although polystyrene has a slightly different surface polarity than PMMA, the Γ_m are comparable, implying a loop-train-tail arrangement for the adsorbed PEO on PMMA.

Comparison with the unmodified PEO suggests that the associative polymers adsorb in crowded layers of stretched chains. From the adsorbed amounts and the molecular weights, the layer thicknesses can be estimated by simple theories for neutral polymer brushes by treating each triblock as two diblock copolymers adsorbed by terminal hydrophobes. Adapting the mean field theories of Alexander (10) and de Gennes (11) for dense brushes of terminally anchored chains in a good solvent to triblocks adsorbed with both hydrophobes on the surface yields the thickness δ on a spherical particle of radius $a >> \delta$, as

$$\delta = \frac{L}{2}\left(\frac{\sigma v}{3\ell}\right)^{1/3} \quad (1)$$

with L (=285 nm) the contour length of the PEO block, ℓ (=0.456 nm) the Kuhn length, and v the excluded volume. The number of chains per unit area σ is related to the measured surface coverage Γ_m by

$$\sigma = \frac{\Gamma_m N_A}{M} \quad (2)$$

where N_A is the Avogadro's number and M is molecular weight of the PEO. The uncertainty in the dimensionless excluded volume, v/ℓ^3=0.12-0.33, reflects variations in the reported radius of gyration of PEO in solution from which it is estimated (4). The latter is accomplished via a mean field theory comparable to (1), on the presumption that uncertainties in the theories themselves should cancel at least partially. The predicted layer thicknesses for ODU and HDU in Table I increase with hydrophobe size and exceed by more than 20% the end-to-end distance of the homopolymer (11-16 nm).

Hydrodynamic diameters of particles with adsorbed layers measured by dynamic light scattering increase rapidly with total polymer concentration c_p and reach full thickness at concentrations below the plateau adsorbed amount Γ_m in Figure 1. On the plateau the layers are significantly thicker for the associative polymers (Table I). For unmodified PEO, δ=10±4 nm is similar to those adsorbed onto polystyrene particles (δ=6-12 nm for 20-40 kg/mol PEO and a=120 nm) (8,9), but smaller than predicted for chains that adsorb terminally. Thus, PEO must contact the particle surface at many points along its backbone. The increase in δ with hydrophobe size is also consistent with predictions from Eqn (2), corroborating the notion that associative polymers adsorb via their hydrophobe endcap, essentially forming dense terminally-anchored brushes. If hydrophobic interactions dictate adsorption, then the bigger hydrophobes adsorb more strongly, form denser layers, and hence have a more stretched configuration. However, the prediction for ODU is considerably smaller than the measured value, whereas that for HDU just falls within the error bars, supporting the argument that ODU adsorbs as micelles or hemi-micelles instead of individual chains. Furthermore, both adsorbed amount (Figure 1) and layer thickness increase smoothly for HDU, while ODU exhibits discontinuities near the *cmc* ($c_s \sim$ 100-200 ppm). Of course, definitive resolution of the actual configuration would require more direct visualization of the adsorbed layer.

Viscosity

The steady shear profiles of PMMA dispersions thickened with HDU (Figure 2) generally exhibit a Newtonian viscosity at sufficiently low shear rates, shear thickening at intermediate shear rates, and thinning at shear rates of 100 s^{-1}, as reported previously for dispersions containing ODU. The dispersion viscosity increases monotonically with polymer concentration and is comparable to that of the neat solution, but can decrease with the addition of particles.

Compared to ODU, HDU enhances the viscosity less and causes shear thinning that moves to lower shear rates (0.04–0.2 s^{-1}) at higher volume fractions and is more gradual. Several observations are unique to PMMA+HDU. For example, at high polymer concentration (c_p>3 wt%) and high particle volume fraction (ϕ=0.17), the dispersion viscosity gradually decreases over all accessible shear rates (Figure 2). At the other extreme for c_p<1.5 wt%, the steady shear viscosity is Newtonian at intermediate shear rates (10-1000 s^{-1}), but can creep up at very low shear rates, producing an apparent dynamic yield stress of 0.02 Pa (Figure 3). The latter can be suppressed by (i) preshearing the dispersion at a moderate shear rate (100 s^{-1}) then measuring the low shear viscosity or (ii) performing the steady shear sweep from high to low shear rates. This behavior is consistent with our visual observation of weak flocculation in these dispersions at low polymer concentrations, apparently due to reversible bridging. The resulting weak percolation network (*3*) increases the apparent viscosity but can be easily destroyed by moderate or high shear. At lower shear rates a transient precedes the steady state, but has not been characterized systematically in this work.

Table II: Apparent intrinsic viscosity and Huggins coefficient of dispersions

Polymer	[η]	k_h
ODU (*4*)	2.3 ±0.6	3.2 ±0.4
HDU	2.4 ±0.6	3.1 ±1.2

Clearly the strength of the hydrophobic interaction affects several aspects of the rheology of dispersions containing associative polymers. Both the viscosity of the associated solution and the amount of adsorbed polymer increase with hydrophobe size, with the former increasing and the latter decreasing the dispersion viscosity. However, we also see a more subtle effect arising from associations between the

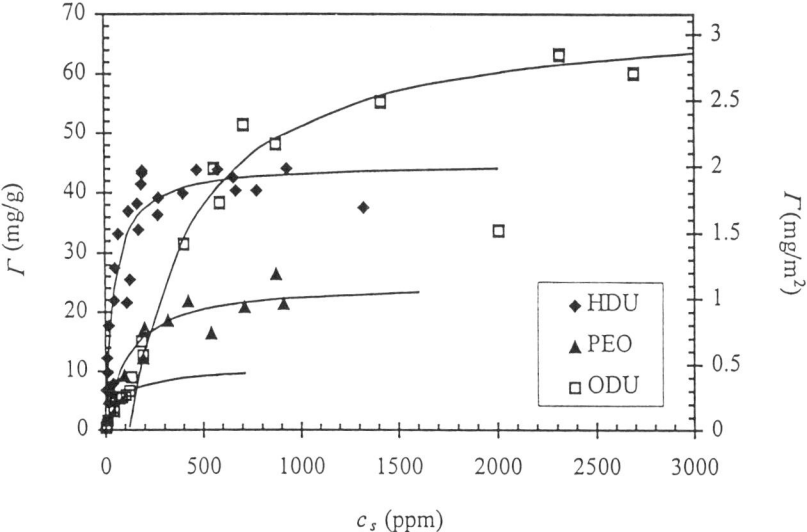

1. Adsorption isotherms for HDU, ODU, and PEO on PMMA latices.

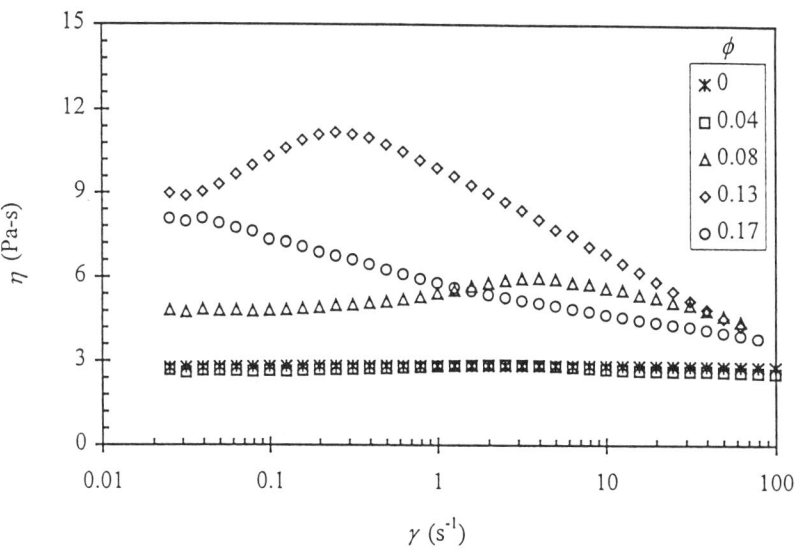

2. Steady shear viscosity of dispersions with 3 wt% HDU.

particles. To understand the situation more fully we start by examining the effect of polymer and particle concentrations on the low shear viscosity through the correlation developed for ODU (4).

The low shear viscosity η_o obtained from the Newtonian plateau or the lowest shear rate (0.025 s^{-1}) increases with c_p from a value at < 1 wt% about the same as with ODU to values about an order of magnitude lower at high c_p. The same trends observed for PMMA+ODU, such as η_o increasing monotonically with ϕ at high c_p and decreasing abruptly at high ϕ, are also apparent for PMMA+HDU.

To correlate the low shear viscosity across the range of particle and polymer concentrations we account for adsorption by rescaling the volume fraction as $\phi_{eff}=(1+\delta/a)^3\phi$ and adjusting the polymer concentration to $c_s=[c_p(1-\phi)-\Gamma_m\phi]/(1-\phi_{eff})$, with $\Gamma_m=1.79$ mg/m^2 and $\delta=18$ nm from the measurements above. The low shear viscosity η_s of HDU solutions is related empirically to the solution concentration c_s (wt%) by (5)

$$\frac{\eta_s}{\mu} = 1 + 9c_s \exp\left(\frac{4.2c_s}{1+0.59c_s}\right) \quad (3)$$

where μ is the water viscosity at 25°C. Then plotting the data according to

$$\frac{\eta_o}{\eta_s(c_s)} = 1 + [\eta]\phi_{eff} \exp(k_h[\eta]\phi_{eff}) \quad (4)$$

produces an intrinsic viscosity $[\eta]=2.4\pm0.6$ and a Huggins coefficient $k_h=3.1\pm1.2$ (Figure 4 and Table II) that lie within experimental error of the values for ODU+PMMA (4), though the scatter here is considerably more. The no-slip limit with $[\eta]\approx2.5$ indicates effective "hydrodynamic" coupling between the coated particles and polymer solution, while the enhanced $k_h>1$ reveals non-hydrodynamic interactions between particles, presumably due to association between adsorbed layers. The comparable $[\eta]$ and k_h for ODU and HDU suggest that the hydrophobe size does not change the nature of the interactions between particles and polymers. The dispersion viscosity is still enhanced primarily by the more viscous associated solution and augmented by particles coupled through their adsorbed layers.

Figure 5 compares the correlation (lines from Eqn (4)) with the full set of data. For $c_p\geq2$ wt%, the high solution viscosity η_s and increased particle interaction k_h maintain a high dispersion viscosity η_o that increases monotonically with ϕ_{eff}. This regime persists to lower c_p than with ODU, since the weaker adsorption depletes the solution polymer less drastically. For $c_p=1.5$ wt%, addition of particles reduces η_o more than an order of magnitude below the original solution viscosity, while for $c_p<1$ wt% η_o becomes independent of the polymer concentration and increases slightly with ϕ_{eff}. Though qualitatively correct, the correlation describes η_o poorly in this regime, perhaps because these concentrations fall within the two-phase regime for the solutions (5). Alternatively, values of η_o lower than predicted for $c_p<1.5$ wt% may suggest adsorption of more polymers at high ϕ than indicated by measurements at dilute ϕ. This is plausible if a significant fraction of the adsorbed chains bridge

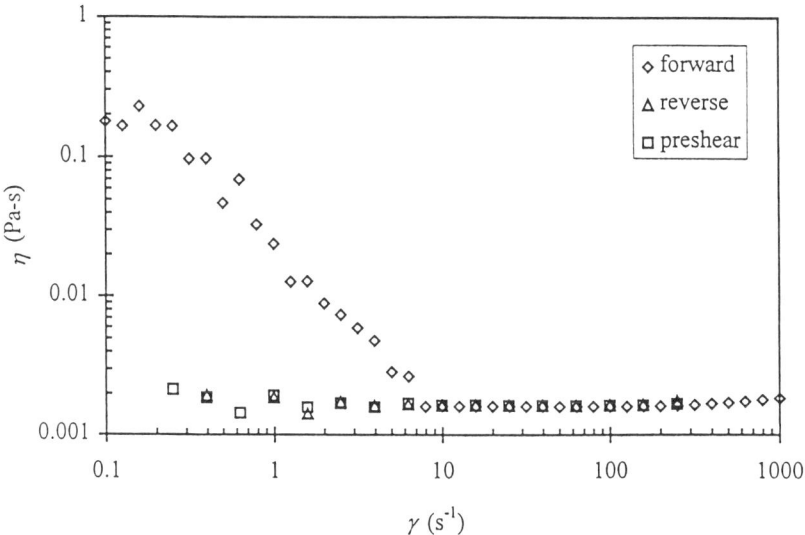

3. Steady shear viscosity of dispersion with $\phi=0.06$ containing 0.5 wt% HDU.

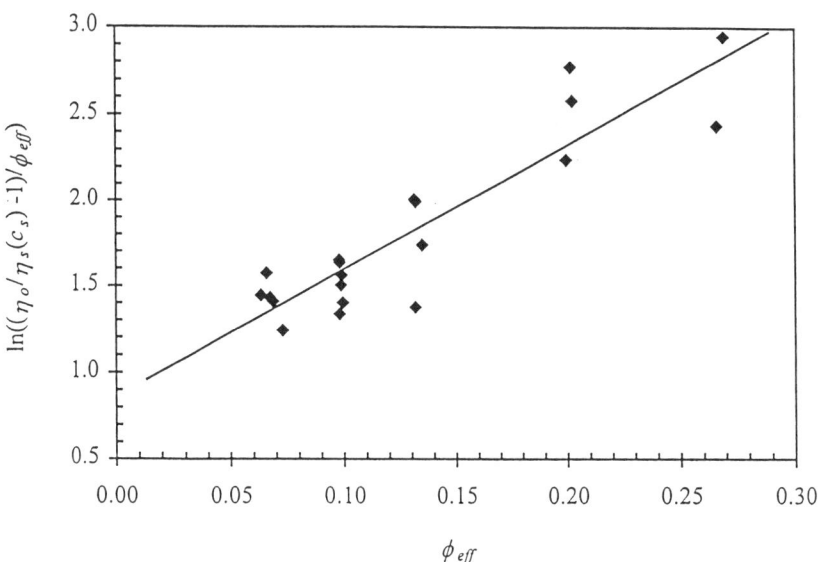

4. Linear regression of low shear viscosities with line indicating least squares fit.

the upturn in the lowest curve, when solution polymer is entirely depleted and particles with adsorbed layers interact in pure water.

Another way to characterize the enhanced viscosity is to define the thickening efficiency as $\eta_o(\phi,c_p)/\eta_o(\phi,0)$, the ratio of the viscosity of the dispersion with associative polymer to the viscosity of a neat dispersion at the same volume fraction, as calculated from Eqn (4) with $k=1$, $[\eta]=2.5$, and $c_p=0$. Figure 6 demonstrates that HDU at $c_p>2$ wt% increases the dispersion viscosity by 3-4 orders of magnitude. The thickening effect is drastically diminished for $c_p=1.5$ wt% and becomes essentially zero for $c_p<1$ wt%, consistent with the correlation and the effect of adsorption.

To distinguish the thickening effects due to the solution from the enhanced particle interactions, we characterize the fraction of the thickening due to the solution viscosity as

$$f = \frac{\frac{\eta_o(\phi,0)}{\mu}}{\frac{\eta_o(\phi,c_p)}{\mu}} \frac{\eta_s(c_s(\phi))}{\mu}. \qquad (5)$$

The remainder (1-f) is due to the increased hydrodynamic size of and direct interactions between the particles. For $c_s>0$, f decreases from 1 to 0.4 for $0.04<\phi<0.2$ (Figure 7) with less influence of polymer concentration. At low ϕ, the solution viscosity dominates and $f\to 1$; but with increasing ϕ, the increased hydrodynamic volume and interactions with $k_h>1$ reduce f to 0.60 at $\phi=0.17$.

Overall, HDU thickens latex dispersions by the same mechanism as ODU. However, the HDU solution has a lower viscosity than ODU, so the dispersion viscosity is lower and stresses generated by the particles seem more important.

Linear Viscoelasticity

In solutions of associative polymers, the placement and content of hydrophobes affects the viscoelasticity with terminal hydrophobes producing the greatest elasticity (2,12). Increase in hydrophobe size extends the relaxation time exponentially, while enhancing the viscoelasticity through an increase in micellar aggregation number (5). In dispersions, longer hydrophobes and/or lower backbone molecular weights impart greater elasticity (2,13).

The viscoelasticity of PMMA+HDU deviates from the simple Maxwellian behavior of the solution (2,5,14) by exhibiting multiple relaxation modes. As expected, the loss G'' and storage G' moduli increase with polymer concentration c_p, whereas increasing the particle concentration ϕ alters profoundly the power law dependence of G' on frequency. Thus at low frequencies $G' \sim \omega^n$, with n falling from 2 to almost 0 as ϕ increases (Figure 8). The key differences for the HDU are (a) the extension of the power law relaxation to intermediate frequencies, (b) a much weaker dependence of the moduli on frequency at low ω, and (c) gelation with $G' \approx G''$ for c_p

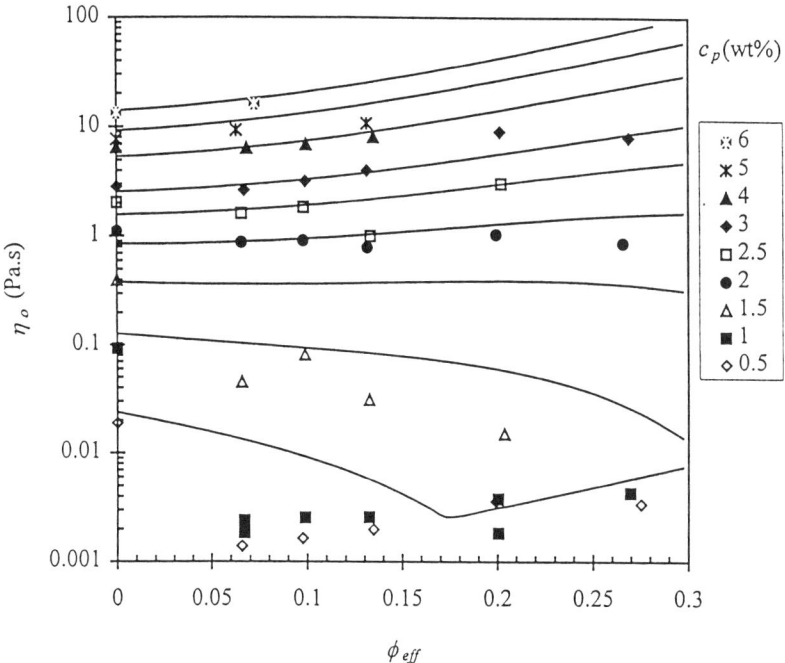

5. Measured low shear viscosities as functions of total polymer concentration and effective particle volume fraction compared with lines from Eqn (5).

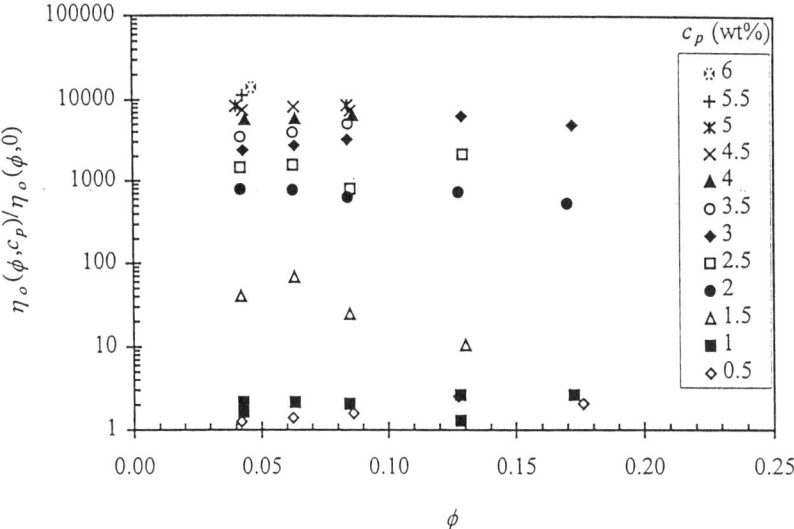

6. Ratio of viscosity of dispersion with associative polymer to that without.

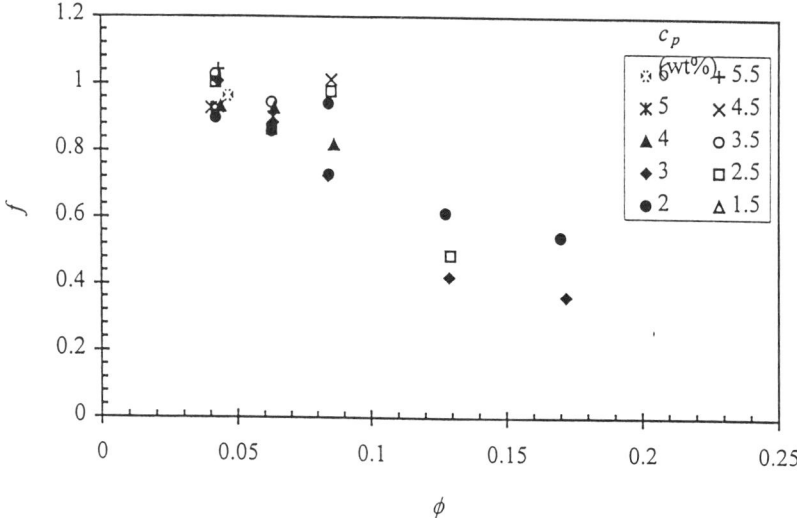

7. *The fractional enhancement of the dispersion viscosities due to the solution.*

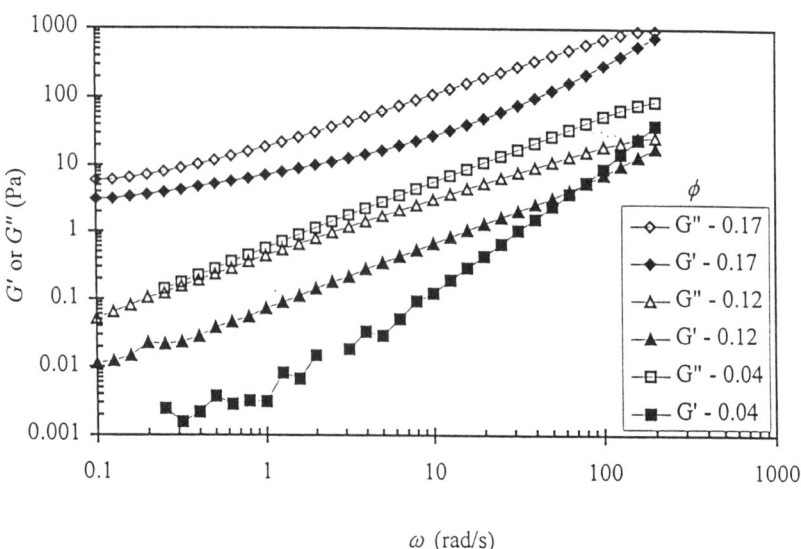

8. *Storage and loss moduli of dispersions containing 3.5 wt% HDU.*

=2-3 wt% and ϕ>0.12. Moreover, G' and G'' approach, but do not reach, the crossover at high frequency, which lies beyond the range (ω >200 rad/s) of the RFSII rheometer.

In the power law region at low frequency where $G'\sim G''\sim \omega^n$, $0<n<1/3$ with $G'\approx G''$ for some concentrations. This behavior strongly resembles the viscoelasticity of crosslinking polymers near the gel point (15,16,17,18), which is frequently defined as the transition from a viscoelastic fluid to a viscoelastic solid as function of time or temperature. For a crosslinked polymer network with no dangling ends, $n=1/2$ and $G''=G'$ at the gel point. Prior to gelation, the viscous component dominates with $G''>G'$ and $n>1/2$; after gelation a crossover emerges at high ω, with $G''<G'$ at low ω (16,18). From the Kramers-Kronig relation, Larson (19) and Chambon and Winter (16) found that $G''>G'$ for $n>1/2$, $G''=G'$ for $n=1/2$, and $G''<G'$ for $n<1/2$. Beyond gelation, the material is an elastic solid with $G'\to G_o$ and $n\to 0$. This behavior matches the observations for PMMA+HDU in Figures 8, though "gelation" of our dispersions is determined by the concentration. Identifying the relaxation modes requires decomposing the frequency dependence of the moduli.

Model

Our interpretation of the viscoelasticity recognizes that the rapid dynamics of the associated solution control at high frequency, while the slower particle dynamics introduce relaxations at low frequency. The shift in the low frequency asymptote of G' suggests a relaxation spectrum of the form

$$H(\lambda) = G'_\infty \delta(\lambda - \lambda_m) + H_o \left(\frac{\lambda_o}{\lambda}\right)^n H(\lambda_o - \lambda) \tag{6}$$

with $\delta(x)$ and $H(x)$ the Kroenecker delta and Heaviside step function, respectively. The moduli are determined as integrals over this sum of a single rapid relaxation and a power law distribution of slower ones as

$$G' = \int H(\lambda) \frac{(\omega\lambda)^2}{1+(\omega\lambda)^2} \frac{d\lambda}{\lambda} \qquad G'' = \int H(\lambda) \frac{\omega\lambda}{1+(\omega\lambda)^2} \frac{d\lambda}{\lambda}. \tag{7}$$

Unfortunately, the high frequency crossover for PMMA+HDU lies outside the range of the instrument, making the shortest relaxation time λ_m somewhat uncertain. Furthermore, although the power law slope approaches unity ($n=1$) in most cases, $n<1$ at high ϕ, indicating a wide frequency range for the power law behavior. Thus, the parameters n, G', H_o, λ_m, λ_l, and λ_o cannot be obtained independently.

We first fit G'' and G' at high frequency to estimate λ_m and G'. Next, this contribution is subtracted from the data, and the residuals are fitted to the power law

Table III: Fitting parameters for Figure 9

ϕ	c_p [wt%]	ϕ_{eff}	n	λ_m [s]	G' [Pa]	λ_l [s]	$H\lambda_o$ [Pa-s]
0.084	4.49	0.13	1	0.0027	1300	0.013	0.08

model with $n=1$ or 0.5, yielding H_o and λ_o. Figures 9a and b illustrate the process with $n=1$ and the parameters in Table III. The high frequency relaxation (bold solid and dashed lines) accounts for G' at high ω and most of G'', but contribute minimally to G' at low ω. The residuals after its subtraction are significant for $\omega<30$ rad/s. The power law is extracted by noting that at ω_c where the residual $G'(\omega_c)=G''(\omega_c)$

$$\int_{\lambda_l\omega_c}^{\infty} \frac{d\bar{\lambda}}{1+\bar{\lambda}^2} = \int_{\lambda_l\omega_c}^{\infty} \frac{d\bar{\lambda}}{\bar{\lambda}(1+\bar{\lambda}^2)}. \tag{8}$$

Numerical solution yields $\lambda_l\omega_c=0.285$, allowing λ_l to be deduced from the crossover. For example, in Figure 9b $\omega_c=21.9$ rad/s determines $\lambda_l=0.013$ s. Then combining the power law and discrete models and adjusting $H_o\lambda_o$ and G' produces quite a good fit of both the residuals and the full spectrum. However, with more strongly associated samples, such as $\phi>0.17$ and $c_p=2.5-3.5$ wt%, setting $n=0.5$ and following the same procedure errs at $O(1)$ for the lowest frequencies. Introduction of a static limit $G' \to G_o$ as $\omega \to 0$ might help but has not been attempted.

Figures 10-13 represent the results as a function of effective volume fraction ϕ_{eff}, normalized where appropriate by the properties of the polymer solution at concentration c_s. The high frequency modulus of the dispersion $G'(c_p,\phi)$ generally increases with c_s but falls below the modulus for the solution $G'(c_s,0)$ except at the highest ϕ_{eff}. Normalized as the ratio of $G'(c_p,\phi)$ to $G'(c_s,0)$, with the latter calculated from curve fitting the measured values (5), shows that $G'(c_p,\phi)/G'(c_s,0)<1$, first decreasing for $0<\phi_{eff}<0.2$, and then increasing for $\phi_{eff}>0.2$ (Figure 10). As expected from the solution behavior, the relaxation time λ_m controlling the dissociation process, is shorter for HDU (4.4-5.1 ms) than for ODU (10-15 ms). Curiously our estimates for the dispersions are significantly shorter than for the neat HDU solution and increase with ϕ_{eff} (Figure 11). These trends for both G' and λ_m are difficult to rationalize, leading us to suspect the extrapolation required at high frequency in the absence of a crossover, which is less accurate with the additional low frequency relaxation, to be at fault.

The other characteristic time λ_l denotes the transition or crossover between the discrete and power law relaxation regimes. In PMMA+HDU, λ_l varies from 0.01 to 0.1 s and is generally shorter for high c_p and ϕ_{eff} (Figure 12). This is about the right

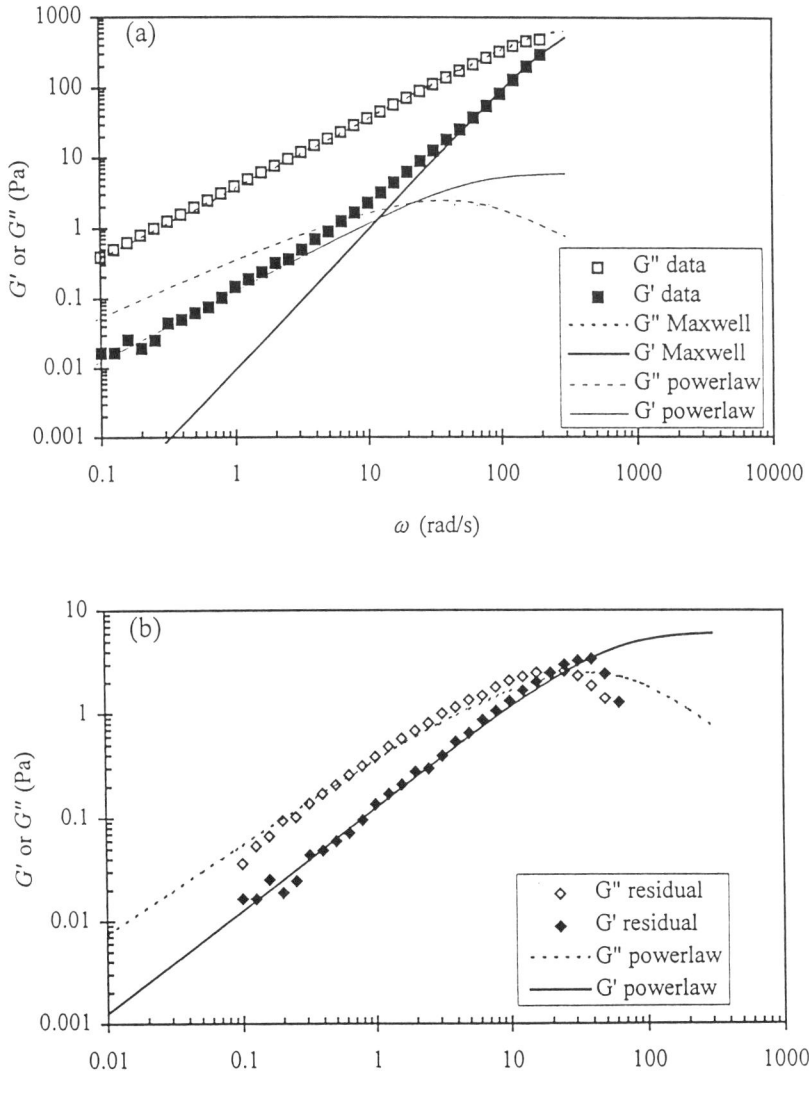

9. Decomposition of viscoelastic spectrum for $\phi=0.08$ and $c_p=4.5$ wt% : (a) full data and (b) residuals after subtraction of high frequency relaxation.

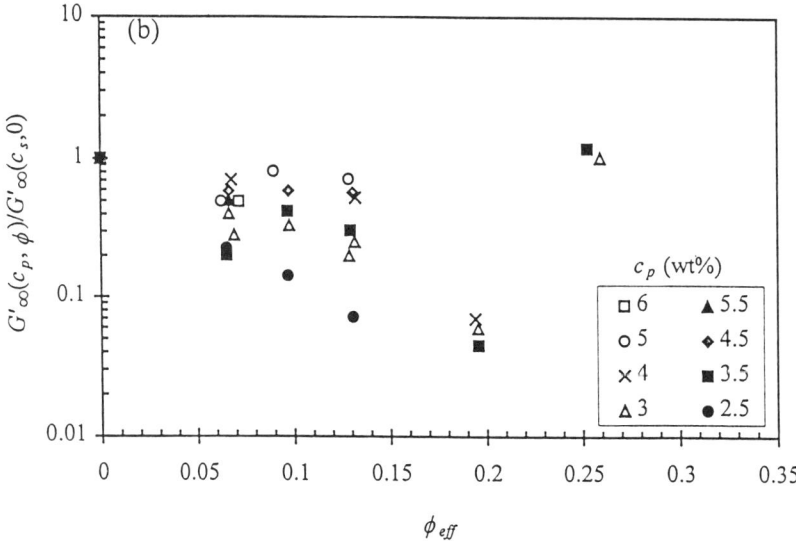

10. *High frequency modulus of dispersion normalized on that of polymer solution*

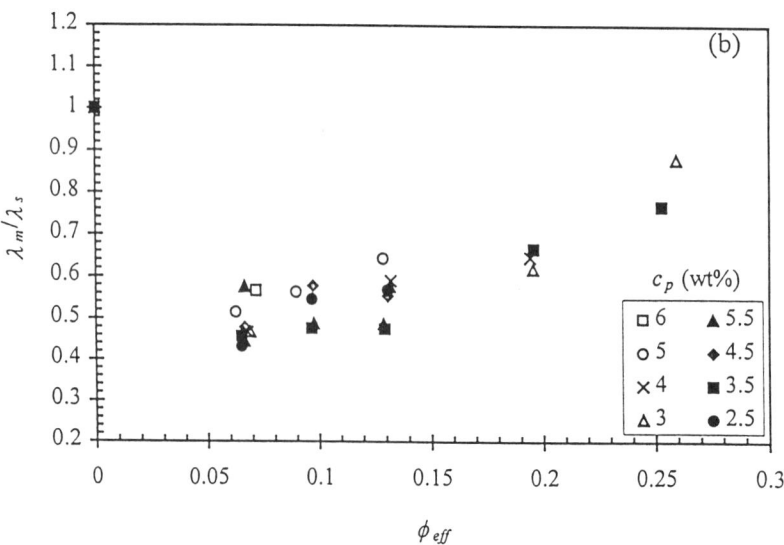

11. *Shortest relaxation time normalized on that of solution.*

diffusional relaxation time for either a single particle (112 nm radius) in pure solvent (.002 Pa-s) or a polymer micelle (15 nm) in the solution (e.g. 5x10^4 Pa-s).

From the power law spectrum, the $H_o\lambda_o$ for $n=1$ (Figure 13), which has units of Pa-s, generally increases with ϕ_{eff} and c_S; whereas the $n=1/2$ regime yields too few data points for $H_o\lambda_o^{1/2}$ to detect any trends. While the trends in $H_o\lambda_o$ are reasonable, the data shows significant scatter and no systematic variations with polymer concentration. The latter may be due to a batch or history dependence of the dispersions, since experiments at 2.5, 3.5, 4.5, and 5.5 wt% HDU belong to one series of samples, while those at 3.0, 4.0, 5.0, and 6.0 wt % represent another set prepared at a different time from the same batch of dialyzed particles but with different polymer solutions. Although the experiments should be repeated to validate these results, the trends we observed and especially the dependence on ϕ_{eff} should still hold.

In crosslinking polymers, the gel strength defined as $S = \pi^{1/2} H_o \lambda_o^{1/2}$ corresponds to the degree of crosslinking and functionality (15). Many dangling ends (imbalanced stoichiometry) produce a soft gel and consequently a small S. As the stoichiometric ratio or extent of crosslinking increases, the polymers interconnect into a stronger network and S increases (17). Hence, S can be viewed as a measure of the mobility of the chain segments in crosslinking polymers or, alternatively, the connectivity of a percolating network (20). For the dispersions, the elasticity with long relaxation times at low frequencies is related to the dynamics of the particles. Thus, $H_o\lambda_o$ increasing with ϕ_{eff} suggests more contacts between particles as the volume fraction increases.

A variety of studies with weakly aggregated dispersions detect elastic moduli that vary as ϕ^n with $n=3-4$. Indeed, $H_o\lambda_o = 30\phi^3$ Pa-s, for example, is consistent with the data, though certainly not unique. This serves to identify a prefactor to test against moduli and relaxations that one might expect for this dispersion. Since the aggregation is weak and the dynamics are slow, the obvious guess is a modulus of roughly kT per particle volume and relaxations controlled by the diffusion time scale for the particles, i.e.

$$H_o \approx \frac{3kT}{4\pi a^3} = 1 \text{ Pa} \qquad \lambda_o = \frac{6\pi \eta_s a^3}{kT} = 44 \text{ s}$$

for $c_S=4.5$ wt%. Thus the magnitudes are quite plausible!

Conclusions

Not surprisingly we find that larger hydrophobes impart greater viscosity and viscoelasticity to an associated solution. Likewise in latex dispersions, the more hydrophobic ODU has a bigger impact on the adsorption and rheology compared to HDU. Both polymers adsorb onto PMMA particles as dense layers that increase the effective hydrodynamic particle volume and reduce the concentration of solution polymer. The larger hydrophobe naturally has denser and thicker adsorbed layers. By comparison with the Alexander-de Gennes theory and unmodified PEO, we

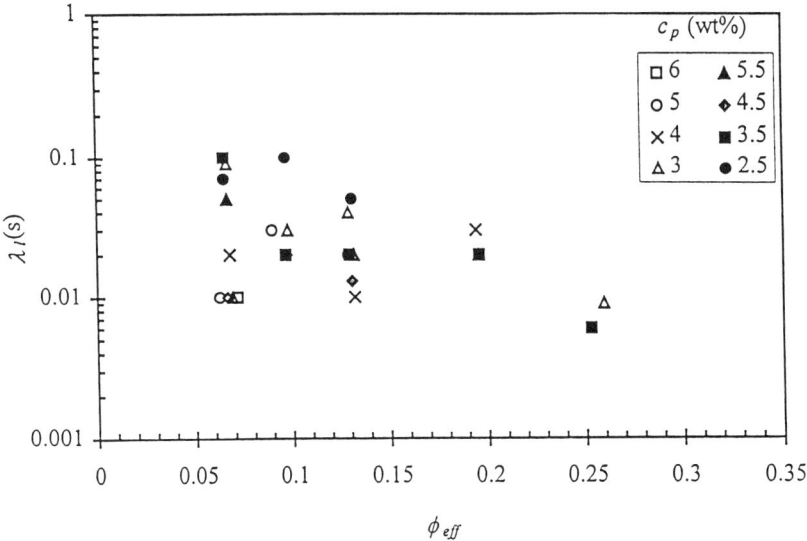

12. *Limiting lower relaxation time for power law spectrum.*

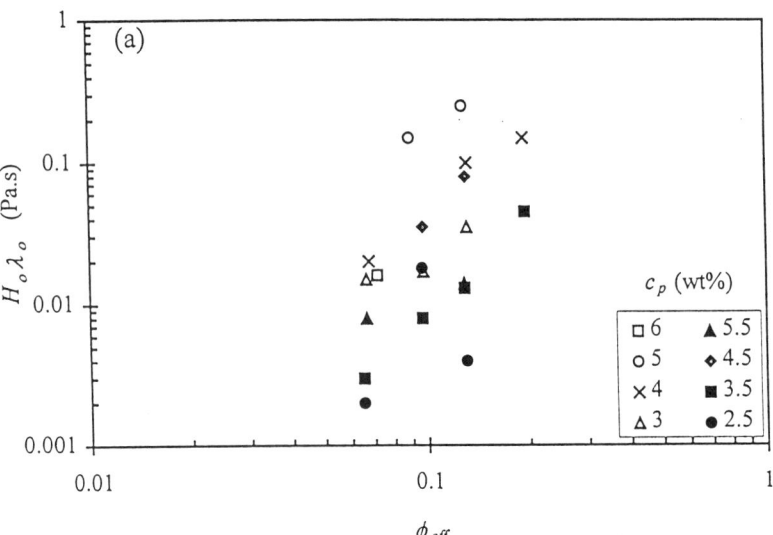

13. *Strength of power law relaxation for n=1.*

hypothesize that HDU adsorbs with both hydrophobes on the particle surface, whereas ODU absorbs as intact micelles beyond the critical micelle/aggregation concentration. In any case, adsorption of associative polymers generates spheres with interactive layers that couple the particles reversibly at modest volume fractions.

The dispersion rheology is then enhanced by the increased particle interactions. The low shear viscosity depends on both the interparticle interactions and the associations in solution. Although the larger adsorbed amount for ODU depletes the solution more than HDU, ODU still produces the higher solution viscosity and, hence, the higher dispersion viscosity. The Huggins coefficient and intrinsic viscosity extracted from the correlation for the low shear viscosity indicate the interactions between particles and with the polymer solution to be similar for ODU and HDU. For dispersions containing HDU, the viscoelasticity represents a superposition of the single rapid relaxation of associations in solution and power law spectrum of slower relaxations arising from reversible couplings between particles. Therefore, the high frequency modulus is larger and the shortest relaxation time longer for ODU reflecting the associations in the micellar solution. At low frequency, the power law spectrum due to the particle dynamics is consistent with the behavior of other weakly flocculated dispersions and our expectations for this system. Overall, the larger hydrophobe imparts a greater enhancement of the dispersion rheology of dispersions. Consequently the couplings between particles are more noticeable in both steady shear and linear viscoelastic measurements with the HDU.

Acknowledgments

The measurements of adsorbed amounts for HDU and PEO on PMMA were performed by Jessica Tucker and the dynamic light scattering experiments for layer thicknesses by Judy Tan. This research was supported by the National Science Foundation through Grant No. CTS 9521662.

References

1. Sperry, P.R.; Thibeault, J.C.; Kostanek, E.C. *Adv. Org. Coatings Sci. Tech.* **1987**, *9*, 1-11.
2. Jenkins, R.D. Ph.D. thesis, Lehigh University, Bethlehem, PA 1990.
3. Santore, M.M.; Russel, W.B.; Prud'homme, R.K. *Faraday Discuss. Chem. Soc.* **1990**, *90*, 323-333.
4. Pham, Q.T.; Lau, W.; Russel, W.B. *J. Rheology* **1998**, *42*, 159-176.
5. Pham, Q.T.; Thibeault, J.C.; Lau, W.; Russel, W.B. *Macromolecules* (submitted).
6. Jenkins, R.D.; Durali, M.; Silebi, C.A.; El-Aasser, M.S. *J. Colloid Interface Sci.* **1992**, *54*, 502-521.

7. Fleer, G.J.; Cohen Stuart, M.A.; Scheutjens, J.M.H.M.; Cosgrove, T.; Vincent, B. *Polymers at Interfaces* Chapman and Hall, London, UK, 1993.
8. Cosgrove, T.; Vincent, B.; Crowley, T.L.; Cohen-Stuart, M.A. *ACS Symposium Series* **1984**, *240*, 147-159.
9. Kato, T.; Nakamura, K.; Kawaguchi, M.; Takahashi, A. *Polymer J.* **1981**, *13*, 1037-1043.
10. Alexander, S., *J. Phys.* **1977**, *38*, 983-987.
11. de Gennes, P.G., *J. Phys. Lett. (Paris)* **1975**, *36*, L55-57.
12. Lundberg, D.J.; Brown, R.G.; Glass, J.E.; Eley, R.R. *Langmuir* **1994**, *10*, 3027-3034.
13. Huldén, M., *Colloids & Surfaces A* **1994**, *88*, 207-221.
14. Annable, T.; Buscall, R.; Ettelaie, R.; Whittlestone, D. *J. Rheol.* **1993**, *37*, 695-726.
15. Winter, H.H.; Chambon, F. *J. Rheol.* **1986**, *30*, 367-382.
16. Chambon, F.; Winter, H.H. *J. Rheol.* **1987**, *31*, 683-697.
17. Winter, H.H.; Morganelli, P.; Chambon, F. *Macromolecules* **1988**, *21*, 532-535.
18. Coviello, T.; Burchard, W. *Macromolecules* **1992**, *25*, 1011-1012.
19. Larson, R.G. *Rheologica Acta* **1985**, *24*, 327-334.
20. Winter, H.H. In *Encyclopedia of Polymer Science and Engineering*, 2nd ed.; Wiley and Sons: New York, NY, 1989, pp. 343-351.

Chapter 14

Force Study of Adsorbed Layers of Hydrophobically Modified Polyacrylamide

P. T. Starkey[1], H. T. Davis[1], M. V. Tirrell[1], J.-F. Argillier[2], A. Audibert[2], and J. Lecourtier[2]

[1]Department of Chemical Engineering and Materials Science, University of Minnesota, Minneapolis, MN 55455
[2]Institut Français du Petrole, Ruell-Malmaison, France 92506

Layers of an associative hydrophilic/hydrophobic random copolymer of acrylamide and nonylmethacrylate (HAPAM) adsorb onto a negatively charged mica substrate. The interactions in aqueous media involving opposed HAPAM layers are measured as a function of separation using the surface force apparatus (SFA). Increasing the salt and polymer concentration in solution increases the adsorbed amount of polymer, indicating multilayer build-up. For adsorbed HAPAM layers in the SFA, the range of the repulsive force decreases irreversibly after the initial compression. The force profile remains constant for subsequent compressions. At lower HAPAM concentrations ($\leq$760ppm), a long-range attraction is observed during separation. At these concentrations, the smaller layer thickness allows for polymer chains to bridge to the other surface. At higher concentrations, though, the chains are unable to bridge between the surfaces due to the increased layer thickness, and no attraction is observed. The addition of the small percent of hydrophobic groups results in a polymer structure in which hydrophobes are either interacting with the surface, or associating in micelle-like structures away from the surface.

Introduction

In this chapter, the hydrophobically modified polymers that are of interest have a hydrophilic backbone that contains a small percentage of randomly distributed hydrophobic monomer (<5%). This type of hydrophobic modification, which leads to association in solution, also results in association with dispersed solids. The polymer adsorbs in layers, which have properties that differ from bulk solution properties, where the conformational freedom of the chain is much higher (1). For many applications, it is very important to understand the association of hydrophobically modified polymers with particles. In some cases, hydrophobic segments associate with the surfaces of latex particles to form a dynamic network structure (2). Polymer adsorption is dependent on the presence of large hydrophobic groups or a sufficient number of smaller anchor groups (3).

Recently, there has been research which focuses on the dependence of the adsorbed amount of modified polymer on polymer concentration (4,5). Argillier *et al* found that modified polyacrylamide exhibits an increase in adsorbed amount onto mineral particles with increasing polymer concentration (6). For the non-modified version, the adsorbed amount reaches a plateau and does not increase with polymer concentration above a critical concentration, indicating that the surface is saturated. The adsorption isotherm is explained by the formation of multiple layers through hydrophobic associations, the adsorbed layer being formed in part by some chains that are not directly in contact with the surface. Similar results are obtained in another study in which a polyacrylamide backbone was modified with different hydrophobic comonomers at various distributions (7). They found that the adsorbed amount increases continuously with concentration, and the adsorption behavior is similar whether the copolymers exhibit strong thickening properties or not.

Studying polymer adsorption onto a flat surface is easier to characterize than particle adsorption, and can be used as a model for it. In a previous study in our group, hydrophobically modified hydroxyethyl cellulose (HMHEC) is adsorbed onto a mica surface and studied in the SFA (8). It was found that there is little affinity between HEC and mica, but HMHEC adsorbs when the salt concentration is at least 0.1M. Theory has been developed that models nonadsorbing polymers with varying hydrophobic groups onto parallel plates (9). Strong hydrophobes hold coils in the gap between surfaces at separations where non-associating polymer would be excluded or expelled. With extremely strong adsorption, a longer range "bridging" attraction appears at plate separations on the order of the end-to-end distance. The bulk polymer concentration and the surface coverage determine the strength of the bridging attraction. The depth of the attraction increases with polymer concentration, but when the surface becomes filled, further increases in concentration reduce the attraction. In conclusion, an associative polymer's affinity towards the surface depends on the chemical make-up of a hydrophobe, its placement within the polymer and the molecular weight of the backbone.

The surface force apparatus (SFA) provides a means to directly measure forces between adsorbed layers at very small separations. The SFA relates the structure of the adsorbed layers to the layer properties. The SFA has been used extensively to study polymers, but there has been relatively less investigation into water-soluble polymers. Some of these studies have dealt with polyelectrolytes (10,11,12,13), biological polymers such as proteins and polypeptides (14,15), and nonionic polymers such as polyethylene oxide (16,17).

In this study, hydrophobically associative polyacrylamide (HAPAM) is the polymer of interest. HAPAM has been chosen because of a collaboration with the

Institut Francais du Petrole, which has already done some investigation on HAPAM's solution behavior and adsorption behavior on mineral particles (6). The purpose of this work is to use surface force measurement to understand what role the hydrophobic interactions are playing in the adsorption of HAPAM onto mica and determine if these interactions are contributing to the build-up of HAPAM layers. The results will then be used to develop a possible picture of what the adsorbed polymer layers look like. The composition of these films is representative of many polymers which are used in latex paint or cosmetic applications. Being able to understand the role of hydrophobic interactions will allow for the potential build-up of multilayer films via these interactions.

Materials and Methods

PAM and HAPAM were specially manufactured by SNF Floerger (Saint-Etienne, France), synthesized by an aqueous phase emulsion polymerization process. The structure of HAPAM is given in Figure 1. The average molecular weight M_w of HAPAM is 7.7×10^6 g/mol. Distilled, deionized water was used in all experiments.

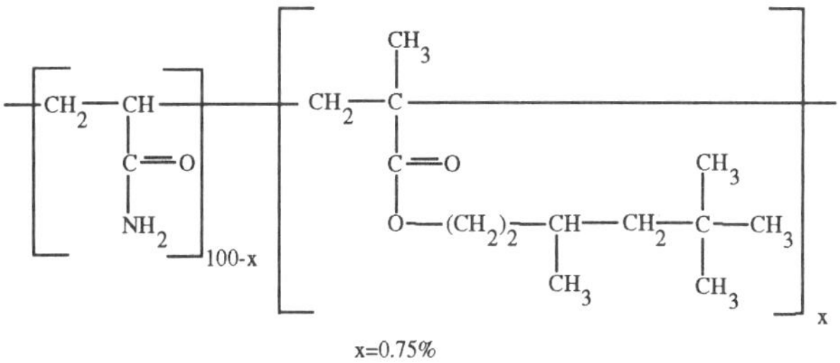

Figure 1. Copolymer of acrylamide and approximately 0.75% nonylmethacrylate.

Force measurements were carried out using the surface force apparatus (SFA), developed by Israelachvili (18,19). A detailed description of the experimental procedure involved in force measurement has been given before (20,21). In the SFA, molecules are adsorbed onto mica substrates that have a semi-reflective silver layer on the opposite side and are glued to quartz lenses. The upper lens is held fixed in place, while movement of the lower lens is controlled in the nanometer range by a lower motor-driven rod operating on a differential spring mechanism. The surface separation can be controlled down to 10Å. The actual separation distance is measured by an optical technique in which discrete wavelengths interfere constructively, are resolved by a spectrometer and viewed as a series of fringes called fringes of equal chromatic order (FECO) (22). The surface separation has been derived as a function of the fringe wavelength and is measured with an accuracy of less than 5Å (20). As the lower lens is moved (reference distance), the actual separation is determined from the fringe movement. The difference between the reference and separation distances gives the deflection of the spring. The force on the surfaces (F) is the force constant of the spring holding the lower lens multiplied by the deflection. The precision in the force measurement is as small as 10^{-7} N. The force is normalized by the mean radius of curvature, R, of the mica surfaces in the region where they are interacting. According to the Derjaguin approximation(23), the ratio $F(D)/R$ equals 2π times the interaction energy per unit area between the two surfaces.

In a standard experiment, where the layers are fully compressed and then separated until there are no forces present (>3000Å), and then recompressed and separated as many times as desired. There is about one minute between each measurement. Usually, a separation is performed ~5 minutes after the compression. The subsequent compressions are performed anywhere from 5 minutes to 24 hours later. For a multistep experiment, the layers are partially compressed, then separated far apart (>3000Å), then recompressed farther, separated and so on until the layers are fully compressed.

In SFA experiments, it is possible to measure the adsorbed amount of polymer. This method involves draining the apparatus and then drying out the polymer layers. The dry surfaces are compressed and a layer thickness is measured. By assuming a polymer density, the adsorbed amount can be estimated from the dry layer thickness. To ensure that the measurement of adsorbed amount is as accurate as possible, the surfaces are left in solution for eight hours, then rinsed with distilled water and immediately dried and compressed.

For rheological experiments, solutions of modified and non-modified polyacrylamide were prepared at 1wt% and 3wt% with millipore water and 0.9M NaCl. The solutions were stirred overnight to ensure maximum solubility. The linear viscoelastic properties of HAPAM solutions were obtained on a Rheometrics fluid spectrometer, RFS-II, with a controlled shear rate. A couette configuration was used, since η_o< 100 Pa·s for the samples, with R_i=16mm, R_o=17mm, and L=33mm. Results were taken at room temperature (23°C). Dynamical strain sweeps were performed at frequencies of 5rad/s and 10rad/s. Dynamical measurements were carried out in the linear viscoelastic region at strains of γ=0.07, 0.1, and 0.4 for 1wt% and γ=0.04 and 0.08 for 3wt%.

Results

Several experiments were performed in which the force-distance profile was obtained for surfaces that were incubated overnight (~8 hours) in electrolyte solution that contained various concentrations of modified PAM or non-modified PAM. The excess polymer was removed and the SFA cell was filled with electrolyte solution. The surfaces were dried and compressed to determine the adsorbed amount at various salt and polymer concentrations. Experiments were done with PAM in electrolyte solution for comparison.

In general, increasing the polymer concentration in the adsorption solution increases the adsorbed amount. The HAPAM concentration is varied from 500ppm to 2000ppm. The adsorbed amount increases from 2mg/m^2 for a 500ppm HAPAM solution in 0.9M NaCl, to 6mg/m^2 for 2000ppm HAPAM in 0.9M NaCl. The plot in Figure 2 compares the increase in adsorbed amount for HAPAM on mica with the adsorption of PAM (6). The HAPAM isotherm shows no apparent plateau in this range of polymer concentrations, while the PAM isotherm plateaus very quickly.

The concentration of salt also affects the adsorbed amount. Increasing the salt concentration simply shifts the adsorption isotherm upwards, with each isotherm increasing at the same slope with the increase in polymer concentration. For example, at a polymer concentration of 2000ppm HAPAM, the adsorbed amount is 4 mg/m^2 for 0.3M NaCl, and 6 mg/m^2 for 0.9M NaCl. The effect of salt concentration is slightly more complicated than for polymer concentration. The presence of electrolyte enhances the water structure organization around hydrophobic groups, thus forming microdomains and enhancing hydrophobic associations (24). Initially, an increase in NaCl concentration increases the adsorbed amount of HAPAM due to this effect. Eventually, at very high salt, the effect is not as noticeable since the solvation of the polymer is poor and aggregates may form.

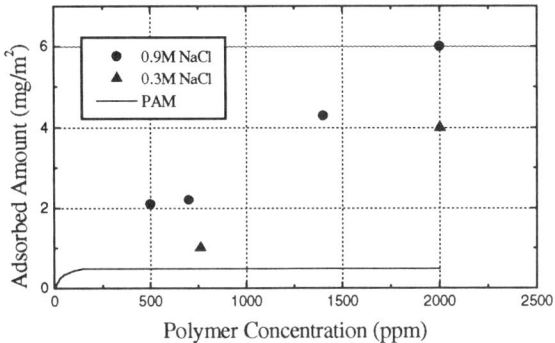

Figure 2. Adsorbed amount of HAPAM with salt and polymer concentration, as well as an adsorption isotherm for non-modified PAM.

Surface force measurements offer a unique way to relate the addition of the hydrophobic moieties to changes in layer-layer interactions. Figure 3 shows the force profiles in 0.3M NaCl solution, following the adsorption of 760ppm HAPAM in 0.9M NaCl solution. This results in a relatively low adsorbed amount. When HAPAM is introduced, no forces are measured down to a separation of ~2000Å for the first compression, and then a repulsion is observed which increases until the surfaces are pressed close together. The profiles that are measured are reproducible down to ~1200Å, but for D<1200Å the compression results in some irreversible changes to the surface layer. Figure 3 demonstrates a feature that is common in all HAPAM experiments at this concentration. The force-distance profile measured on the first compression is not reversible. Upon separation, the force drops sharply and an attraction is even observed. The second compression shows a profile shorter in range of interaction than the first compression, and subsequent separations and compressions follow the profile of the first separation and second compression, respectively. This effect has been seen in other experiments involving uncharged polymer (16,17), but they usually saw a separation profile more similar to the compression profile. In some experiments the compression/separation cycles are done immediately after one another, while in others the surfaces are left overnight before the next compression. The repulsive interaction does not vary with time, and no desorption is observed during the time of the experiment.

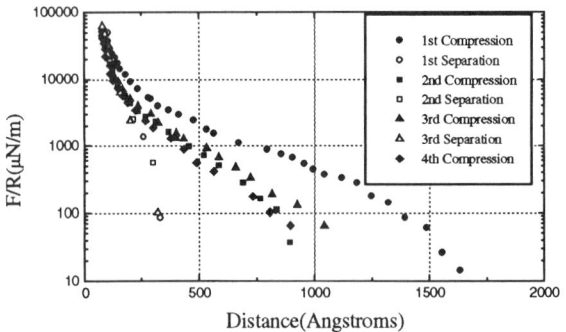

Figure 3. Force-distance profile for 760ppm HAPAM adsorbed in 0.9M NaCl, results taken in 0.3M NaCl.

In order to further investigate this feature, experiments with multi-step compression/separation cycles were carried out, as described in the methods section. The results are shown in Figure 4. When the compression is done in steps, the irreversibility observed for a one step approach is broken down into components, eventually resulting in a reversible path. Also, each partial compression is irreversible and closely follows the previous separation. Figure 4 also contains a reversible compression curve from a force profile at a different spot on the surface under the same conditions. As you can see, it is comparable to the fourth compression of the multi-step experiment, indicating that the multi-step experiment is now a reversible profile. As previously mentioned, the separation cycle does not follow the same path as the compression profile due to an attractive force between the surfaces.

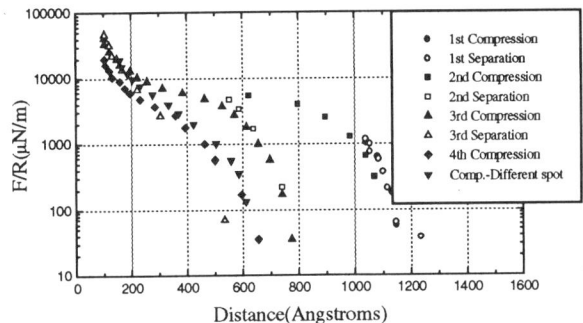

Figure 4. Multi-step compression force-distance profile for 760ppm HAPAM adsorbed in 0.9M LiCl, results taken in 0.3M LiCl. The inverted triangles represent a reversible compression from a different spot on the surface.

Figure 5 shows a close-up of a separation cycle for HAPAM at the same conditions as Figure 3. The attractive force is very long range (~2000Å) and is unlike van der Waals attraction (19). This type of hysteresis in the closer region is in small part due to the relaxation of the polymer over time (16), but is more likely due to hydrophobic interactions. These interactions may be between hydrophobic groups bridging from one surface to the other or hydrophobic groups at the interface.

Surface force measurements were taken on PAM in order to compare the force profiles with those of the modified version, to prove that the attraction and irreversibility in the first compression are due to the addition of the 1% hydrophobic groups. When PAM is adsorbed onto the mica surface at a concentration of 760ppm and forces are measured in 0.3M NaCl, the profile is very different than for HAPAM. The repulsion is shorter range and the compression is reversible. The separation curves follow the same path as the compression curves, showing no signs of attraction as seen in HAPAM experiments. Similar results are also found for PAM adsorbed at a higher polymer concentrations.

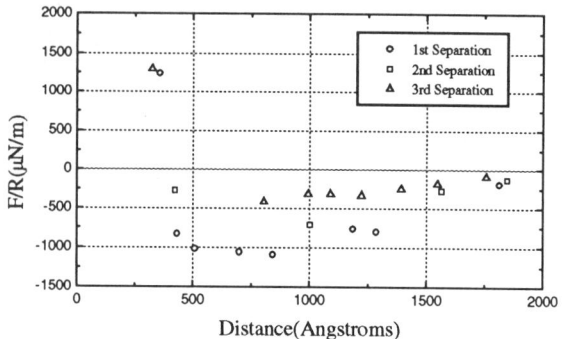

Figure 5. Close-up of separation force-distance profile for 760ppm HAPAM.

One way to examine the role of the hydrophobic moieties is to examine how changes in polymer concentration in solution affect the adsorption behavior. As the polymer concentration increases, the adsorbed amount increases. Force measurements were taken for HAPAM concentrations ranging from 400ppm to 2000ppm. The force profile for 400ppm does not have any obvious differences from the force profile for 760ppm. When the polymer concentration is increased to 2000ppm HAPAM in 0.3M NaCl, though, the force profile changes as seen in Figure 6. In this case, the compression curve has a similar shape, but there is not an attraction observed during separation as in the profiles at lower concentrations.

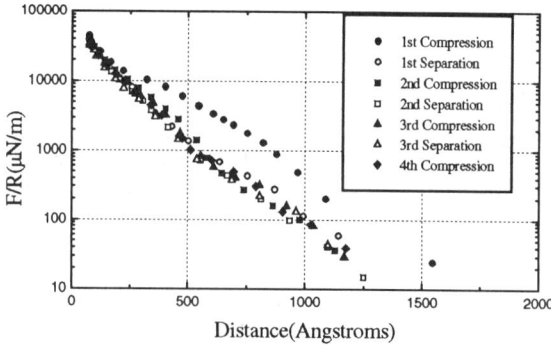

Figure 6. Force-distance profile for 2000ppm HAPAM adsorbed in 0.3M NaCl.

In order to understand the nature of the attraction observed upon separation, a series of experiments were performed in which the HAPAM was only adsorbed onto one surface. In one of the experiments, 760ppm HAPAM in 0.9M NaCl was adsorbed on the mica surfaces and then one surface was replaced with a bare mica surface in 0.3M NaCl. During the first compression, the repulsion increases as the surfaces are brought closer together and the force profile is not reproducible, as in a 2-sided experiment. The range of interaction for a 1-sided experiment is less than half of the range for a 2-sided experiment. The most interesting aspect of this experiment, though, is the separation profile. As shown in Figure 7, the first separation results in an attraction on the order of -3000μN/m, and is very long-range in nature (~2500Å). The second separation exhibits an attraction of smaller magnitude, and the third shows only a small attraction. Since the successive separations show less attraction, it is concluded that some of the chains are remaining attached to the bare surface after the first compression/separation cycle. Another one-sided experiment involved the adsorption of 2000ppm HAPAM in 0.9M NaCl, which results in the largest adsorbed amount. By using the thickest layer, polymer chains may be prevented from bridging from surface to surface. The results are shown in Figure 8. There is not an attraction observed, as is observed with other one-sided experiments at lower adsorbed amounts.

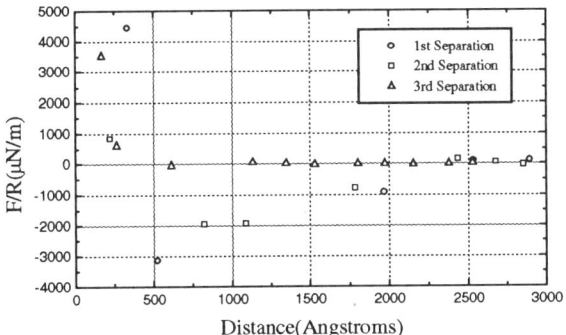

Figure 7. Force-distance profile for a bare mica surface and 760ppm HAPAM adsorbed in 0.9M NaCl.

Rheological experiments were performed in order to understand the effectivenes of the hydrophobic interactions within the adsorbed layer that are contributing to multilayer build-up. Solutions of 1wt% and 3wt% HAPAM and PAM were prepared. These concentrations are representative of the density of polymer in the region away from the substrate. For these polymers, measurements are taken in the linear viscoelastic regime. A characteristic relaxation time can be estimated form the frequency at which the storage and loss moduli cross. At 1wt% HAPAM, the longest relaxation time was 0.01s, which is extremely fast. At 3wt% the longest relaxation time was 0.33s. Comparing the complex viscosity for HAPAM solutions and PAM solutions, the addition of the hydrophobes does not enhance the viscosity as might be expected. The fast relaxation times and minimal viscosity changes indicate that any inter-chain interactions are short lived and do not greatly affect the rheology of the solution.

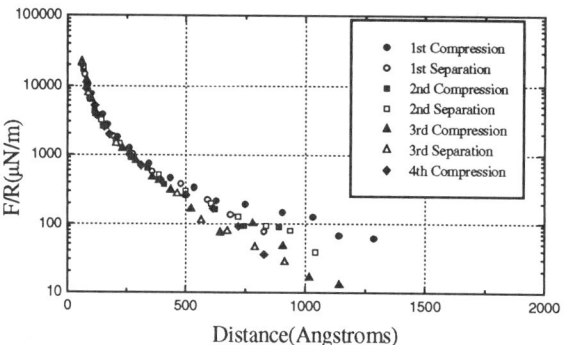

Figure 8. Force-distance profile for a bare mica surface and 2000ppm HAPAM adsorbed in 0.9M NaCl.

Discussion

Enhancement of adsorption of hydrophobically modified polymers versus the non-modified version has been seen in polyacrylamide systems and hydroxyethyl cellulose (HEC) systems (5,6,7). The adsorbed layer has some segments in contact with the surface and others that are linked through hydrophobic interactions to chains that are adsorbed on the surface. For HEC polymers, though, the adsorption is enhanced with the hydrophobic modification, but there is not a correlation between polymer concentration and adsorbed amount, using lower polymer concentrations, though. Obviously, the effects of polymer concentration on adsorbed amount and the structure of the copolymer layers are not completely clear.

The present results are useful in addressing the disparity in the literature over how the hydrophobic modification actually leads to multilayer build-up and whether adsorbed amount is dependent on polymer concentration. Looking closely at the adsorbed amounts in Figure 2, overall there is an increase in adsorbed amount with polymer concentration, but between 500ppm and 760ppm in 0.9M NaCl, there is only a very small increase, similar to the case of HMHEC in this concentration range (5). This is similar to what is observed for HEUR adsorbed on polystyrene latex where the adsorbed amount increases rapidly, followed by a plateau region and then another increase (25). This indicates that the polymer layer may go through a conformational rearrangement on the surface, due to the strong affinity of the hydrophobic groups for the surface, until the surface is saturated. Figure 9a is a schematic of the proposed structure of the modified polymer adsorbing onto the surface. At lower polymer concentrations, the chains initially adsorb in loops and tails and are loosely extended into solution. The hydrophobic interactions with the surface are interspersed with polymer backbone interacting with the surface, though they are not densely packed. As the polymer concentration in solution is increased and the layers are compressed, more hydrophobes find their way to the surface, adding additional surface interactions as well as replacing some of the polymer backbone that is at the surface. The chains will organize in such a way to maximize the surface interactions.

One point that has not been addressed in previous research is exactly how the hydrophobic moieties interact to form multilayers. Not all of the hydrophobes can be interacting with the surface because some are in the polymer loops and tails interacting with each other. Hydrophobes that are not in contact with the surface do not want to be left dangling in water, so they will tend to cluster with other hydrophobes in micelle-like formation, where the polymer backbone protects the hydrophobic core, as shown in Figure 9a. During this phase, the actual layer thickness does not drastically increase, but the layer's interaction with the surface becomes stronger.

Examining the force profile gives insight on the interactions at the mica surface as well as the layer/layer hydrophobic interactions. In the initial compression, there is a long-range repulsive interaction that increases steadily as the separation decreases. The multi-step compression in Figure 4 of the extended copolymer gives more detail on the irreversibility of the first compression. The first approach is similar to other first compression profiles. The first partial compression does not compress far enough to actually change the layer structure. Therefore, the first separation follows the same path of the first compression, as well as the second partial compression. After the second compression, for every separation and recompression the onset of repulsion moves progressively inward, and each partial compression produces irreversible changes in the layer until the layer is fully compressed. Subsequent separations and compressions are shorter-range interactions. Since the range of interactions does not change with time after the first compression in the span of our experiments, it is suggested that the extended layers have been irreversibly distorted by compression.

The repulsive force is initially due to polymer-polymer overlap and increases with decreasing distance. When the polymer layers encounter each other, there is a reduction in conformational entropy of the coils, which results in an increase in free energy and thus a repulsive force between the surfaces (26). The second part of the repulsive interaction is steric in nature as the configuration of the layer actually changes due to the other layer. This is the region where each successive compression starts closer in, it would seem that it is the steric component of the interaction that is irreversible, and that the overlap forces are essentially reversible.

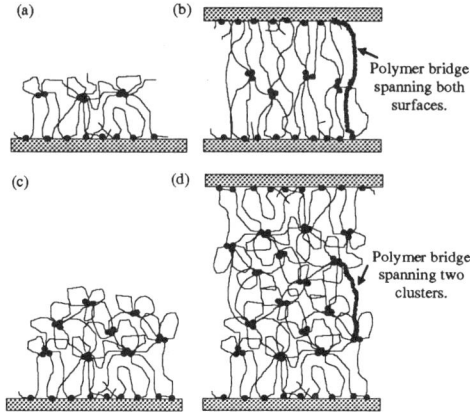

Figure 9. Schematic representation of the adsorbed layers formed by HAPAM at (a), (b) low polymer concentration and (c), (d) high polymer concentration.

One way to explain the irreversibility of the first compression of the HAPAM layers has to do with the possibility of unoccupied binding sites on the substrate upon initial adsorption, or sites occupied by polymer backbone. The extended polymer presents an energy barrier which prevents further adsorption. On compression, this barrier is overcome and hydrophobes are able to reach the unoccupied binding sites and also replace polymer backbone at other sites. Since the HAPAM profile has an irreversible first compression while the PAM profile does not, the interaction between the hydrophobe and the surface is more favorable than between the polymer backbone and the surface. It may be surprising that the hydrophobes would interact strongly with the mica and replace polymer backbone at the surface since mica is hydrophilic. The surface, however, consists of Si-O-Si bonds that have a certain hydrophobic character. Also, the relative penalty of having a hydrophobe in the aqueous environment may drive it to replace backbone at the surface. Hydrophobic interactions within the copolymer also contribute to the stability of the final conformation of the layer. Since only a small percentage of HAPAM is hydrophobic, the initial extended form provides less opportunity for the hydrophobic moieties to interact. Compression increases the effective density of hydrophobic groups, which enhances the environment for hydrophobic attraction to occur.

One feature that is interesting in this system is the weak, long-range attraction that is seen upon separation for some of the systems. Assuming that not all of the hydrophobes are in contact with the surface, there will be some hydrophobes or hydrophobe clusters, which will be away from the substrate near the layer/layer interface. Upon compression, the two surfaces approach each other, as in Figure 9b, and hydrophobes are able to bridge to the other surface and bind to the substrate. The hydrophobe clusters will go through some rearrangement to permit the bridging. Hydrophobic interactions would mainly be seen during a separation cycle because the statistical density of hydrophobic groups will be higher in the compressed state. For PAM, the separation curve follows the compression curve. This gives assurance that the attraction in the separation curves is really a result of hydrophobes bridging, not just polymer relaxation.

As the polymer concentration is increased, as shown in Figure 9c, hydrophobes from polymer chains attached to the surface interact with hydrophobes from other polymer chains in solution to form a network of clusters. The clusters are micelle-like in the upper portion of the layer, where the hydrophobes form the core and the polymer backbone forms loops around the core, protecting it from the water. This change in the layer affects the separation profiles. At higher concentrations, corresponding to adsorbed amounts greater than 2.5mg/m^2, the separation curve follows the compression curve relatively closely, and there is not any attraction observed. At lower concentrations corresponding to adsorbed amounts of 2.5mg/m^2 or less, the separation curves follow a different path from the compression curves and there is at least a small attraction observed. The magnitude of the attraction varies as the adsorbed amount continues to decrease. As the thicker layers are compressed in Figure 9d, the chains are unable to span from one surface to the other. This result suggests that the observed attraction is due to polymer bridging from surface to surface only, and the ability of a polymer chain to bridge depends on the thickness of the layers.

Examining the separation profiles for the 1-sided experiments, results also suggest that the observed attraction is due to polymer bridging, and is dependent on the layer thickness. There is no attraction when a large amount is adsorbed onto one surface, which indicates that polymer chains are not able to bridge from one surface to the other. When attraction is observed for a 1-sided experiment, it decreases after each compression, indicating that some of the polymer chains have remained attached to the bare mica surface. If this occurs at the lower adsorbed amount, it is also occurring at the highest adsorbed amount, though no attraction is observed in the force profile. We know that the adsorbed amount increases due to hydrophobes interacting to form multilayers, yet at high adsorbed amounts, there are not obvious interactions between hydrophobes at the layer/layer interface. This means that the polymer chains that are being pulled away from the layer are not attached to the surface, but are being pulled away from a hydrophobe cluster.

The hydrophobes that interact with the surface do not desorb in the time-scale of the experiment, indicating a relatively long relaxation time of these interactions. In the SFA force profiles, we do not see an attraction due to hydrophobe/hydrophobe interactions near the interface, or bridging from layer to layer. In these experiments, there is approximately a minute between measurements. Therefore, the longest relaxation time of the hydrophobe/hydrophobe interactions determined by rheological measurements, is very fast relative to the time-scale of the experiment. This means that if the hydrophobes within the layer are interacting in clusters, they can rearrange themselves very quickly and pulling these interactions apart during separation will not result in a visible attraction in the separation force profile. This is similar to what was found by Volpert *et al* for one of their hydrophobes, where no viscosifying effect is found and it is attributed to the short lifetime of the hydrophobic associations (7). They have shown that the associative properties of a hydrophobically modified polyacrylamide depend strongly on the bulkiness of the hydrophobic comonomer (27). Therefore, attraction seen in the force profiles is due only to polymer chains bridging from surface to surface.

From this discussion it would seem that initially the polymer would adsorb in such a way as to maximize the number of surface contacts. The layer thickness would increase with polymer concentration as the number of chains interacting with the surface increased, and the chains formed a more extended conformation. In this conformation, there would be many hydrophobes available in clusters near the interface to either bridge to the other surface or interact with other hydrophobes in solution, depending on the polymer concentration. At higher polymer concentrations, the adsorbed amount continues to increase as the hydrophobes of free chains associate with the extended hydrophobes of chains that are connected to the surface, creating multilayers.

Conclusion

Surface force measurements have been used to study the interactions in a system of hydrophobically modified polyacrylamide. The adsorbed amount of HAPAM increases with increasing polymer and salt concentrations, which indicates multilayer build-up. The first compressions under any adsorption conditions are irreversible due to extended polymer layer being compressed to a more stable state in which there are more hydrophobic interactions at the mica surface and within the layers. Subsequent compressions are reproducible and the shorter range of repulsion reflects the reduction of extended polymer after the first compression. At lower adsorbed amounts, the separation profiles show an attraction between the HAPAM layers, which is due to polymer chains bridging from one surface to the other. At the higher adsorbed amounts, the polymer chains are unable to bridge from one surface to the other, so there is not an observed attraction. Polymer chains may bridge from layer to layer, but the attraction is not visible in the SFA due to the fast relaxation of the hydrophobe/hydrophobe interactions. This indicates that the interactions between the hydrophobes and the surface are more effective than interactions within polymer layers. This feature is dependent on the size and type of hydrophobe, as well as the amount, as demonstrated by the research that has been performed on these types of polymers in solution and in particle dispersions. Therefore, by changing the size and type of hydrophobe, the stability of the multilayer structure can be controlled.

References

1. Kientz, E.; Holl, Y. *Colloid & Polym. Sci.* **1994**, *272*, 141.
2. Evani, S.; Rose, G.D. *Polym. Mater. Sci. Eng.* **1987**, *57*, 477.
3. Buscall, R.; Corner, T. *Colloids and Surfaces* **1986**, *17*, 39.
4. Chen, M.; Wetzel, W.H.; Ma, Z.; Glass, J.E. *J. of Coatings Tech.* **1997**, *69*,73.
5. Tanaka, R.; Williams, P.A.; Phillips, G.O. *Colloids & Surfaces* **1992**, *66*, 63.
6. Argillier, J.-F.; Audibert, A.; Lecourtier, J.; Moan, M.; Rousseau, L. *Colloids & Surfaces A - Physicochemical & Engineering Aspects* **1996**, *113*, 247.
7. Volpert, E.; Selb, J.; Candau, F.; Green, N.; Argillier, J.-F.; Audibert, A. *Langmuir* **1998**, *14*, 1870.
8. Argillier, J-F.; Ramachandran, R.; Harris, W.C.; Tirrell, M. *J. of Colloid & Interface Sci.* **1991**, *146*, 242.
9. Santore, M.M.; Russel, W.B.; Prud'homme, R.K. *Macromolecules* **1990**, *23*, 3821.
10. Ananthapadmanabhan, K.P.; Mao, G.-Z.; Goddard, E.D.; Tirrell, M. *Colloids & Surfaces* **1991**, *61*, 167.
11. Dhoot, S.; Goddard, E.D.; Murphy, D.S.; Tirrell, M. *Colloids & Surfaces* **1992**, *66*, 91.
12. Luckham, P.F.; Klein, J. *J. Chem. Soc., Faraday Trans. 1* **1984**, *80*, 865.
13. Marra, J.; Hair, M.L. *J. Phys. Chem.* **1988**, *21*, 6044.
14. Klein, J.; Luckham, P.F. *Colloids Surf.* **1984**, *10*, 65.
15. Afshar-Rad, T.; Bailey, A.I.; Luckham, P.F.; McNaughton, W.; Chapman, D. *Colloids Surf.* **1987**, *25*, 263.
16. Klein, J.; Luckham, P.F. *Macromolecules* **1984**, *17*, 1041.
17. Israelachvili, J.N.; Tandon, R.K.; White, L.R. *J. Colloid Interface Sci.* **1980**, *78*, 430.
18. Israelachvili, J.N.; Tabor, D. *Proc. R. Soc. London Ser. A* **1972**, *331*, 19.
19. Israelachvili, J.N.; Adams, G.E. *J. Chem. Soc., Faraday Trans. 1* **1978**, *74*, 975.
20. Israelachvili, J.N. *J. Colloid Interface Sci.* **1973**, *44*, 259.
21. Hadziioannou, G.; Patel, S.; Granick, S.; Tirrell, M. *J. Am. Chem. Soc.* **1986**, *108*, 2869.
22. Tolansky, S. *An Introduction to Interferometry;*, Longmans Green & Co.: London, England, 1955.
23. Derjaguin, B.V. *Kolloid Zh.* **1934**, *69*, 155.
24. Israelachvili, J.N. *Intermolecular and Surface Forces;* Academic Press: London, England, 1985.
25. Jenkins, R.D.; Durali, M.; Silebi, C.A.; El-Asser, M.S. *J. Colloid Interface Sci.* **1992**, *154*, 502.
26. Evans, D.F.; Wennerstrom, H. *The Colloidal Domain;* VCH publishers: New York, NY, 1994.
27. Volpert, E.; Selb, J.; Candau, F. *Macromolecules* **1996**, *29*, 1452.

Chapter 15

Complexations of Beta-Cyclodextrin with Surfactants and Hydrophobically Modified Ethoxylated Urethanes

Analytical Application in Adsorption Measurements

Zeying Ma[1] and J. Edward Glass

Polymers and Coatings Department,
North Dakota State University, Fargo, ND 58105

Complexations of beta-cyclodextrin with surfactants and model associative thickeners of the **H**ydrophobically-modified, **E**thoxylated **UR**ethane (HEUR) type have been investigated by a visible spectral displacement method. In this method, phenolphthalein is displaced from the beta-cyclodextrin cavity by the hydrocarbon moiety of the surface-active agent, resulting in the enhancement of the 554 nm band of the basic form of phenolphthalein. The results are correlated with the hydrophobe size and geometry of the compounds. The application of this complex technique to the determination of adsorption of HEUR associative thickeners, containing no chromophores, on polymeric latices is discussed.

[1]Current address: R&D Center, Hewlett-Packard Corp., San Diego Site, 16399 West Bernando Drive, San Diego, CA 92127-1899.

Introduction

Cyclodextrins (CDs) are non-reducing cyclic oligosaccharides produced by bacterial degradation of starch.[1] The molecular geometry may be described as toroidal shape; the narrower opening contains the O(6)H groups, whereas the wider opening is occupied by the O(2)H and O(3)H groups. Crystal structure analysis[2] reveals that the glucose units always have a C1-chair conformation, and the C6-O(6) bonds preferentially direct away from the center of the ring. With the steric positioning of the hydroxyl groups on the exterior, the interior surface of the cavity is hydrophobic (apolar), and the external surface of the molecule is hydrophilic (polar). The structural features of CDs enable the formation of inclusion complexes with a variety of compounds. The main criterion for complex formation is the spatial fit between a CD host cavity and guest molecule. The dominant driving force is the displacement of water molecules from the cavity by hydrophobic guest molecules to attain apolar-apolar association, producing a more stable, lower energy state. Upon inclusion, water molecules, expelled from the cavity and stripped from the hydration sphere of guest molecules, gain degrees of freedom, resulting in a large increase in entropy. The complexation property of CDs has provided the molecules with a wide range of applications in the food, cosmetic and pharmaceutical industries, as well as in chemical analysis and reaction chemistry.[3,4] An illustration of the CD structure is given in the chapter that will follow.

A number of studies on the associations of CDs with surfactants have been reported. Since neither CDs nor most surfactants bear a chromophore, most published results were obtained by conductance measurements with ionic surfactants. Adding CDs to ionic surfactant solutions dramatically affects their conductivities when inclusion complexes are formed.[5] The critical micelle concentrations of sodium dodecyl sulfate (SDS) and cetyltrimethylammonium bromide (CTABr) increase upon the addition of alpha-CD and beta-CD in aqueous media, indicating a portion of surfactant monomers in the micellization process is removed by complexation. Aqueous solutions of CDs do not exhibit surface activity. The surface tension of a CTABr solution increases with addition of beta-CD. For surfactant-CD complexes, a 1:1 stoichiometry is usually presumed.[5,6,7] The association constants of alpha-CD and beta-CD with a homologous series of sodium 1-alkanesulfonates and alkylsulphate increase monotonously with increasing carbon atoms (n) and become abruptly constant at n=10.[6,7] The association constants of CD with ionic surfactants of the same alkyl chain length are similar to one another, regardless of the nature of the charge of the ionic head group.[6] In general, the association constants with alpha-CD and beta-CD are of the same order of magnitude. The formation of 2:1 type alpha-CD-alkylsulphate anions is also reported.[8] The 2:1 beta-CD-surfactant complexes are formed with alkanesulfonates of n>10 and alkyl sulfates of n>8, based on fluorescence-probe studies.[9] Literature data on gamma-CD are scarce.[10] At a given chain length, fluorocarbon surfactants bind much stronger with beta-CD than the corresponding hydrocarbon surfactants.[11]

Surfactant-modified water-soluble polymers contain multiple hydrophobes and are often used concurrently with surfactants in a variety of formulations. From a "model" thickener viewpoint the most important of these compounds is the

Hydrophobically-modified Ethoxylated URethane (HEUR). This investigation examines the associations of beta-CD with anionic surfactants (alkanesulfonates), nonionic surfactants (ethoxylated alkyls and aryls), and surfactant-modified water-soluble polymers of the HEUR type, using a visible spectral displacement technique, recently reported by Sasaki et al.[12,13] The results are correlated to the hydrophobe size and geometry of the compounds. The spectral displacement technique was examined for its analytical application in the determination of HEUR adsorption on polymeric latices.

Experimental Section

Beta-CD monohydrate, phenolphthalein, and sodium carbonate were purchased from Aldrich. Surfactants studied were provided or purchased from a variety of manufacturers. Model associative thickeners studied were made in this laboratory.[14] The structures discussed in the first part of this chapter were prepared by a step-growth synthesis; the adsorption studies in the latter part of this study were narrow molecular weight HEURs. Experimental procedures for complexation analysis follow that by Sasaki et al.[13], with only a slight variation.

Preparation of stock solutions: (1) A 1.5 mM phenolphthalein stock solution was prepared by dissolving 0.0477 g of the compound with 95 v/v % ethanol to a total volume of 100 ml; (2) A 1 mM beta-CD stock solution was made by dissolving 0.2838 g of the solid with DDI water to a total volume of 250 ml; (3) A phenolphthalein aqueous solution was prepared by placing 0.2120 g of sodium carbonate powder and 10 g of the phenolphthalein ethanol solution, prepared in (1), into a 100 ml volumetric flask, then diluting with DDI water to 100 ml; (4) Solutions of surfactants and associative thickeners were prepared by dissolving the solids into DDI water by stirring gently for one day.

Preparation of individual samples: Four grams of the 1 mM beta-CD solution, 4 g of the aqueous phenolphthalein solution, and 12 g of known concentration surfactant or thickener solution were mixed in a vial. All the samples in this study contain a constant amount of Beta-CD monohydrate, phenolphthalein, and sodium carbonate. Their concentrations in each sample are 0.2 mM, 0.03 mM, and 4.0 mM, respectively. The only variation in each sample is the surfactant or thickener concentration.

UV measurement: Spectra were taken on a Hewlett-Packard 8451A diode array UV-visible spectrophotometer equipped with a deuterium lamp. All the measurements were made within two hours after the preparation of the phenolphthalein aqueous solution. In determining the concentrations of surfactants and associative thickeners in the supernatants of latex dispersions, standard solutions were prepared and analyzed concurrently with the unknowns.

Latex: Three monodisperse poly(methyl methacrylate)[PMMA] latices were used as substrates in this adsorption study: one was free of methacrylic acid oligomeric surface stabilizers, 2 and 4 wt.% methacrylic acid were charged in the latter stages of a starved emulsion synthesis for the remaining substrates. The preparation and characterization of the latices have been described[15]. The latices were dialyzed free of their synthesis surfactant prior to the adsorption studies. The adsorption sample is prepared by adding a certain amount of stock latex dispersion to a measured volume of thickener solution of known concentration. After adsorption reaches equilibrium, latex solid is removed by centrifugation. The supernatant is analyzed for thickener concentration. A series of thickener aqueous solutions of varying concentrations have to be made concurrently each time with adsorption samples to construct a calibration curve.

Results and Discussion

In studying the adsorption by the disperse phase, it is necessary to know the equilibrium concentration of the adsorbate in the medium after adsorption has occurred. There are numerous ways of determining the concentration of water-soluble polymers in aqueous medium. These include UV,[16,17] FTIR,[18] and fluorescence[19] spectrophotometric analysis, differential refractometric method,[20] size exclusion chromatographic,[21] surface enhanced resonance Raman scattering technique,[22,23] electron spin resonance study,[24,25] turbidity measurement,[26] and the use of radioisotopes.[27] With all of these techniques, UV and visible spectroscopy are the most convenient and accurate methods for quantitative analysis. However, most associative thickener molecules do not contain a chromophore in the accessible ultraviolet or visible regions of the spectrum, at least not at sufficient strength to quantify adsorption. The above-discussed spectral displacement study of the beta-CD/thickener complexation provides a new tool to approach this problem.

Beta-CD/Surfactants Complexations

Phenolphthalein and beta-CD can form, at pH 10.5, a 1:1 complex with a formation constant of 2.1×10^4 liters mol^{-1}; this 1:1 complex has an absorbance practically equal to zero in the region of the 550 nm band of the basic form of phenolphthalein.[28,29] Thus, the only species having an absorbance in the visible region is due to the colored form of phenolphthalein uncomplexed by beta-CD.

When sodium dodecyl sulfate (SDS) or N,N-dimethyl-N-dodecylamine oxide (DDAO) is added to the beta-CD and phenolphthalein solution, the alkyl moiety of the surfactant displaces[15] some of the phenolphthalein at pH=10.5, resulting in a significant increase in absorbance at 550 nm. The two surfactants previously studied[13] have the same dodecyl hydrophobe. Our interest is in using this displacement method to examine the complex formations of CD with nonionic surfactants having alkyl- and aryl-hydrocarbon groups of various sizes and geometry and with the hydrophobes of HEUR thickeners.

Effect of Surfactant Hydrophobe Size

The absorbance of the 554 nm phenolphthalein band as a function of total concentration of four added anionic surfactants, sodium C_8H_{17}- to $C_{14}H_{29}$- alkyl sulfonates, is illustrated in Figure 1.

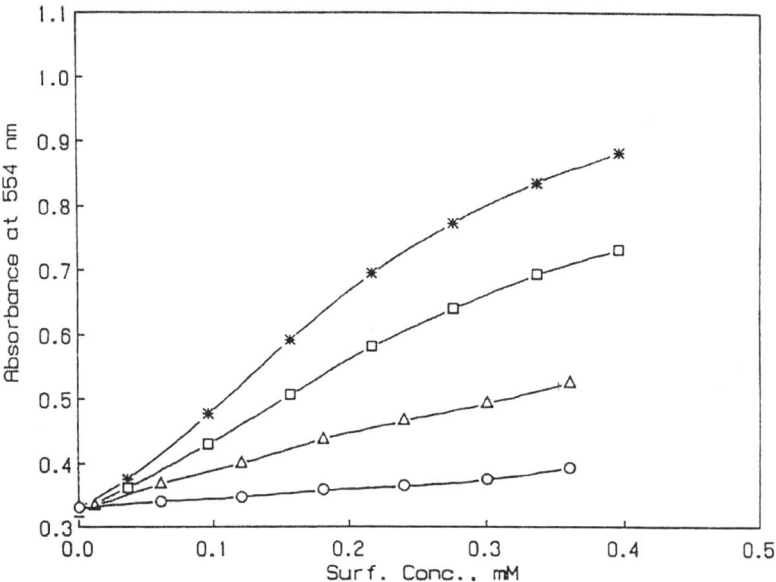

Figure 1. Dependence of absorbance at 554 nm on surfactant concentration. [Solutions contain 0.2 mM β-cyclodextrin, 0.03 mM phenolphthalein, and 4 mM Na_2CO_3.] Symbols: O, $CH_3(CH_2)_7SO_3Na$; Δ, $CH_3(CH_2)_9SO_3Na$; □, $CH_3(CH_2)_{11}SO_3Na$; *, $CH_3(CH_2)_{13}SO_3Na$

All the solutions were made having a specific and constant total concentration of beta-CD and phenolphthalein as described in the experimental part. Obviously, the formation constant of the complexes increases as the hydrophobe size of the surfactant increases from C_8H_{17}- to $C_{14}H_{29}$-. The absorbance of the same band versus the concentration of four nonionic surfactants with variable size hydrophobes is illustrated in Figure 2. They are non-branched $C_{16}H_{33}$- and $C_{18}H_{37}$-, octyl phenyl (OP), and nonyl phenyl (NP). The addition of $C_{16}H_{33}$- and $C_{18}H_{37}$- nonionic surfactants has no effect on the absorbance of the 554 nm band. The highest sensitivity of the absorbance change as a function of surfactant concentration is observed with the surfactants containing octyl phenyl or nonyl phenyl hydrophobes. The results suggest that the central griddle of beta-CD is too small to accept a hydrophobe equal to or larger than an aliphatic $C_{16}H_{33}$-. The best match occurs with a linear $C_{14}H_{29}$- hydrophobe. The aliphatic $C_{12}H_{25}$- and $C_{10}H_{21}$- hydrophobes appear to contact the interior partially, but the likely configurations leave several holes for water molecules. Hydrophobes equal to and smaller than linear C_8H_{17}- are too small to satisfy the size criterion of host-guest complex. Of course what one is measuring here is the competition with phenolphthalein for the beta-CD cavity. The striking increases in absorbance for octyl phenyl and nonyl phenyl are attributed to the presence of the aromatic ring, which interacts strongly with beta-CD.[2]

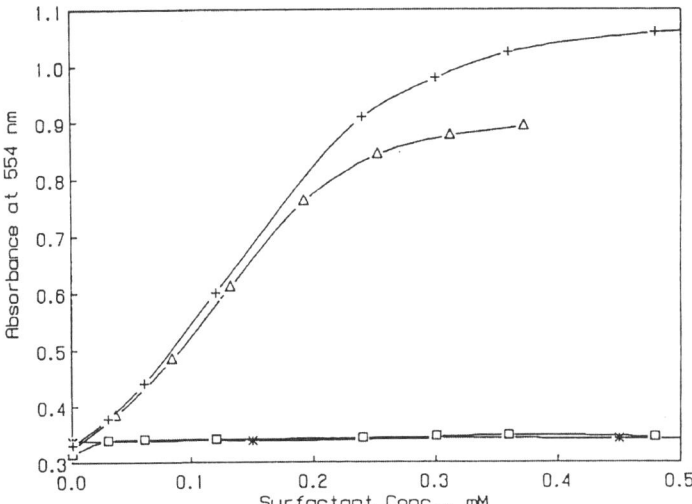

Figure 2. Dependence of absorbance at 554 nm on surfactant concentration. Solutions contain 0.2 mM β-cyclodextrin, 0.03 mM phenolphthalein, and 4 mM Na_2CO_3. Symbols:
+, $C_8H_{17}C_6H_4(OCH_2CH_2)_{10}OH$; Δ, $C_9H_{19}C_6H_4(OCH_2CH_2)_{12}OH$;
□, $CH_3(CH_2)_{15}(OCH_2CH_2)_{20}OH$; *, $CH_3(CH_2)_{17}(OCH_2CH_2)_{20}OH$.

Effect of Polar Head Groups of Surfactant

Figure 3 illustrates the dependence of the absorbance of the phenolphthalein band on the molar concentration of the surfactants having the same dodecyl hydrophobic tail, but with anionic, cationic, and nonionic hydrophilic groups. The data superimpose when the surfactant concentration is below 0.2 mM, indicating that at low concentrations, the complexation is independent on the chemical and electrical properties of the hydrophilic segment. A similar result is observed with the surfactants having a nonyl phenyl hydrophobe but different polyethylene oxide chain length (Figure 4). In this case, the concentration at which the data become divergent

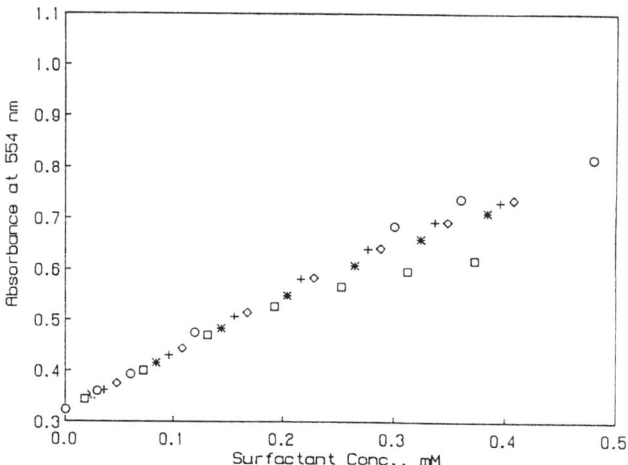

Figure 3. Dependence of absorbance at 554 nm on surfactant concentration. Solutions contain 0.2 mM β-cyclodextrin, 0.03 mM phenolphthalein, and 4 mM Na_2CO_3. Symbols: O, $CH_3(CH_2)_{11}SO_4Na$; +, $CH_3(CH_2)_{11}SO_3Na$; ◊, $CH_3(CH_2)_{11}N(CH_3)_3Br$; *, $CH_3(CH_2)_{11}N(CH_3)_2(O)$; □, $CH_3(CH_2)_{11}(OCH_2CH_2)_{23}OH$.

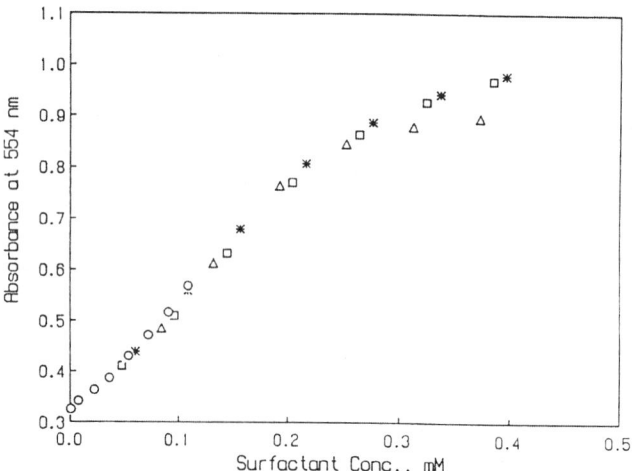

Figure 4. Dependence of absorbance at 554 nm on surfactant concentration. Solutions contain 0.2 mM β-cyclodextrin, 0.03 mM phenolphthalein, an 4 mM Na_2CO_3. Symbols:
O, $C_9CH_{19}C_6H_4(OCH_2CH_2)_9OSO_3NH_4$; Δ, $C_9CH_{19}C_6H_4(OCH_2CH_2)_{12}OH$; □, $C_9CH_{19}C_6H_4(OCH_2CH_2)_{40}OH$; *, $C_9H_{19}C_6H_4(OCH_2CH_2)_{100}OH$.

increases to 0.25 mM. This is due to the similarity between the polar head groups. The separations of the curves, i. e., the differentiation in their complex formation constants in the higher surfactant concentration region, may be related to the solubilities of the surfactants and their critical micelle concentrations (CMC). As illustrated in Figures 3 and 4, the surfactants with lower CMC ($CH_3(CH_2)_{11}(OCH_2CH_2)_{23}OH$ in Figure 3 and NP-O(EtO)$_{12}$H in Figure 4) tend to deviate from the common curve at a lower surfactant molar concentration than those having higher CMC values. The formation of micelles reduces the free surfactant in aqueous phase, and thus the absorbance of the 554-nm band of the displaced phenol-phenolphthalein.

Complexation of Beta-CD with HEURS

Three types of Hydrophobically-modified, Ethoxylated URethanes (HEURs) are examined in these adsorption studies. The first type is illustrative of commercial thickeners made by a step-growth process[14]. Although a general chemical composition based on what is expected from the stoichiometry of reactants is given in Figure 5, the products studied are in fact a cesspool mixture of materials to be expected from a step-growth synthesis. In these syntheses the diisocyanate structure can be varied, as long as it is aliphatic in structure; an aromatic linked urethanes hydrolyze in alkaline media. In the second synthesis type a $C_{18}H_{37}$ monoisocyanate is added directly to a low molecular weight polyethylene glycol. The third synthesis involved the addition of a 100 mole ethoxylate of nonylphenol (NP) to meta-tetramethylxylene diisocyanate (TMXDI)[30].

Employing a small excess of diisocyanate in the step-growth synthesis, the compositions formed have terminal isocyanates units that can be reacted with alcohols or amines, to provide terminal hydrophobes. The terminal isocyanates contribute to the effective terminal hydrophobe size and for this reason the hydrophobe capped H$_{12}$MDI compositions, that lead to the greatest viscosity build, are examined below. Four model associative thickeners with H$_{12}$MDI internal hydrophobes but different terminal hydrophobes (aliphatic C_6H_{13}-, C_8H_{17}-, $C_{10}H_{21}$-, and $C_{12}H_{25}$-, Figure 5) were studied.

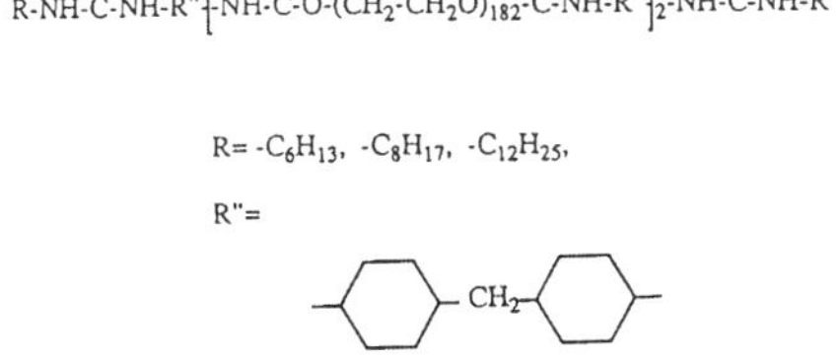

$$R\text{-NH-}\overset{O}{C}\text{-NH-R''-}[\text{-NH-}\overset{O}{C}\text{-O-}(CH_2\text{-}CH_2O)_{182}\text{-}\overset{O}{C}\text{-NH-R''-}]_2\text{-NH-}\overset{O}{C}\text{-NH-R}$$

R= -C_6H_{13}, -C_8H_{17}, -$C_{12}H_{25}$,

R''= ⟨⟩-CH_2-⟨⟩

Figure 5. Chemical structures of model associative thickeners with H$_{12}$MDI internal hydrophobes and different aliphatic terminal hydrophobes.

From the surfactant study noted above (Figures 3 and 4), a stronger interaction between beta-CD and thickeners was expected as the terminal hydrophobe size increases from C_6H_{13}- to $C_{12}H_{25}$-. This however does not account for the $H_{12}MDI$ contribution to the effective terminal hydrophobe size. It is observed (Figure 6) that C_6H_{13}-$H_{12}MDI$ and C_8H_{17}-$H_{12}MDI$-HEURs are more effective in displacing the phenolphthalein (i.e., give a greater enhancement in absorbance) than $C_{10}H_{21}$-$H_{12}MDI$ and $C_{12}H_{25}$-$H_{12}MDI$-HEURs.

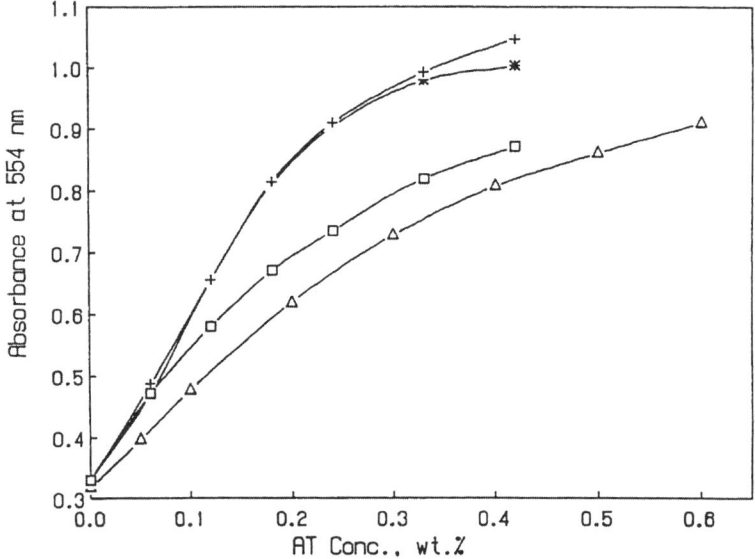

*Figure 6. Dependence of absorbance at 554 nm on model associative thickener concentration. Solutions contain 0.2 mM β-cyclodextrin, 0.03 mM phenolphthalein, and 4mM Na_2CO_3. Symbols: *, C_6H_{13}-$H_{12}MDI$; +, C_8H_{17}-$H_{12}MDI$; □, $C_{10}H_{12}$-$H_{12}MDI$;), Δ $C_{12}H_{25}$-$H_{12}MDI$*

As noted with the $C_{18}H_{37}$-surfactants (Figure 2) there is a size limitation for displacement of phenolphthalein from the beta-CD cavity. If a $C_{18}H_{37}$-monoisocyanate is reacted directly with a polyetherdiol, a narrower molecular weight with large hydrophobes is produced. This HEUR with a smaller $(EO)_{182}$ spacing would participate in dimer formation. Perhaps both this and/or the size of the $C_{18}H_{37}$ hydrophobe inhibits displacement of phenolphthalein from the beta-CD cavity (Figure 7).

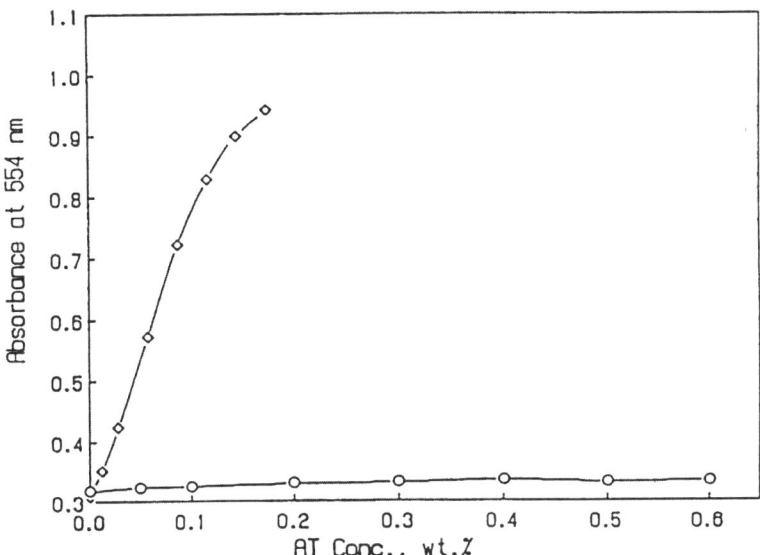

Figure 7. Dependence of absorbance at 554 nm on model associative thickener concentration. Solutions contain 0.2 mM β-cyclodextrin, 0.03 mM phenolphthalein, and 4mM Na_2CO_3. Symbols: ◊, $(NP-EtO_{100})_2TMXDI$; O, $(C_{18}H_{37})_{1.6}EtO_{182}$.

The third method of synthesis is to couple two nonyl-phenol (NP) surfactant with an average 100 mole of oxyethylene units by reaction with TMXDI[30]. This HEUR contains three aromatic groups, but its UV absorbance is not strong enough to be useful in adsorption studies. Although it has the same spacing as the $C_{18}H_{37}$-HEUR, there are no urethane linkages next to the NP hydrophobic groups and the strength of interaction among the smaller NP groups is less. The enhanced interaction of aromatics with the CD cavity also appears to contribute to the displacement of phenolphthalein from the beta-CD cavity. It is obvious that even in simple solutions, the beta-CD analysis would not be universally applicable for analysis in HEUR concentration in aqueous solutions.

HEUR Adsorption on Latices

When the HEUR Hydrophobes are Plentiful and UV Active

Determining the aqueous solution concentration of model associative thickeners without a chromophore is difficult. A colorimetric method[31], originally developed for nonionic polyoxyethylene surfactants, has been utilized in previous associative thickener adsorption studies. This method is based on the complex formation of iodine with oxyethylene units. Large experimental errors were encountered in our attempt to use this method due to the necessary multiple dilution. As noted in the earlier discussion of the $NPEO_{100}$ HEUR coupled with TMXDI, three aromatic rings in an HEUR do not provide enough UV sensitivity for quantifying adsorption. The utilization of nonylphenol units in a comb architecture (discussed in Chapters 6 and 14 of this text) can supply a higher concentration of NP units for quantifying adsorption. The adsorption isotherms of a NP-comb HEUR on four latices at pH=6 are illustrated in Figure 8.

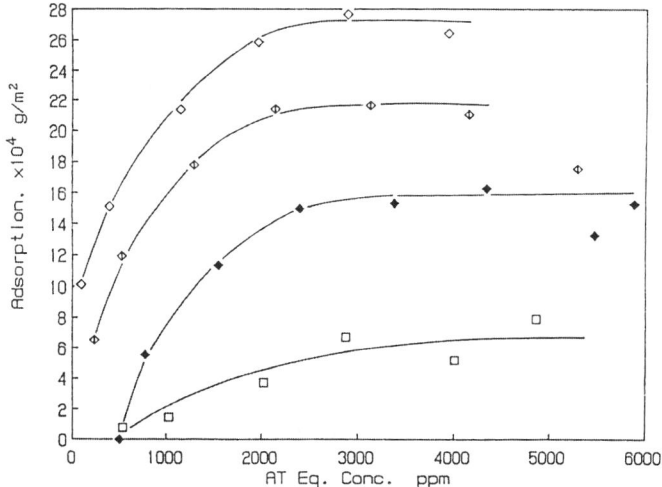

Figure 8. Adsorption isotherms of Nonyl phenol comb HEUR on P(MMA-MAA) and VA latices at pH 6, determined by the UV absorbance of aromatic ring at 276 nm. Symbols: ♦, *MMA, MAA (100/0);* ♦, *MMA/MAA (98/2); ◊, MMA/MAA (96/4); □, poly (vinyl acetate).*

The adsorption isotherms exhibit a rapid increase in the amount adsorbed at low concentrations and reach a plateau at high concentrations, in agreement with Langmuir-type behavior. The greater adsorption in proportion to the amount of surface acid stabilizer is attributable to the POE ether linkages hydrogen bonding with the surface acid units. The lower adsorption on the poly(vinyl acetate) (PVA) latex is primarily related to the lower interfacial tension at the PVA latex/water interface than at the acrylic latex /water interface.[32,33] In this study the vinyl acetate latex is also slightly larger than the 130 nm PMAA latices and therefore has a lower surface area for adsorption.

To examine the validity of this displacement technique, adsorption of NP-$(EtO)_{12}H$ on a PMMA latex has been determined at the same time by two methods:

one is based on the absorbance of the 276 nm band of the chromophore of NP-O(EtO)$_{12}$H, and the other is on the absorbance of the 554 nm band of the phenolphthalein displaced from the beta-CD cavity by the hydrophobes of the surfactant (Figure 9). Both methods give the same type of monolayer adsorption isotherms, consistent with published results.[34] However, the plateau value obtained by the 554 nm band is two units greater than that by the 276 nm band.

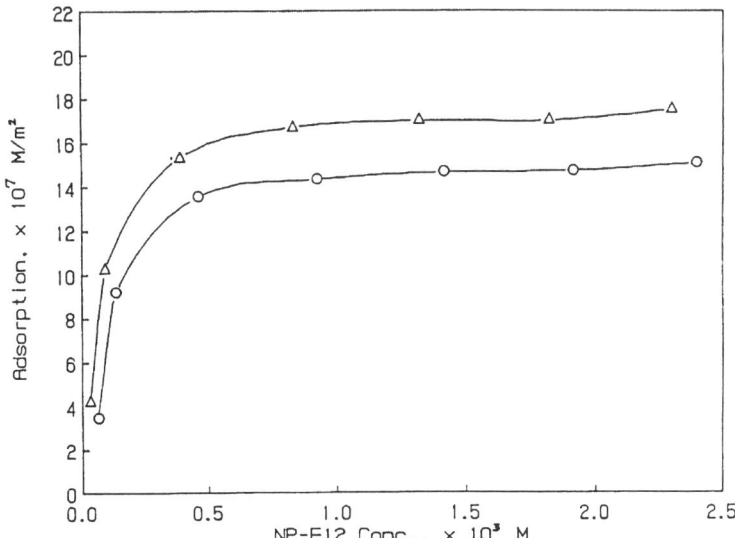

Figure 9. Adsorption isotherms of nonionic surfactant NP-O (EtO)$_{12}$H on PMMA latex with 0.785x10^{-3} meq/m^2 surface acid at pH 6. Symbols: O, determined by absorbance at 276 nm; Δ, determined by absorbance at 554 nm.

With this knowledge, we turn to the adsorption of the step-growth HEUR with H$_{12}$MDI urethane linkages on PMMA latices. With the smaller terminal C$_6$H$_{13}$H$_{12}$MDI- and C$_8$H$_{17}$H$_{12}$MDI-HEUR hydrophobes, the displacement absorbance follows the expected results (Figure 10). What is surprising is the unusual absorbance behavior noted with the C$_{12}$H$_{25}$-H$_{12}$MDI-HEUR (Figure 11). There was nothing in the clear solution study (Figure 6) to suggest such a deviation. The data in Figure 11 do however bear a marked resemblance to the trend in adsorption data reported with a similar step-growth HEUR using the iodine complex analysis.[35] We will forego an interbridging interpretation of larger hydrophobes, for the real interest in a practical application lies in the adsorption behavior of associative thickeners in the presence of surfactants. This can not be done with the beta-CD approach because of the competition for the CD cavity. The competitive analysis can be done when the surfactant has an aromatic

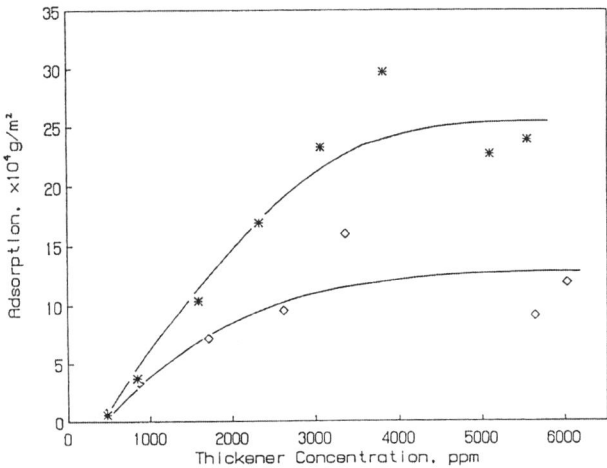

Figure 10. Adsorption isotherms of C_8H_{17}-$H_{12}MDI$ and $C_6H_{13}H_{12}MDI$ on the P(MMA-MAA) latex with $5.582x10^{-3}$ meq/m^2 surface acid at pH 9.5, obtained by β-cyclodextrin complex method. Symbols: *, C_8H_{17}-$H_{12}MDI$; ◊, C_6H_{13}-$H_{12}MDI$

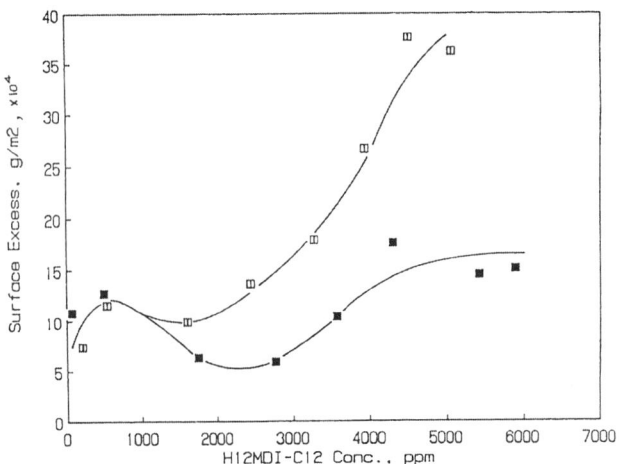

Figure 11. Adsorption isotherms of $C_{12}H_{25}$-$H_{12}MDI$ on P(MMA-MAA) latices at pH 9.5, obtained by β-cyclodextrin complex method. Symbols: ■, MMA/MAA (100/0); ▫, MMA/MAA (98/2).

chromophore, nonylphenol (NP), and the HEUR does not contain NP groups in a comb architecture. The technique involves dialyzing the latex free of synthesis surfactant, and then adding sufficient surfactant to just saturate the surface[36]. The extent of the step-growth HEUR thickener adsorbed can be quantified by appropriate calibration of the surfactant liberated.

As a complementary check, HEUR adsorption can also be studied with Size Exclusion Chromatography. This approach can be used with the large $C_{18}H_{37}$-uniHEUR, for which the beta-CD approach could not be used (Figure 7); it is reported in Figure 12. With the large hydrophobe-HEUR the surfactant can not compete, particularly with increasing HEUR concentration. There also appears to be a spacer length effect on the competitive adsorption behavior (Figure 13). This would be expected due to the decreasing solubility with decreasing EO length; it might also be correlated to a greater intrahydrophobic bonding effect associated with the shorter EO spacing. A proper interpretation of the data, however, is marred by the variation in the degree of hydrophobe substitution among the samples with different EO spacer lengths. This was a consequence of residual water in the different POE molecular weights in their synthesis[37].

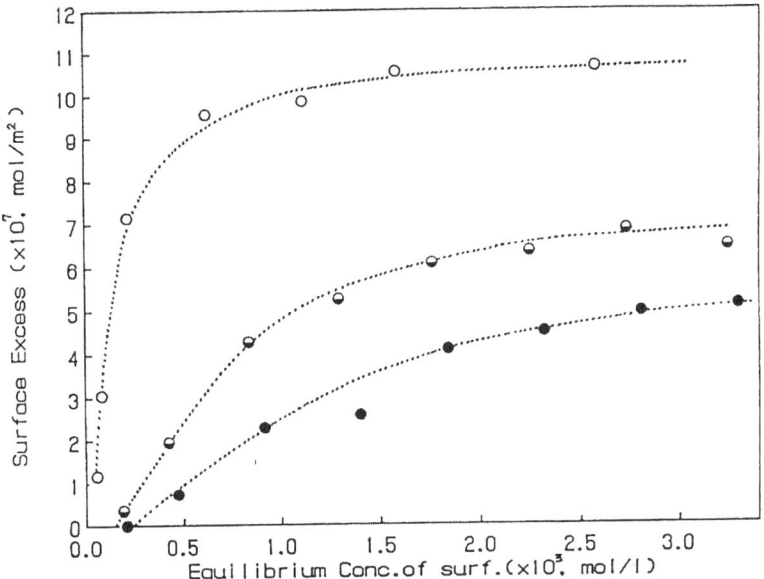

Figure 12. Adsorption isotherms of NP-$O(EtO)_{12}H$ on MMA/MAA (96/4) latex in the presence of $(C_{18}H_{37}HN$-COO-$)_{1.6}(EtO)_{182}$ at pH 9.5. Symbols: ○, no thickener; ◐, 0.2 wt% $(C_{18}H_{37}HN$-CO-$)_{1.6}(EtO)_{182}$; ●, 0.4 wt% $(C_{16}H_{37}HN$-CO-$)_{1.6}(EtO)_{182}$.

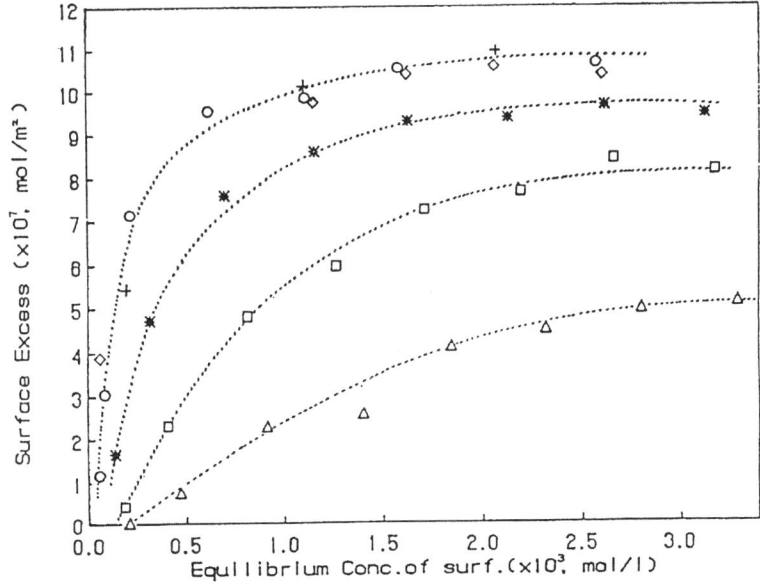

Figure 13. Adsorption isotherms of NP-O(EtO)$_{12}$H on MMA/MAA (96/4) latex in the presence of model associative thickeners with different POE chain lengths at pH 9.5. Symbols: O, no associative thickener; +, 0.4 wt% POE of molecular weight 8000; ◊, 0.4 wt% ($C_8H_{17}HN$-CO-$)_2$-EtO_{182}; *, 0.4 wt% ($C_{18}H_{37}HN$-CO-$)_2$-EtO_{663}; □, 0.4 wt% ($C_{18}H_{37}HN$-CO-$)_{1.4}$-EtO_{331}; Δ, 0.4 wt% ($C_{18}H_{37}HN$-CO-$)_{1.6}$-EtO_{182}.

Conclusions

Phenolphthalein and beta-CycloDextrin (CD) can form, at pH 10.5, a 1:1 complex with a formation constant of 2.1×10^4 liters mol^{-1}; this 1:1 complex has an absorbance practically equal to zero in the region of the 550 nm band of the basic form of phenolphthalein. We have used this observation to quantify the adsorption of surfactants and HEUR thickeners on latices. It is observed that anionic and nonionic surfactants can displace phenolphthalein from the CD cavity, followed in alkaline solutions by the 554 nm absorbance band of the dye. Considering that the diisocyanate coupling the hydrophobes to the polyetherdiol chain has considerable hydrophobic character itself, it is observed that the C_6H_{13}-H_{12}MDI and C_8H_{17}-H_{12}MDI-HEURs are more effective in displacing the phenolphthalein (i.e., give a greater enhancement in absorbance) than $C_{10}H_{21}$-H_{12}MDI and $C_{12}H_{25}$-H_{12}MDI-HEURs. As noted with the $C_{18}H_{37}$-surfactants there is a size limitation for displacement of phenolphthalein from the beta-CD cavity. It is obvious that even in simple solutions, the beta-CD analysis would not be universally applicable for analysis of HEUR concentrations in aqueous solutions.

Turning to the real objective of this study, the adsorption of NP-(EtO)$_{12}$H and HEURs on a PMMA latex were considered. The adsorption of the nonionic surfactant was determined by two methods: one is based on the absorbance of the 276 nm band of the chromophore of NP-O(EtO)$_{12}$H, and the other is on the absorbance of the 554 nm band of the phenolphthalein displaced from the beta-CD cavity. The two different methods of determining adsorption are complementary. The beta-CD technique also can be used to measure the adsorption of HEURs, but the results should be coordinated with Size Exclusion Chromatography. To address the real goal, competitive adsorption of associative thickeners in the presence of surfactants, more extensive studies are needed, keeping in mind that what one is really measuring is the competition with phenolphthalein for the beta-CD cavity.

References

1. Starnes, R. L., *Cereal Foods World*, **1990**, *35*, 1094-1099.
2. Saenger, W., *Angew. Chem. Int. Ed. Engl.*, **1980**, 19, 344-362.
3. Haggin, J., *C&EN*, May 18, **1992**, 25.
4. Parrish, M. A., *Specialty Chemicals*, **1987**, *7*, 366.
5. Okubo, T.; Kitano, H.; Ise, N., *J. Phys. Chem.*, **1976**, *80*, 2661.
6. Satake, I.; Ikenoue, T.; Takeshita, T.; Hayakawa, K.; Maeda, T., *Bull. Chem. Soc. Jpn.*, **1985**, *58*, 2746.
7. Satake, I.; Yoshida, S.; Hayakawa, K.; Maeda, T.; Kusumoto, Y., *Bull. Chem. Soc. Jpn.*, **1986**, *59*, 3991.
8. Hersey, A.; Robinson, B. H.; Kelly, H. C, *J. Chem. Soc. Faraday Trans.*, **1986**, *82*, 1271.
9. Park, J. W.; Song, H. J., *J. Phys. Chem.*, **1989**, *93*, 6454.
10. Saenger, W.; Muller-Fahrnow, A., *Angew. Chem., Int. Ed. Engl.*, **1988**, *27*, 393.
11. Guo, W.; Fung, B. M.; Christian, S. D., *Langmuir*, **1992**, *8*, 446-451.
12. Sasaki, K. J,; Christian, S. D.; Tucker, E. E., *Fluid Phase Equilibria*, **1989**, *49*, 281.
13. Sasaki, K. J,; Christian, S. D.; Tucker, E. E., *J. Colloid Interface Sci.*, **1990**, *134*, 412.
14. Kaczmarski, J. Philip and Glass, J. Edward, *Langmuir*, **1994**, *10(9)*, 3035.
15. Karunasena, A., Glass, J. E., *Progress in Organic Coatings*, **1989**, *17*, 301
16. Hulden, M.; Sjoblom, E., *Progr. Colloid Polym. Sci.*, **1990**, *82*, 28.
17. Ma, C.; Li, C., *J. Colloid Interface Sci.*, **1989**, *131*, 485.
18. Kawaguchi, M.; Kawaguchi, H.; Takahashi, A., *J. Colloid Interface Sci.*, **1988**, *124*, 57.
19. Richey, B.; Kirk, A. B.; Eisenhart, E. K.; Fitzwater, S.; Hook, J., *J. Coatings Tech.*, **1991**, *63*, *No.798*, 31.

20. Parnas, R. S.; Chaimberg, M.; Taepaisitphongse, V.; Cohen, Y., *J. Colloid Interface Sci.,* **1989**, *129*, 441.
21. Ramachandran, R.; Somasundaran, P., *J. Colloid Interface Sci.*, **1987**, *120*, 184.
22. Kim, J.; Cotton, T. M.; Uphaus, R. A.; Mobius, D., *J. Phys. Chem.*, **1989**, *93*, 3717.
23. Somasundaran, P.; Kunjappu, J. T., *Colloid & Surface*, **1989**, *37*, 245.
24. Meadows, J.; Williams, P. A.; Garvey, M. J.; Harrop, R.; Phillips, G. O., *J. Colloid Interface Sci.*, **1989**, *132*, 319.
25. Robb, I. D.; Sharples, M., *J. Colloid Interface Sci.*, **1982**, *89*, 301.
26. Kronberg, B.; Kuortti, J; Stenius, P., *Colloid & Surface*, **1986**, *18*, 411.
27. Pefferkorn, E.; Jean-Chronberg, A. C.; Chauveteau, G; Varoqui, R., *J. Colloid Interface Sci.*, **1990**, *137,* 66.
28. Buvari, A. and Barcza, L., *Inorg. Chim. Acta*, **1979**, *33*, L179.
29. Vikmon, M., in *Proceedings of the First International Symposium on Cyclodextrins,* Szetli, J., Ed, Reidel Publishing Co., Boston, **1982**, p.69.
30. Lundberg, David J.; Brown, Richard G; Glass, J. Edward; and Eley, Richard R., *Langmuir* **1994, 10**(9) 3027.
31. Baleux, B., *C. R. Acad. Sci.*, **1972**, *Ser. C 274*, 1617.
32. Bartell, F. E.; Davis, J. K., *J. Phys. Chem.*, **1941**, *45*, 1321.
33. Vijayendran, B. R., *Polymer Colloids II*, Fitch, R. M., Ed., Plenum, New York, **1980**, 209.
34. Kronberg, B.; Stenius, P., *J. Colloid Interface Sci.*, **1984**, *102*, 410.
35. J. C. Thibeault, P.R. Sperry, E. J. Schaler, *Advances in Chemistry Series 213: Water-Soluble Polymers: Beauty with Performance*, ed., Glass, J.E., American Chemical Society, Washington, D. C., Chapter 20, **1986**.
36. Ma, Zeying; Chen, Mao and Glass, J. Edward, *Colloids and Surfaces*, **1996**, *112(2/3)*.
37. Kaczmarski, J. Philip; Glass, J. Edward, *Macromolecules*, **1993**, *26*, 5149.

Chapter 16

Association Thickener by Host–Guest Interaction of β-Cyclodextrin Polymers and Guest Polymers

Gerhard Wenz[1], Meik Weickenmeier[1], and Jürgen Huff[2]

[1]Polymer-Institut der Universität Karlsruhe (TH),
Hertzstrasse 16, D–76187 Karlsruhe, Germany
[2]BASF Aktiengesellschaft, D–67056 Ludwigshafen, Germany

Host polymers with pendant β-cyclodextrin side groups and guest polymers with pendant hydrophobic 4-t-butyl anilide side groups were synthesized by polymer analogous reactions starting from copolymers of maleic anhydride. The host guest interaction of these polymers with suitable monomeric counterparts was proven by titration microcalorimetry. The interaction of the host polymer and the guest polymer in aqueous solution leads to a tremendous increase in viscosity. Polymer chains appear to form a physical network due to host guest interactions. This interaction can be switched off either by dilution or by addition of a competitive monomeric host or guest.

Introduction

Cyclodextrins **1** are a unique group of water-soluble host molecules. They are cyclic oligomers of glucose produced by enzymatic degradation of starch*(1)*. The pure homologous compounds consisting of 6, 7 or 8 glucose rings, called α-, β-, or γ-cyclodextrin (**1a, 1b, 1c**) are produced on a large scale. They are potentially useful for the formation of host guest complexes, so-called inclusion compounds, with a great variety of hydrophobic molecules*(2)*.

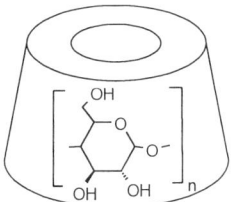

1a, n = 6
1b, n = 7
1c, n = 8

α-Cyclodextrin **1a** with an internal diameter of about 5 Å is able to accommodate linear alkyl chains*(3)*. β-Cyclodextrin **1b** with an internal diameter of about 6 Å can complex benzene and naphthalene groups*(4)* and γ-cyclodextrin **1c** with about 8 Å diameter can hold polycyclic aromatics*(5)* and even buckminster fullerene*(6)*. The hydrophobic interaction between host and guest is assumed to be the major driving force for the binding ability of cyclodextrins*(2)*.

Polymers also can be bound by cyclodextrins. If the hydrophobic binding sites are within the polymer backbone, cyclodextrins become threaded like beads on a string. Linear aliphatic polyethers, like polyethylene glycol, form insoluble inclusion compounds with **1a** *(7,8)*, while poly(imino-oligomethylenes) form soluble ones *(9,10)*. The attachment of bulky substituents at the chain ends or along the chain leads to stable polyrotaxanes*(9,11,12)*.

Hydrophobic binding sites have also been attached as side groups to a polymer backbone. Side chain guest polymers have already been synthesized suitable for cyclodextrins **1a** – **1c** *(13-15)*. In this regard cyclodextrins are applied to control the association of so-called associative thickeners*(16)*. Associative thickeners are water-soluble polymers with a few hydrophobic side groups*(17-20)*. They are used for the viscosification of water based on the hydrophobic interactions of the side groups. These hydrophobic groups can be masked by cyclodextrins to ease the dissolution process of the polymers in water*(16)*.

The synthesis of cyclodextrin polymers allows the design of two component associative thickeners (s. Scheme 1). The specific interaction between host and guest moieties leads to a three dimensional physical network that exhibits a high viscosity. We have synthesized cyclodextrin polymers by polymer analogous reaction of poly(maleic anhydride-*alt*-isobutene) with **1b** *(15)* and guest polymers by reaction of the same polymer with t-butyl aniline. A tremendous increase of the viscosity occurred after the two aqueous solutions of the host and the guest polymers were mixed*(21,22)*.

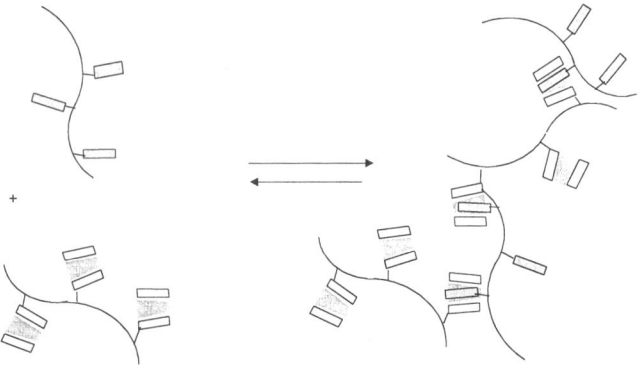

Scheme 1. Two component associative thickener by interaction of a host polymer with a guest polymer.

Very recently, Sebille, Amiel and coworkers reported also on two component associative thickeners based on β-cyclodextrin epichlorohydrin polymers(23) with guest polymers based on polyethylene glycol or hydroxyethyl methacrylate (24-26).

In the following sections we describe our thickening system in more detail. Special emphasis is given to the effects of the stoichiometry of host and guest moieties, of the molecular weights of the interacting polymers and of low molecular weight additives on the viscosification.

Experimental

Materials and methods. β-Cyclodextrin **1b** (Wacker-Chemie GmbH, Burghausen, Germany) was dried under vacuum at 95°C. 4-t-butyl aniline, 4-t-butyl benzoic acid, maleic anhydride and LiH (Aldrich) were used without further purification. N,N-Dimethylformamide (DMF) was distilled over CaH_2. Poly[(maleic anhydride)-*alt*-(isobutene)] (M_w = 3700, 25000, 60000, and 160000 g/mol) and N-vinyl-pyrrolidone were obtained from BASF Aktiengesellschaft, Ludwigshafen, Germany.

NMR spectra were measured with a 400 MHz spectrometer (Bruker) in D_2O with TMS propionate-d_4 (δ = 0.00 ppm) as the reference. An OMEGA titration calorimeter (MicroCal, Inc. Northampton, USA) was used for microcalorimetry at 25°C and the data were fitted with the computer program Origin for ITC (version 2.9) as described elsewhere (22,27). The viscosities of the polymers dissolved in 0.1 M phosphate buffer of pH 7.0 were measured with an Ubbelohde viscometer, the viscosities of the highly viscous mixtures by a rotation viscometer with cone/plate geometry (Rotovisko RV1, Haake).

Synthesis of poly[(maleic anhydride)-*co*-(N-vinyl-pyrrolidone)] (28). 4.6 g (47 mmol) maleic acid, 5.2 g (47 mmol) N-vinyl-pyrrolidone and 50 mg (0.3 mmol AIBN) were dissolved under N_2 in 50 ml benzene and heated to 60°C. After 20 h the precipitate was filtered and dried under vacuum. Yield 4.29 g (44%); IR (DRIFT, KBr): 2951 m (C-H), 1847 s (C=O anhydride), 1773 s (C=O anhydride), 1673 s (C=O lactam), 1491 w, 1462 m (C-H), 1418 s (C-H), 1373 w, 1288 s, 1225 s, 1089 m (C-O-C anhydride), 932 s (C-O-C anhydride), 738 w, 685 w cm^{-1}; viscometry: (DMSO/Acetone 35:65 v/v, 25°C) [η] = 8.49 ml/g.

Synthesis of the cyclodextrin polymers 2. In accordance with Scheme 2 the cyclodextrin polymers **2** were synthesized by reaction of 2-O-lithium-β-cyclodextrinate with a stoichiometric amount of the reactive polymer poly[(maleic acid)-*alt*-(isobutene)] or poly[(maleic anhydride)-*co*-(N-vinyl-pyrrolidone)] in DMF solution as described elsewhere(22). The DMF was then removed under reduced pressure. The solid residue was dissolved in water and ultrafiltered through a cellulose membrane (nominal exclusion limit 20,000 g/mol, UF-BM Berghoff) with water until free cyclodextrin could no longer be detected by SEC (PSS HEMA 1000 equipped with a polarimeter detector (IBZ-Messtechnik, Hannover, Germany). The retentate was freeze-dried and the polymers **2** were obtained as white powders.

2a – 2c: Characterization is described elsewhere*(22)*.

2d: ^{1}H-NMR (D$_2$O): δ = 1.4-3.1 (sh, 19.5 H, CH$_2$, CH main chain), 3.4-4.2 (sh, 23.3 H, H-2, H-3, H-4, H-5, H-6), 4.9-5.2 (bs, 2 H, H-1) ppm; ^{13}C-NMR (D$_2$O): δ = 19.2 (CH$_2$ NVP), 32.6 (CH$_2$), 36.0 (CH), 43.10 (CH$_2$), 44.7 (CH), 61.6 (C-6), 73-75 (C-3, C-2, C-5), 82.5 (C-4), 103.5 (C-1), 178-179 (C=0) ppm; IR (DRIFT, KBr): 3342 s (O-H), 2934 m (C-H), 1724 m (-COO), 1654 s (amide), 1584 s (COO$^-$), 1492 m, 1463 m (C-H),1400 m, 1292 m, 1155 s (C-O), 1079 s (C-O/C-C), 1031 s (C-O/C-C), 947 m, 755 m, 578 s cm^{-1}.

Synthesis of the guest polymers. According to Scheme 3 a series of samples of poly[(maleic acid)-*co*-(isobutene)-*co*-(4-t-butyl maleic anilide)] **3** was synthesized as shown in Scheme 3 *(22)*. The typical procedure was to dissolve poly[(maleic acid)-*alt*-(isobutene)] in DMF and then add a solution of 4-t-butyl aniline in DMF. The mixture was stirred for 2 days at 40°C and for a further 3 days at 20°C. The solvent was subsequently removed under reduced pressure and the solid residue treated with NaOH (1.10 wt.-%) at RT for 1d, then ultrafiltered through a cellulose membrane (nominal exclusion limit 20,000 g/mol) with water. The retentate was freeze-dried. The guest polymers **3** were obtained as white powders. The content of guest moieties was determined from the integrals of the Ar-H signals vs. the signals of the polymer chain by ^{1}H-NMR spectroscopy.

3a – 3d: Characterization is described elsewhere*(22)*.

Results and Discussion

Synthesis of the Host Polymers

The host polymers **3** were synthesized by polymer analogous reactions of 2-O-lithium-β-cyclodextrinate with reactive copolymers containing maleic anhydride moieties (s. Scheme 2). Similar to the work of D'Souza*(29)* 2-O-lithium-β-cyclodextrinate was formed by titration of **1b** with LiH in anhydrous DMF solution. The Li salts are preferable to the Na salts of **1b** as they have greater solubility in DMF. Alternating copolymers of maleic anhydride and isobutene or statistical copolymers of maleic anhydride and N-vinyl-pyrrolidone were used. The molecular weights were in the range of 3,000 to 60,000 g/mol. Crosslinking of these polymers by the polyfunctional cyclodextrins was prevented by stopping the reaction at low conversions. Therefore only moderate yields of **3** (6-16% based on **1b**) could be reached. Unreacted **1b** was recovered by ultrafiltration. Residual anhydride groups had to be hydrolyzed to maleic acid groups in acetate buffer at pH 4.7. The polymers **3** are highly soluble in water but insoluble in most organic solvents. As the products **3** in polymers **2**. Close to one hundred guest groups could be attached to one polymer chain (s. Table 2). The guest polymers **3** form clear solutions in water with low viscosities. No indication was found for any association of the hydrophobic groups. Possibly their hydrophobic interaction is not strong enough to overcome the repulsion of the anionic polymer backbones. They therefore lend themselves well to investigate their interactions with monomeric or polymeric cyclodextrins.

Interaction of the Host Polymers 2 with Monomeric Guests

A single chain of host polymer **2** contains from 4 to 30 cyclodextrin groups. We were interested in discovering whether all of these are accessible to guest molecules. Some of the cyclodextrin groups could be hidden inside of the coil of the polymer chain. We chose 4-t-butyl-benzoic acid as a probe to investigate the binding constants K_S of the host polymers **2** as it forms a very stable complex with monomeric **1b***(30)*. The binding constant K_S was measured by titration microcalorimetry, because this method is very accurate and also provides the stoichiometry ratio *n*, the enthalpy $\Delta H°_S$ and the entropy of binding $\Delta S°_S$ *(31)*. The results are presented in Table 3. The stoichiometric ratio *n* was in all cases $n = 1 \pm 0.19$, which shows that all cyclodextrin rings are accessible for binding. The binding constants K_S of the cyclodextrin polymers are of the order of 3,000 M^{-1}, which is significantly lower than the value of $K_S = 18,000$ M^{-1} for the monomeric cyclodextrin **1b**. The lower binding ability of the cyclodextrin polymers is mainly due to a significant loss in the entropy term $T\Delta S°_S$. This loss of entropy might be caused by some unfavorable changes in the conformations of the polymer backbone of **2** upon inclusion of the anionic guest.

Table 3: Interaction of 4-t-Butyl-benzoic Acid with Host polymers 2 as Measured by Titration Microcalorimetry*

Host	n	K_S [M^{-1}]	$\Delta H°_S$[kJ/mol]	$T\Delta S°_S$ [kJ/mol]
2a	0.98	3140 ± 130	-24.08 ± 0.16	-4.12
2b	1.16	2900 ± 150	-23.91 ± 0.27	-4.15
2c	1.19	4040 ± 260	-25.27 ± 0.31	-4.68
1b	0.99	17900 ± 650	-20.10 ± 0.09	4.18

*0.1 M Phosphate buffer pH 7.2, 25°C

Interaction of the Guest Polymers 3 with Monomeric Cyclodextrins

The interactions of the polymeric guests **3** with the monomeric host **1b** were also studied by titration microcalorimetry. In all cases the stoichiometric ratios *n* were close to 1 showing that all 4-t-phenylanilide side groups are readily accessible to monomeric cyclodextrins. The guest polymers **3** showed binding constants of around

K_S = 23,000 M^{-1} with β-cyclodextrin. These were not dependent on the distance X between two neighboring binding sites. Also for a monomeric guest, 4-t-butylaniline, a similar value K_S = 25,000 M^{-1}, was found. It appears that the attachment of the guest to the polymer chain does not affect formation of its inclusion compounds.

Interaction of the Cyclodextrins Polymers 2 with the Guest Polymers 3

It was shown above that the cyclodextrin polymers **2** and the guest polymers **3** each form stable inclusion compounds with their monomeric counterparts in aqueous solution. We would now like to discuss the interaction of two complementary polymers. Both host and guest are restricted by their attachment to a polymer chain. The chains must therefore change their conformation if an host guest interaction is to occur. Such an interaction should lead to the formation of a three dimensional network; and as the interaction is reversible, this network should be mobile and responsive to changes in its environment.

A tremendous increase in viscosity was in fact observed immediately after 2 wt.-% aqueous solutions of **2** and **3** were mixed (s. Figure 1). While the individual polymers showed viscosities in the order of η = 10 mPas, the 1:1 mixtures had much higher values of η = 100 – 4,000 mPas, depending on the shear rate D. The viscosity of the mixture declined with increasing shear rate. This shear thinning was attributed to the rupture of inclusion complexes by mechanical forces. At a polymer concentration of 5% the mixture was so stiff that it could no longer flow. On the other hand, for polymer concentrations of 1 wt.-% or less no thickening effect was found at all. All of the mixtures remained perfectly clear and there was no indication of any phase separation. The interaction of host polymer **2b** and guest polymer **3c** was also measured by titration microcalorimetry. The observed interaction enthalpy, around $\Delta H°_S$ = -20 kJ/mol, was simular to those values found for the monomer/polymer systems (s. Table 3). The binding constant K_S could not be calculated due to the complexity of the system. K_S appears to decrease with conversion because of the restricted mobility of the host and guest moieties.

If the specific interaction between host and guest moieties of polymers **2** and **3** is behind the observed gelation, the largest effect should be seen at a 1:1 molar ratio of host and guest: that is, a maximum should be observed at a mol fraction of 0.5 in the so-called continuous variation plot or Job's plot(32) of the viscosity. This was indeed the case. The dependency of the viscosity η on the mol fraction of the guest polymer showed a pronounced maximum at a mol fraction of 0.5 and had almost mirror symmetry (s. Figure 2). The shape of the curve remains unaffected by the shear rate D. Therefore we conclude that specific host guest interactions are responsible for the thickening effect.

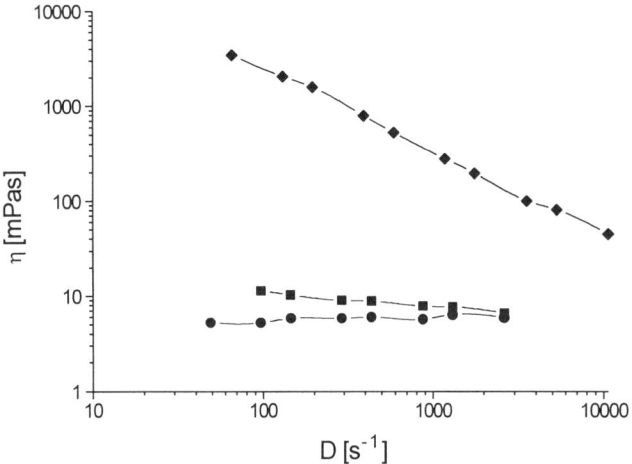

Figure 1. Viscosity η of (■) a 2 wt.-% solution of the host polymer 2c, (●) a 2 wt.-% solution of the guest polymer 3c, and (♦) a 1:1 mixture of both as a function of the shear rate D.

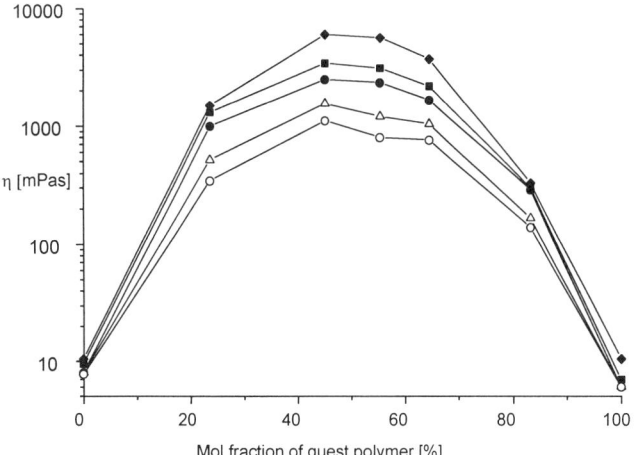

Figure 2. Viscosity η of aqueous mixtures of host polymer 2c and guest polymer 3c at several shear rates [D = 66 (♦), 131 (■), 196 (●), 393 (Δ) and 590 s^{-1} (O)] vs. the mol fraction of 3c at a constant total polymer concentration of 2 wt.-%.

In addition to the stoichiometry, the chain lengths of both polymers should have a strong influence on the gelation effect. As the absolute molecular weights of the polymers **2** and **3** were not available due to the charge of the polymer backbone, we chose their intrinsic viscosities [η] as a measure of their chain lengths. The viscosity η of the 1:1 mixture was determined as a function of the intrinsic viscosity [η] of the guest polymer **3** for the 4 different host polymers **2a - 2d** (s. Figure 3). The concentrations of both polymers were 2 wt.-% before mixing. Obviously, the thickening effect strongly increases with the intrinsic viscosity of the guest polymer **3**. The same happens for the host polymer **2**: whereas almost no thickening was achieved for **2a** with the lowest intrinsic viscosity of [η] = 5.8 ml/g, strongest thickening occured for the host polymer **2c** with the highest intrinsic viscosity [η] = 72 ml/g. The structure of the polymer backbone had no pronounced influence on the thickening effect. The isobutene copolymer **2b** and the N-vinyl-pyrrolidone copolymer **2d** behaved very similarly (s. Figure 3). Thus the chain length appears to be a crucial factor in the thickening. The steep increase in the viscosity of the mixture at a guest polymer intrinsic viscosity of around 50 ml/g (s. Figure 3) might be interpreted as a critical overlapping phenomenon. At [η] = 50 ml/g the corresponding critical overlapping concentration of polymer **3** amounts to $c^* \cong 2$ wt.-%. The same is true for the host polymer **2**. Hence the solvent volume is completely filled with polymer coils at a total polymer concentration of 2 wt.-%. It seems that at this critical overlapping concentration a macroscopic array of coils is formed which causes the large increase in viscosity. Therefore the thickening effect might be described by percolation theory.

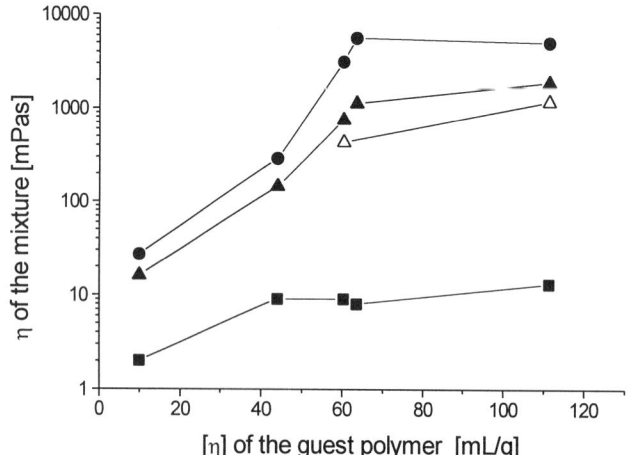

*Figure 3. Viscosity η of aqueous 1:1 mixtures of 2 wt.-% solutions of host polymers **2** and 2 wt.-% solutions of guest polymer **3** at a shear rate of $D = 66\ s^{-1}$ vs. the intrinsic viscosity [η] of the guest polymers, for host polymers: (■) **2a**, (▲) **2b**, (●) **2c**, and (△) **2d**.*

Switching the Thickening Effect

The formation of the associative network between polymers **2** and **3** is a fully reversible process. On dilution of the gel with water the viscosity is immediately reduced. There is no temporary formation of gel particles. The addition of a competitive monomeric guest or host also leads to a strong reduction in the thickening effect. As an example, the influence of hydroxypropyl β-cyclodextrin on the viscosity of the mixture of polymers **2c** and **3c** was investigated in some more detail (s. Figure 4). An equimolar quantity of the monomeric host is sufficient to switch off the association entirely. The monomeric host exhibits a much higher affinity to the guest moieties of polymer **3c** than the polymeric host **2c** (s. Table 3). Therefore the polymer-bound host sites are displaced by the monomeric ones. The addition of the monomeric guest 4-t-butyl benzoic acid to the 1:1 mixture of polymers **2c** and **3c** also led to a total loss of the extra viscosity. The amount required, 4 equivalents of 4-t-butyl benzoic acid, exceeded the amount of hydroxypropyl β-cyclodextrin. This finding is understandable, since the binding constant K_S of 4-t-butyl benzoic acid and the host polymer **2c** is lower than the K_S of hydroxypropyl β-cyclodextrin and guest polymer **3c**. In both cases the thickening ceased immediately the monomeric host or guest was added. This might be useful for applications where only a temporary viscosification is needed.

On the other hand, the thickening effect in our host guest system is scarcely affected by the pH and the ionic strength of the solvent. This contrasts with the behavior of conventional thickeners (e.g. those based on poly acrylic acid).

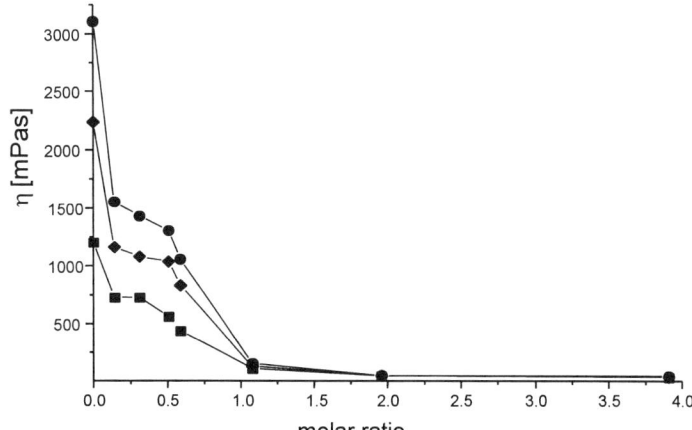

*Figure 4. Viscosity η of a mixture of 2 wt.-% solutions of the host polymer **2c** and guest polymer **3c** vs. the molar ratio of added hydroxypropyl-β-cyclodextrin at a shear rate of D = (■) 393, (♦) 13, and (●) 66 s^{-1}.*

Conclusion

A new principle for the formation of reversible hydrogels based on specific host guest interactions has been demonstrated. Clear homogenous hydrogels can be formed immediately by mixing solutions of a host and a guest polymer. Intrinsic viscosities of at least $[\eta]$ = 50 ml/g are necessary to form a gel at a minimum polymer concentration of 2 wt.-%. The thickening effect is practically independent of the pH and the ionic strength of the solvent, but it can be switched off at will by dilution or by the addition of monomeric cyclodextrins or guests. For real applications lower thickening concentrations and greater stability of the cyclodextrin polymer would be desirable. Further word in this direction is currently underway.

Acknowledgements. The authors thank BASF Aktiengesellschaft, Ludwigshafen, Germany for financial support, the donation of the reactive polymer (Dr. A. Kistenmacher), and for helpful discussions (Dr. E. Winkler). We further thank C. Gruber for the synthesis of polymer **2d** and S. Wehrle for performing the microcalorimetric titrations.

References

1. Wenz, G. *Angew. Chem. Int. Ed. Engl.* **1994**, *33*, 803-822.
2. Connors, K. A. *Chem. Rev.* **1997**, *97*, 1325-1357.
3. Bastos, M.; Briggner, L. E.; Shehatta, I.; Wadsö, I. *J. Chem. Thermodyn.* **1990**, *22*, 1181-1190.
4. Fekharsky, M. V.; Inouc, Y. *Chem. Rev.* **1998**, *98*, 1875-1917.
5. la Pena, A. M. d.; Ndou, T.; Zung, J. B.; Warner, I. M. *J. Phys. Chem.* **1991**, *95*, 3330-3334.
6. Andersson, T.; Nilsson, K.; Sundahl, M.; Westman, G.; Wennerstroem, O. *J. Chem. Soc., Chem. Commun.* **1992**, *8*, 604-606.
7. Harada, A.; Kamchi, M. *Macromol.* **1990**, *23*, 2821-2823.
8. Harada, A. *Acta Polym.* **1998**, *49*, 3-17.
9. Wenz, G.; Keller, B. *Angew. Chem. Int. Ed. Engl.* **1992**, *31*, 197-199.
10. Herrmann, W.; Keller, B.; Wenz, G. *Macromolecules* **1997**, *30*, 4966-4972.
11. Harada, A.; Li, J.; Kamachi, M. *Nature* **1992**, *356*, 325-327.
12. Herrmann, W.; Schneider, M.; Wenz, G. *Angew. Chem. Int. Ed. Engl.* **1997**, *36*, 2511-2514.
13. Holt, P.; Thadani, C. *Die Makromolekulare Chemie* **1973**, *169*, 55-58.
14. Harada, A.; Adachi, H.; Kawaguchi, Y.; Kamachi, M. *Macromolecules* **1997**, *30*, 5181-5182.
15. Weickenmeier, M.; Wenz, G. *Macromol. Rapid Commun.* **1996**, *17*, 731-736.
16. Eisenhart, E. K.; Merritt, R. F.; Johnson, E. A. US Patent 533,148, 1990.
17. Tarng, M.-R.; Kaczmarski, J. P.; Lundberg, D. J.; Glass, J. E. *Adv. Chem. Ser.* **1996**, *248*, 305-41.

18. Glass, J. E.; Schulz, D. N.; Zukoski, C. F. "Polymers as Rheology Modifiers. An Overview". In: *Polymers as Rheology Modifiers, Developed from a Symposium Sponsored by the Division of Polymeric Materials: Science and Engineering at the 198th National Meeting of the American Chemical Society, Miami Beach, Florida, September 10-15, 1989;* Schulz, D. N.; Glass, J. E. Eds.; American Chemical Society: Washington, D. C., 1991, ACS Symp. Ser., Vol. 462, pp. 2-17.
19. Dubin, P.; Bock, J.; Davies, R.; Schulz, D. N.; Thies, C. *Macromolecular Complexes in Chemistry and Biology*; Springer, 1996.
20. Selb, J.; Biggs, S.; Renoux, D.; Candau, F. *Polym. Mater. Sci. Eng.* **1993**, *69*, 128-129.
21. Huff, J.; Kistenmacher, A.; Schornick, G.; Weickenmeier, M.; Wenz, G. German Patent DE 19612768A1, 1996.
22. Weickenmeier, M.; Wenz, G.; Huff, J. *Macromol. Rapid Commun.* **1997**, *18*, 1117-1123.
23. Renard, E.; Deratani, A.; Volet, G.; Sebille, B. *Eur. Polym. J.* **1997**, *33*, 49-57.
24. Moine, L.; Cammas, S.; Amiel, C.; Renard, E. *Macromol. Symp.* **1998**, *130*, 45-52.
25. Gosselet, N. M.; Borie, C.; Amiel, C.; Sebille, B. *J. Dispersion Sci. Technol.* **1998**, *19*, 805-820.
26. Amiel, C.; Sebille, B. *J. Inclusion Phenom. Mol. Recognit. Chem.* **1996**, *25*, 61-67.
27. Wiseman, T.; Williston, S.; Brandts, J. F.; Lin, L. N. *Anal. Biochem.* **1989**, *179*, 131-137.
28. Rios, H.; Gargallo, L.; Radic, D. *J. Polym. Sci., Polym. Phys. Ed.* **1986**, *24*, 2421-2431.
29. Rong, D.; D'Souza, V. T. *Tetrahedr. Lett.* **1990**, *31*, 4275-4278.
30. Höfler, T.; Wenz, G. *J. Incl. Phenom.* **1996**, *25*, 81-84.
31. Brandts, J.; Lin, L. N. *Biochemistry* **1990**, *29*, 6927-6940.
32. Job, A. *Ann. de Chim. Ser. 10* **1928**, *9*, 113.

HMHEC and HMEHEC Polymers

Chapter 17

Fluorescent Labels: Versatile Tools for Studying the Association of Amphiphilic Polymers in Water

Françoise M. Winnik[1], Sudarshi T. A. Regismond[1], and Dan F. Anghel[2]

[1]Department of Chemistry, McMaster University,
1280 Main Street West, Hamilton, Ontario L8S 4M1, Canada
[2]Department of Colloids, Institute of Physical Chemistry,
Spl. Independentei 202, 79611 Bucharest, Romania

Fluorescence techniques are used to monitor the solution properties of two types of polyelectrolytes: poly(acrylic acid) and cationic cellulose ethers. A commercial sample of poly(acrylic acid) has been labeled with low levels of pyrene, either alone, or together with naphthalene. The pH-induced expansion of the labeled polymers was monitored by changes in the ratio I_E/I_M of pyrene monomer to pyrene excimer emission intensities and by changes in the extent of non-radiative energy transfer between naphthalene and pyrene. The cationic hydrophobically-modified hydroxyethyl cellulose ether LM200 was labeled with pyrene. Its interactions with cationic surfactants were evaluated by measuring changes in the ratio I_E/I_M of the pyrene emission. Association of the polymer with a series of pyridinium surfactants was detected by fluorescence quenching experiments.

Introduction

Compounds capable of self-aggregation can alter dramatically the properties of aqueous solutions. Amphiphilic polymers, which are constituted of well-defined hydrophilic and hydrophobic parts, often undergo aggregation in water. There are two broad classes of amphiphilic polymers. They can have either a high content of hydrophobic groups, as in the case of alternating copolymers of ionic and hydrophobic monomers (polysoaps), or a low content of hydrophobic groups, as for example in hydrophobically-modified cellulose ethers or poly(acrylic acids). The latter polymers exhibit rheological features markedly different from those of the unmodified polymers. Interchain aggregation of the hydrophobic groups leads to an enhancement in viscosity and reversible shear sensitivity. Fundamental questions related to their aggregation mechanisms, conformation, and association with surfactants need to be answered to help in designing new polymers and formulating materials of improved performance.

In this chapter we describe applications of fluorescence spectroscopy to investigate the solution properties of amphiphilic polymers. Photophysical studies of amphiphilic polymers require that one adds to the system under scrutiny a luminescent dye, which may be free to diffuse in solution (a probe), or attached covalently to the polymer (a label). This investigation focuses on the use of labeled polymers as a means to elucidate, on a molecular level, the events that control macroscopic rheological phenomena. Two classes of polyelectrolytes, poly(acrylic acid) and cationic cellulose ethers were chosen for this study. Commercially available polymers were labeled either with pyrene or with two chromophores, pyrene and naphthalene, to produce the series of labeled polymers depicted in Figure 1. They are the pyrene-labeled poly(acrylic acid) PAA-Py, the di-labeled poly(acrylic acid) PAA-Py-Np, the pyrene-labeled cationic (hydroxyethyl)cellulose, JR400-Py and its hydrophobically-modified analog, LM200-Py. After a brief outline of the fluorescence techniques employed, we describe their application in the study of the pH-induced conformation changes of poly(acrylic acids) and of the interactions of polyelectrolytes with like-charge surfactants.

Experimental Details

Materials

Water was deionized using a NANOpure water purification system. The polymers Quatrisoft LM200 and JR400 were gifts from Amerchol and were used without further purification. Quatrisoft LM200 is the chloride salt of an N,N-dimethyl-N-dodecyl ammonium derivative of (hydroxyethyl)cellulose (MS_{EO} ca 2.5) with a molar mass of approximately 100,000 daltons. The degree of substitution is reported to be 2.0×10^{-4} mol of $C_{12}H_{25}$ per gram of polymer or on average 1 dodecyl group per 19 glucose units (*1*). JR400 is the chloride salt of an N,N,N-trimethyl ammonium derivative of (hydroxyethyl)cellulose with a molecular mass of approximately 400,000 daltons and a degree of substitution of 5.4 mol % (1). In both polymers, the ethoxy groups are distributed statistically along the cellulose backbone (2, 3). Pyrene-labeled Quatrisoft LM200 (LM200-Py) and JR400 (JR400-Py) were prepared by reaction of the respective polymers with 4-(1-pyrenyl)butyl tosylate in N,N-dimethylformamide in the presence of sodium hydride (4). The amount of pyrene incorporated, determined by UV spectroscopy, was 2.3×10^{-5} mol Py g^{-1} polymer (LM200-Py) and 2.6×10^{-5} mol Py g^{-1} polymer (JR400-Py). Poly(acrylic acid) (PAA, nominal molecular weight 150,000 daltons, 25 % solution in water) was purchased from Wako Chemicals, Japan. It was labeled with pyrene, and both pyrene and naphthlene, as previously described (5). The pyrene modified poly(acrylic acid),

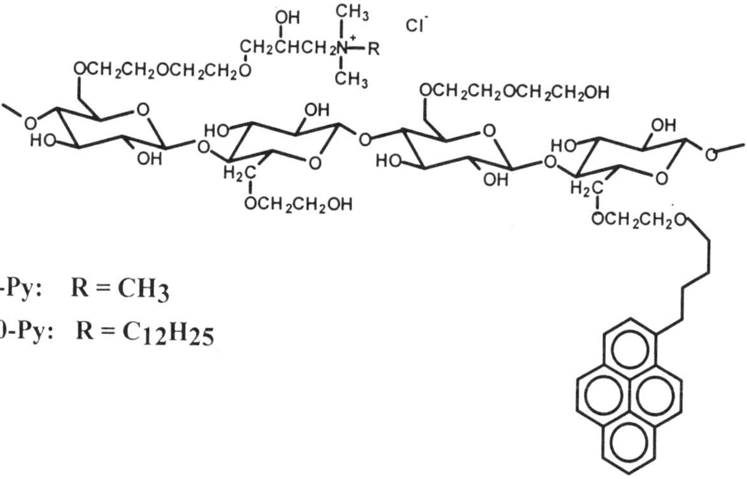

JR400-Py: R = CH$_3$
LM200-Py: R = C$_{12}$H$_{25}$

PAA-Py-Np
x = 98, y = 0.5, z = 1.5

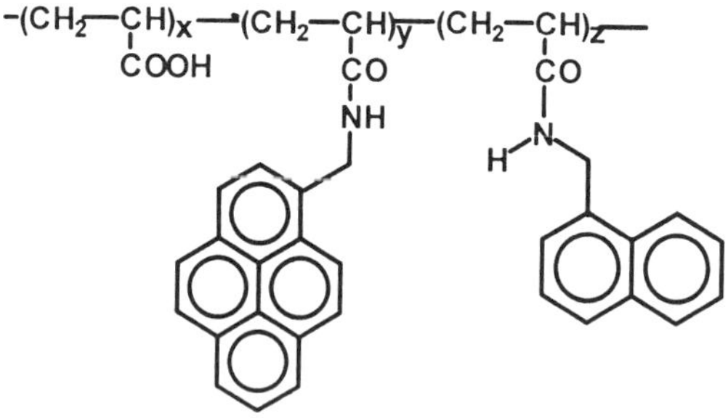

PAA-Py
x = 97, y = 3

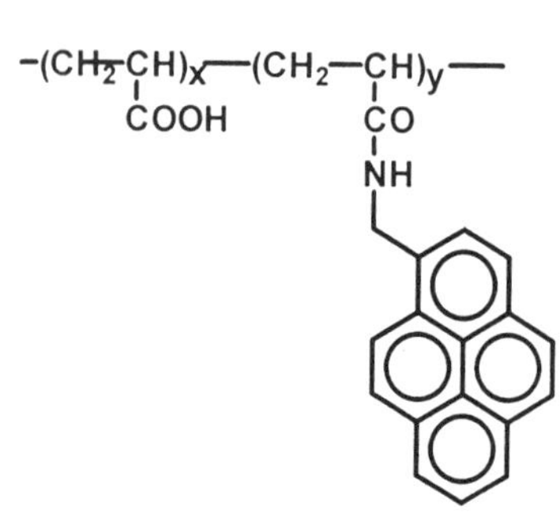

Figure 1. Chemical structure and composition of the polymers used in this study.

PAA-Py, contains 1.9 x 10^{-4} mol Py g^{-1} polymer. The dilabeled poly(acrylic acid), PAA-Py-Np, contains 5.7 x 10^{-5} mol Py g^{-1} polymer and 2.4 x 10^{-4} mol Np g^{-1} polymer. Sodium dodecylsulfate (SDS) was obtained from Sigma (purity > 99 %). Cetylpyridinium chloride (CPC) (purity > 95 %) was obtained from Aldrich Chemicals Corp. Dodecylpyridinium chloride (DPC) (purity > 95 %) was purchased from TCI America. Ethylpyridinium bromide (EPB) was purchased form Acros Organics (>98 %). N,N-dimethylformamide (DMF) was distilled over calcium hydride. All other chemicals and solvents were of reagent grade and used without further purification.

Methods

Fluorescence spectra were recorded at room temperature on a SPEX Fluorolog 212 spectrometer equipped with a DM3000F data system. Emission spectra were not corrected. For pyrene emission spectra the excitation wavelength was set at 344 nm. For naphthalene emission spectra the excitation wavelength was 290 nm. The excimer to monomer intensity ratio (I_E/I_M) was calculated by taking the ratio of the intensity at 485 nm to the half-sum of the emission intensities at 377 nm and 398 nm. The ratio I_{Py}/I_{Np} was calculated as the ratio of the half-sum of the intensities at 378 nm and 398 nm to the intensity at 340 nm. Polymer solutions were prepared by dissolving the polymer into water (1.0 or 0.1 g L^{-1}). They were allowed to equilibrate at room temperature for 24 h. The labeled poly(acrylic acids) were dissolved either in water or in a 0.1 M NaCl solution. Their pH was adjusted by addition of NaOH (5N) or HCl (5N). Surfactant solutions were prepared by dilutions of 1.0 mol L^{-1} stock solutions. Solutions for spectroscopic analysis were prepared by mixing aliquots of surfactant solutions (0.3 mL) with aliquots of polymer solutions (2.7 mL). They were allowed to equilibrate overnight in the dark.

Results and Discussion

Materials and Spectroscopy

Synthesis and Structure of the Polymers.
All the labeled polymers were obtained by post-modification of commercial polymers. Attachment of the labels to the PAA backbone was achieved by reacting PAA with chromophores bearing a short amino-terminated alkyl chain (5). The reactions were carried out in an aprotic solvent, 1-methylpyrrolidinone, in the presence of 1,3-dicyclohexylcarbodiimide. Under these conditions, random attachment of the hydrophobic groups along the chain is favored over the formation of discreet blocks of contiguous chromophores. In the case of cellulose ethers, such as

Quatrisoft LM200 and JR400, the label was attached by ether formation between 4-(1-pyrenyl)butyl tosylate and hydroxyl groups present along the macromolecules (4, 6). It is important to note that the polymer LM200 offers two different sites for pyrene incorporation, namely primary hydroxyl groups of the hydroxyethyl substituents on the glucopyranose ring (site a, Figure 2) and hydroxyl groups of the hydrophobic substituents (site b, Figure 2). Pyrene groups linked to sites a and sites b are expected to experience different environments, hence may exhibit different photophysical properties. In particular the pyrenes attached to sites b, in close proximity to the cationic charge and the hydrophobic group can report on changes in polymer conformation and association upon addition of salts or surfactants. In the case of JR400-Py, which does not have hydrophobic substituents, all chromophore attachment sites are equivalent in terms of pyrene environment. The physical properties and compositions of the labeled polymers are listed in Table I.

Spectroscopy of the Polymers in Solution

The emission of pyrene and naphthalene groups attached to the polymers are sensitive to small changes in the chromophore separation distances. A short separation distance (ca 4 to 5 Å) can be monitored with pyrene emission, and a longer scale (ca 15 to 50 Å) by measuring the extent of non-radiative energy transfer between the two chromophores.

Pyrene monomer and excimer emission. The emission of locally isolated excited pyrenes ('monomer' emission, intensity I_M) is characterized by a well-resolved spectrum with the [0,0] band at 378 nm. The emission of pyrene excimers (intensity I_E) centered at 480 nm is broad and featureless. Excimer formation requires that an excited pyrene (Py*) and a pyrene in its ground state come into close proximity within the Py* lifetime (7). The process is predominant in concentrated pyrene solutions or under circumstances where microdomains of high local pyrene concentration form, even though the total pyrene concentration is very low. This effect is shown for example by comparing the spectra of solutions in water of JR400-Py and its hydrophobically modified analog, LM200-Py (Figure 3). The strong excimer emission from aqueous solutions of LM200-Py (I_E/I_M = 0.86, polymer concentration: 0.1 g L^{-1}) vouches for the presence of hydrophobic microdomains in which the pyrene groups come in close contact. By contrast the emission of pyrene in aqueous solutions of JR400-Py displays mostly pyrene monomer emission.

Non-radiative energy transfer (NRET). This process originates in dipole-dipole interactions between an energy donor in its excited state and an energy acceptor in its ground state. The probability of energy transfer between the two chromophores depends sensitively on their separation distance and to a lesser extent on their relative orientation (8, 9). In solutions of polymers carrying both chromophores along the same chain, variations in NRET provide information on changes in the conformation of the chain, such as coil collapse or coil expansion. The pyrene-naphthalene pair of chromophores is known to interact as energy donor (naphthalene, Np) and energy acceptor (pyrene) by NRET with a characteristic distance, R_o = 29 Å (10), R_o being defined as the interchromophoric distance for which half of the donor molecules decay by energy transfer. With λ_{exc} = 290 nm, a wavelength at which most of the light is

291

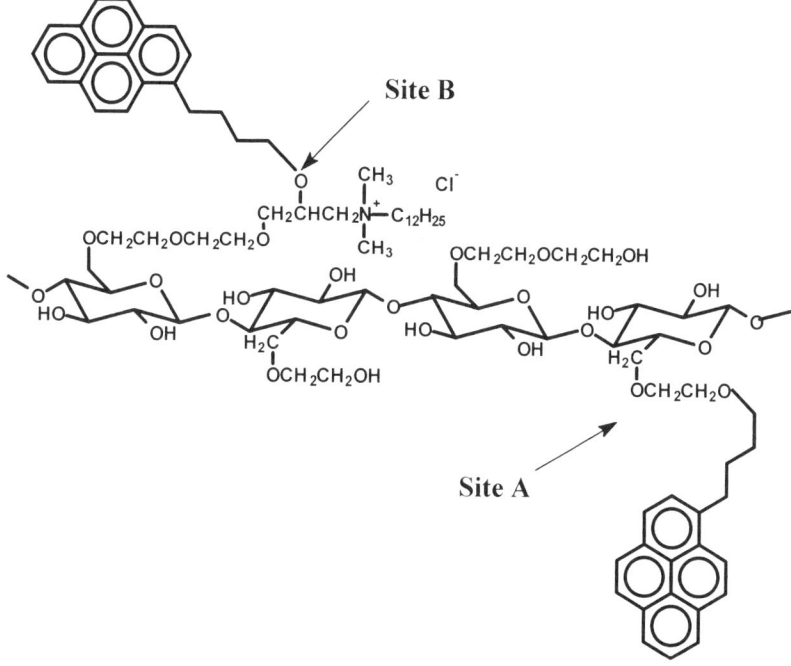

Figure 2. Chemical structure of LM200-Py indicating the two sites of pyrene incorporation.

Table I. Physical Properties of the Polymer used in this Study

Polymer Sample	Molecular weight[a]	Pyrene content[b]	Naphthalene content[b]
PAA-Py	150,000	3.1 mol % (4.4×10^{-4} mol g^{-1})	
PAA-Py-Np	150,000	0.4 mol % (5.64×10^{-5} mol g^{-1})	1.7 mol % (2.40×10^{-4} mol g^{-1})
JR400-Py	400,000	2.6×10^{-5} mol g^{-1}	
LM200-Py	100,000	2.3×10^{-5} mol g^{-1}	

[a] from manufacturers

[b] from UV data

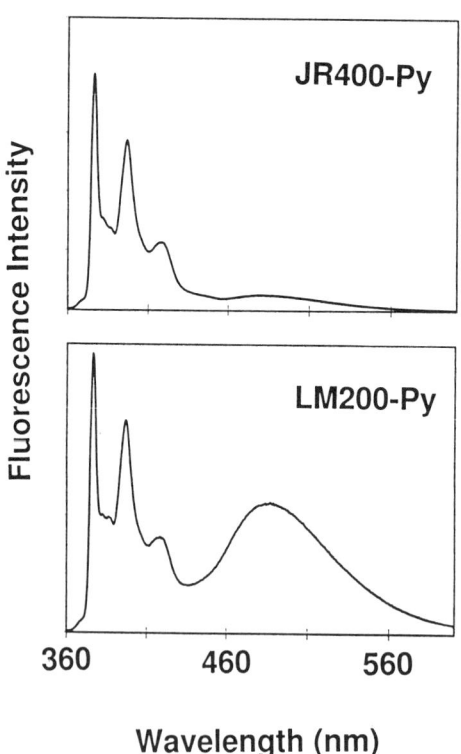

Figure 3. Fluorescence spectra of JR400-Py (top) and LM200-Py (bottom) in water; polymer concentration: 1.0 g L^{-1}; λ_{exc} = 346 nm.

absorbed by Np, one can detect both the direct emission from excited Np and the emission of pyrene excited through NRET from Np*. The emission of solutions of fully neutralized PAA-Py-Np in water (pH 3) presents a contribution from Np and Py upon excitation at 290 nm, indicating the occurrence of intrapolymeric NRET between Np* and Py placed in close enough proximity to satisfy the requirements of NRET.

We will describe next applications of photophysical methods in the study of hydrophobically-modified polymers and their interactions with surfactants. Changes in pyrene excimer emission and in the extent of NRET will be used to monitor the pH-induced expansion of hydrophobically-modified poly(acrylic acid). Quenching of pyrene fluorescence by a series of pyridinium salts will be employed to assess the extent of interactions between the cellulosic polyelectrolytes and surfactants of same charge.

Photophysical Tools as a Measure of the pH-induced Coil Expansion of Labeled Poly(acrylic acids).

Pyrene Excimer Emission

Emission spectra of solutions of the pyrene-labeled polymer PAA-Py in several organic solvents consist of a contribution of pyrene monomer (intensity I_M) and pyrene excimer (intensity I_E). The excimer emission increases with decreasing solvent viscosity at the expense of the monomer emission. A plot of the changes in I_E/I_M as a function of the inverse viscosity follows a linear relationship, as anticipated for dynamic excimer formation, following the mechanism originally proposed by Birks (7).

The emission spectrum of the polymer was recorded also for aqueous solutions of pH 3.0 and 6.0, in the absence of electrolytes and at constant ionic strength (0.1 M NaCl) (Figure 4). The fluorescence quantum yield of the pyrene emission (0.20) (5) is significantly lower for aqueous polymer solutions of pH 3.0, compared to those measured in organic solvents (0.51-0.60) (5), suggesting extensive self-quenching of pyrene emission. Also, the Py emission presents an increased contribution from the excimer, compared to the emission recorded from solutions in organic solvents. Moreover, when recording the excitation spectra of PAA-Py solutions in water, we observed that the spectrum recorded for the monomer emission is different from that recorded for the excimer emission. The general features of the excitation spectra are similar, but the former is blue-shifted by about 3 nm. All these features led us to conclude that the pyrene groups exist predominantly in the form of aggregates in solutions of pH 3.0, where the polymers are fully protonated. The formation of pyrene aggregates is driven by hydrophobic association within the polymer coils, which are expected to assume a compact conformation (11). In solutions of pH 6.0 the polymer is fully ionized and adopts a more expanded conformation, as a result of the electrostatic repulsion between neighboring carboxylate groups (12). This change in conformation is easily detected in the fluorescence spectrum of ionized PAA-Py:

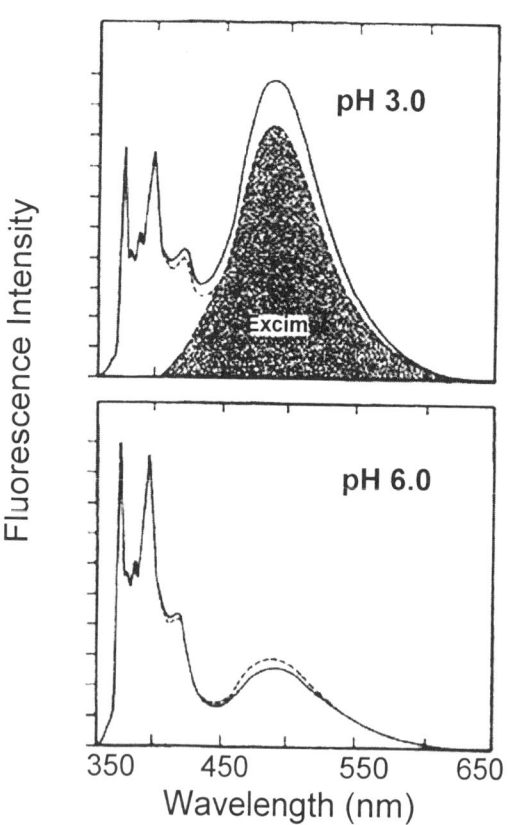

Figure 4. Emission spectra of PAA-Py in water (dotted line) and in 0.1 M aqueous NaCl (full line), pH 3.0 (top) and pH 6.0 (bottom); polymer concentration: 0.1g L^{-1}; λ_{exc} = 344 nm (Reproduced with permission from reference 5; copyright 1998, Elsevier).

the excimer is significantly weaker, compared to solutions of pH 3.0, in agreement with the expected increase in the average separation distance of the pyrene groups in the extended chains.

Non-Radiative Energy Transfer Measurements

Figure 5a shows the emission spectra, upon excitation at 290 nm, of PAA-Py-Np in aqueous NaCl (0.1 M) solutions of pH 3.0 and 6.0. Both spectra exhibit a contribution from Np* (300-360 nm) and a contribution from Py* (360-450 nm). A qualitative measure of the relative extent of energy transfer can be obtained by taking the ratio I_{Py}/I_{Np} of the intensity of the pyrene emission to that at the naphthalene emission (see Methods section). In this scale, a larger value reflects an increase in the efficiency of energy transfer. The largest value was recorded for aqueous solutions of the fully protonated polymer (I_{Py}/I_{Np} = 14.0). It was much lower for solutions of the polymer, either at neutral pH (I_{Py}/I_{Np} = 4.1) or in an organic solvent, such as dimethylformamide (DMF, I_{Py}/I_{Np} = 5.8, Figure 5b) (5). These values confirm that the polymer adopts a rather open conformation in aqueous neutral solution or in DMF, and that the polymer collapses in acidic pH, rendering the NRET process more efficient.

Fluorescence spectra of the doubly-labeled polymer PAA-Py-Np were monitored as a function of pH, from 3.0 to 12.0. The changes with pH of the efficiency of NRET are shown in Figure 6 in terms of the ratio I_{Py}/I_{Np}. Also presented in Figure 6 are the changes in the ratio I_E/I_M recorded from the same PAA-Py-Np solutions, but in this case an excitation wavelength of 344 nm was selected. Under these conditions only emission from directly excited pyrene is detectable. It is noteworthy that the changes in energy transfer efficiency as a function of pH do not occur within the pH range surrounding the pKa of the polymer, as do the changes in I_E/I_M, but they take place for pH values slightly lower than the pK_a and over a much wider range. The transition in the curve (mid-point pH 3.8) is diffuse and resembles a typical titration curve of PAA (13). The leveling-off (pH 4.7) of the curve I_{Py}/I_{Np} is identical to the mid-point in the I_E/I_M transition and correlates reasonably well with the pK_a of the doubly-labeled polymer determined by potentiometric titration (5). Thus the conformational reorganization that accompanies the deprotonation of the PAA carboxylic acid groups takes place over a wide pH range. When the solution reaches the critical pH corresponding to the sharp transition in I_E/I_M, the electrostatic repulsive forces between the carboxylate groups overcome the hydrophobic interactions that, until this pH point is reached in the neutralization process, keep a fraction of the pyrene groups within excimer formation distance.

Photophysical Tools as Measure of the Interactions between Cationic Surfactants and Cationic Cellulose Ethers

While simple polycations do not interact with cationic surfactants, hydrophobically-modified polycations have been shown to interact with surfactants of the same charge and sufficiently long chain length (14, 15). We describe here an

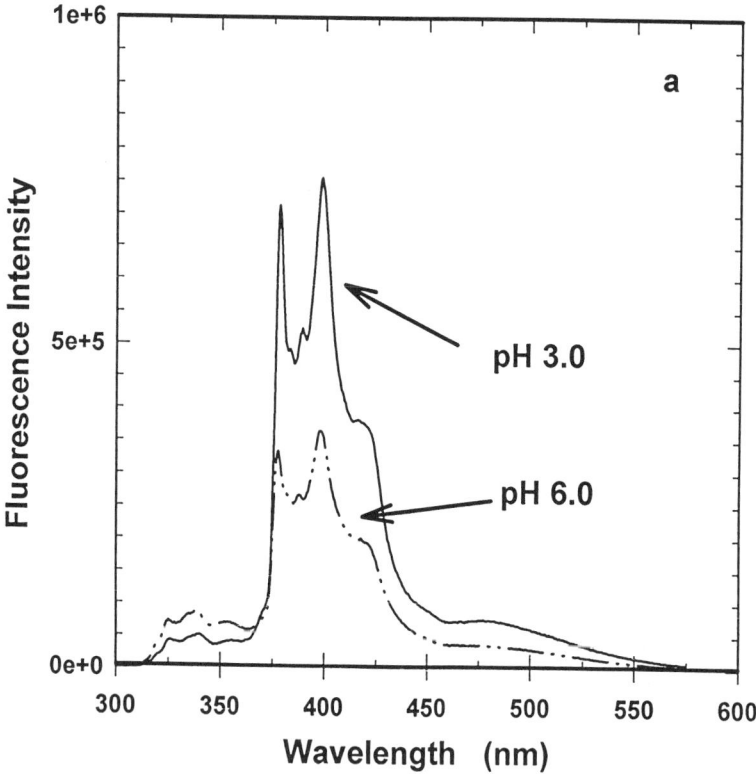

Figure 5. Fluorescence spectra of PAA-Py-Np in (a) 0.1 M aqueous NaCl, pH 3.0 and 6.0; (b) dimethylformamide; polymer concentration: 0.1 g L^{-1}; λ_{exc} = 290 nm. (Reproduced with permission from reference 5; copyright 1998, Elsevier).

Continued on next page.

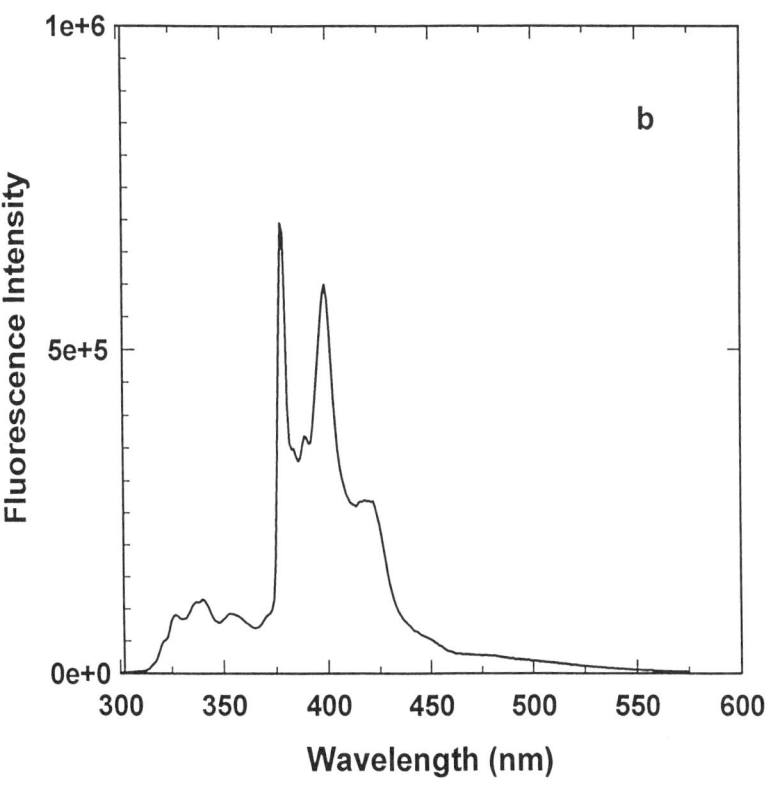

Figure 5. *Continued.*

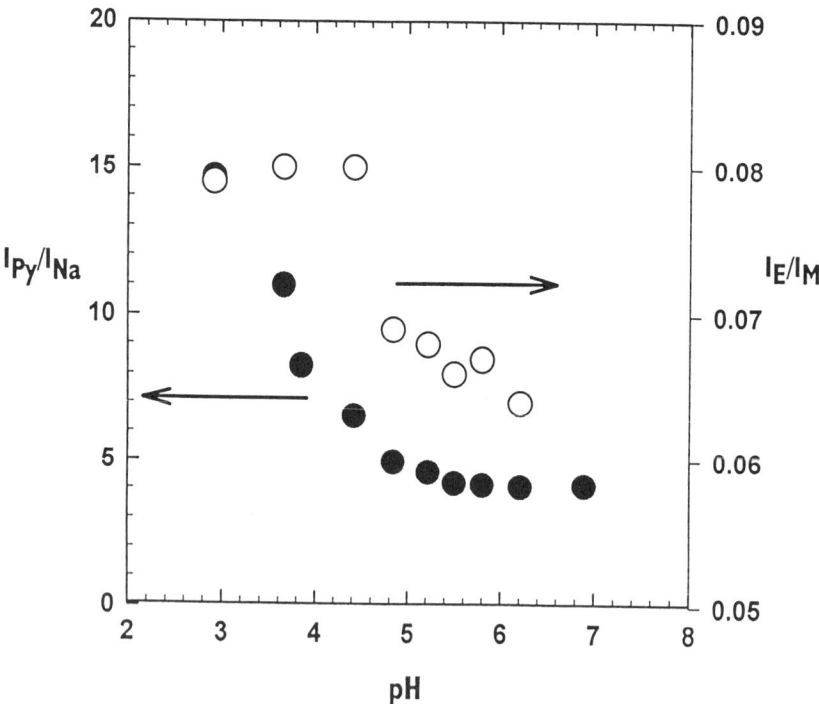

Figure 6. Plot of the ratio of pyrene to naphthalene emission intensities (I_{Py}/I_{Np}) and of the ratio of pyrene excimer to monomer emission (I_E/I_M) for 0.1 M aqueous NaCl solutions of PAA-Py-Np as a function of pH; polymer concentration: 0.1 g L^{-1}; λ_{exc} = 290 nm (for I_{Py}/I_{Np}) and λ_{exc} = 343 nm (for I_E/I_M)

example of such a system, consisting of cationic surfactants and the pyrene-labeled hydrophobically-modified cationic hydroxyethylcellulose, LM200-Py (Figure 1). To assess the role of the hydrophobic substituents in guiding the interactions, the spectroscopy of JR400-Py (Figure 1) was monitored under identical conditions. Association of LM200-Py with cetyltrimethylammonium chloride (CTAC) or dodecyltrimethylammonium chloride (DTAC) was detected by a decrease in the pyrene excimer emission intensity with a concomitant increase in the pyrene monomer emission intensity, upon addition of increasing amounts of surfactant to a dilute solution of LM200-Py (0.1 to 1.0 g L^{-1}). The association takes place at a critical surfactant concentration, slightly lower than the respective critical micelle concentrations (cmc) of the surfactants. The results are shown in Figure 7 in terms of the changes in the ratio I_E/I_M of the pyrene emission as a function of surfactant concentration.

Next, we conducted a series of experiments using cationic surfactants, such as cetylpyridinium chloride (CPC) and dodecylpyridinium chloride (DPC), that possess a head group capable of quenching the fluorescence of pyrene. In this case, formation of complexes between surfactants and polymer should result in a decrease of the overall fluorescence intensity, in addition to changes in the relative intensities of pyrene monomer and excimer emissions. As a control experiment, we monitored first the effect of added ethylpyridinium bromide, EPB, a water-soluble quencher but not a surfactant, on the fluorescence of LM200-Py and JR400-Py in water. This compound is known to quench pyrene fluorescence at a diffusion-controlled rate by an electron transfer mechanism. Quenching results are reported in terms of the Stern-Volmer model (16, 17). In this treatment the fluorescence intensities I_0 and I, in the absence and presence of quencher, respectively, are related to the quencher concentration [Q] by equation 1, where K_{SV} is the Stern-Volmer quenching constant, k_q is the bimolecular quenching constant, and τ_o is the lifetime of the fluorophore in the absence of quencher:

$$I_0/I = 1 + K_{SV}[Q] = 1 + k_q\tau_o[Q] \qquad (1)$$

Linear Stern-Volmer plots were obtained for quenching of pyrene monomer and excimer emissions in aqueous solutions of JR400-Py. Stern-Volmer parameters are reported in Table II. It is interesting to note that the cationic quencher actually interacts with chromophores linked to a cationic polymer. This confirms that in JR400-Py the chromophores are linked to the glucopyranose groups at sites removed from the quaternary ammonium group. In contrast, the quenching efficiency of EPB for aqueous solutions of LM200-Py is much lower, compared to JR400-Py, indicating that the hydrophobic groups protect the chromophores against quencher molecules. Moreover, since the hydrophobic groups are linked to the quaternary ammonium group, encounter of EPB and pyrene groups is prevented through electrostatic repulsion.

Addition of increasing amounts of cetylpyridinium chloride to solutions of LM200-Py resulted in efficient quenching of pyrene monomer and excimer emission. The Stern-Volmer plots were linear for low surfactant concentration ([CPC] < 10^{-4}

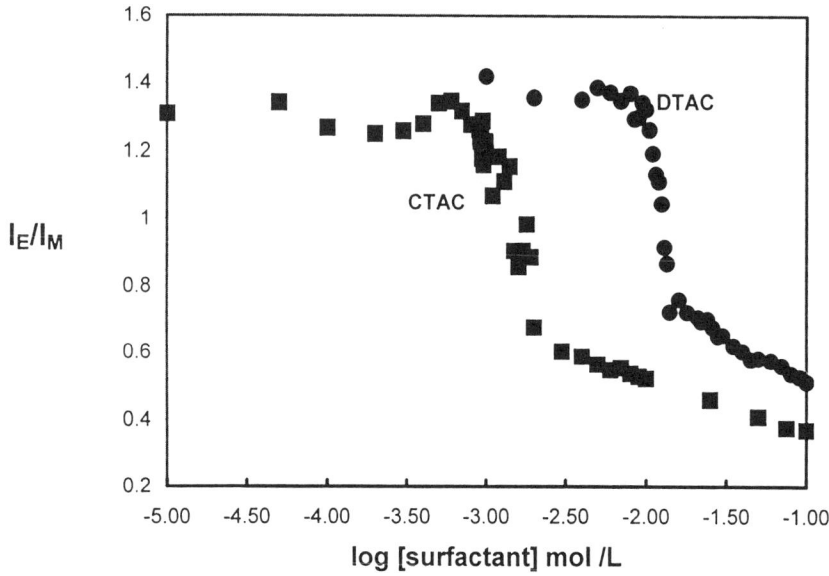

Figure 7. Plot of the ratio I_E/I_M of pyrene monomer to excimer emission intensities for aqueous solutions of LM200-Py in water as a function of added surfactant full squares: cetyltrimethylammonium chloride (CTAC); full circles: dodecyltrimethylammonium chloride (DTAC); polymer concentration: 0.1 g L^{-1}; λ_{exc} = 346 nm.

Table II. Stern-Volmer Constants (K_{SV}) Obtained for Quenching of Pyrene-labeled Polymers with Various Pyridinium Chloride Derivatives

Quencher	JR400-Py monomer	LM200-Py monomer	LM200-Py excimer
EPB	70 ± 3 M^{-1}	---	---
DPC	128 ± 5 M^{-1}	348 ± 10 M^{-1}	436 ± 20 M^{-1}
CPC	83 ± 5 M^{-1}	6780 ± 50 M^{-1}	$36{,}730 \pm 50$ M^{-1}

NOTE: Values obtained from linear part of the plots (low surfactant concentrations, see text)

M). For higher CPC concentrations, the plots exhibited substantial upward curvature, indicating that quenching takes place by a static mechanism. Addition of the shorter chain surfactant, dodecylpyridium chloride, to LM200-Py solutions had the same overall effect on pyrene emission, but the quenching efficiency was significantly lower than in the case of CPC. Since both surfactants were able to quench pyrene fluorescence, whereas EPB was ineffective, this set of measurements confirms the occurrence of association of like-charged species when the association is aided by hydrophobic attraction. Further support to this conclusion is given by the fact that neither CPC nor DPC had any significant effect on the fluorescence of the JR400-Py, the analog of LM200-Py lacking hydrophobic substituents.

Conclusions

Fluorescent groups have been attached to commercial polymers by simple chemical reactions. The modification slightly alters the solution properties of these polymers in water, as a result of the hydrophobicity of the chromophores (18). However, the labeled polymers become excellent models for the study of hydrophobically-modified polymers, an important class of commercial polymers. Several photophysical properties of a single chromophore, pyrene, were employed: a) changes in the ratio of pyrene monomer to excimer emission intensities; b) changes in the extent of non-radiative energy transfer between naphthalene and pyrene; and c) quenching of fluorescence. Each technique reveals a different facet of the rich physical chemistry of the polymers in aqueous solutions, either alone or in the presence of surfactants.

Acknowledgments

F.M.W. and S.T.A.R. thank the National Science and Engineering Research Council of Canada for financial support. D.F.A. acknowledges the Ontario Center Materials Research for a visiting scientist fellowship. Thanks are also due to Michael Berg for help in the fluorescence measurements.

Literature Cited

1 Goddard, E. D.; Leung, P. S. *Colloids Surf.* **1992**, *65*, 211.
2 Glass, J. E.; Buettner, A. M.; Lowther, R. G.; Young, S.; Cosby, L. A. *Carbohydr. Res.* **1980**, *84*, 245.
3 Glass, J. E.; Shah, S.; Lu; D.-L.; Seneker, S.D. in *Polymer Adsorption and Dispersion Stability*, ACS Symposium Series 240, E.D. Goddard; B. Vincent; Eds, American Chemical Society, Washington DC, 1984, Chapter 7.

4 Winnik, F. M.; Regismond, S. T. A.; Goddard, E. D. *Langmuir* **1997**, *13*, 111-114.
5 Anghel, D. F.; Alderson, V.; Winnik, F. M.; Mizusaki, M.; Morishima, Y. *Polymer* **1998**, *39*, 3035-3044.
6 Winnik, F. M.; Winnik, A. M.; Tazuke, S.; Ober, C. K. *Macromolecules* **1987**, *20*, 38-44.
7 Birks, J. B. *Rep. Prog. Phys.* **1975**, *38*, 903.
8 Förster, T. *Discuss. Faraday Soc.* **1959**, *27*, 7.
9 Lakowicz, J. R. *Principles of Fluorescence Spectroscopy;* Plenum Press: New York, NY, 1983, Chapter 10.
10 Winnik, F. M. *Polymer* **1990**, *31*, 2125.
11 Winnik, F. M. *Chem. Rev.* **1993**, *93*, 587.
12 Morawetz, H. *Macromolecules in Solution,* 2nd edition; Wiley: New York, NY, 1975, Chapter 7.
13 Mandel, M. *Eur. Polym. J.* **1970**, *6*, 807.
14 Bakeev, K. N.; Ponomarenko, E. A.; Sishkanova, T. V.; Tirrell, D. A.; Zezin, A. B.; Kabanov, V. A. *Macromolecules* **1995**, *28*, 2886.
15 Winnik, F. M.; Regismond, S. T. A.; Goddard, E. D. *Colloids Surfaces A* **1996**, *106*, 243.
16 Ref 7, Chapter 8.
17 Webber, S. E. *Photochem. Photobiol.* **1997**, *65*, 33.
18 Morawetz, H. *Macromolecules* **1996**, *29*, 2689.

Chapter 18

The Adsorption and Surface Dilatational Rheology of Unmodified and Hydrophobically Modified EHEC, Measured by Means of Axisymmetric Drop Shape Analysis

Rolf Myrvold[1,3], Finn Knut Hansen[1,4], and Björn Lindman[2]

[1]Department of Chemistry, University of Oslo,
P.O. Box 1033, Blindern, N–0315 Oslo, Norway
[2]Physical Chemistry 1, Chemical Center, University of Lund,
P.O. Box 124, S–22100 Lund, Sweden

Studies of adsorbed and spread layers of ethylated hydroxyethyl cellulose (EHEC) and a hydrophobically modified analogue (HM-EHEC) at the air-water interface have been performed by means of four different experimental routes. The adsorption characteristics visualized by means of dynamic surface tension measurements, dilatational elastic behavior measured by means of a oscillating sessile bubble method, and the surface pressure response to surface area reductions measured by means of a Langmuir surface balance all give the same result. Relatively strong interactions can be observed when adding sodium dodecyl sulfate (SDS) to the polymer solutions. The SDS concentration for optimal enhancement of surface properties is 5-6 mM, indicating the formation of the most surface-active polymer/SDS complex. Despite the fact that the bulk properties of these two polymer/SDS systems are quite different, both the dynamic surface tension and the surface rheological properties are practically the same. This emphasizes the different effect of hydrophobic modification on bulk and surface behavior.

Introduction

Several types of water-soluble polymers, and in particular biopolymers, play an important role in different types of applications. One class of water-soluble

---

[3]Current address: Pronova Biopolymer, P.O. Box 494, N–3002 Drammen, Norway.
[4]Corresponding author.

biopolymers is made up of the different types of cellulose derivatives. These polymers may serve as stabilizers and as a part of the retard system in control drug delivery [1,2].

The equilibrium values of surface tension and adsorption are often used to describe the effect of surface-active materials or phenomena connected to colloidal stability. In many instances, these properties alone have been shown to be insufficient for correlation of the macroscopic behavior to fundamental microscopic phenomena, such as emulsification and foaming. In these systems it is often the dynamic properties of the interface that are important as a result of surface tension gradients and surface mobility. The Marangoni effect and the surface rheological parameters are used as a measure of the surface's properties under dynamic conditions. The surface shear viscosity and elasticity, the 2 dimensional equivalents to ordinary bulk rheology, have been shown to be closely correlated with the stability of emulsions and foams [27]. However, probably more important for these processes are the surface dilatational viscosity and elasticity that are often several orders of magnitude higher than the shear parameters [12]. These properties are, however, less readily measurable, except for the Gibbs elasticity that is the static limit of the latter. Several methods have been developed in order to measure the dynamics of adsorption and surface rheological properties, among these are the oscillating jet, oscillating bubble, surface waves etc. The interested reader is referred to ref. [12] for a further description of these methods and references.

In this group we have developed an instrument for axisymmetric drop shape analysis [6] that can also measure dynamic properties of surface films by means of the oscillating bubble technique [7]. In this paper we will report the results from a study of the interfacial properties of ethylated hydroxyethyl cellulose (EHEC) and a hydrophobic modified analogue (HM-EHEC) and their synergies with sodium dodecyl sulfate (SDS). It has been shown that above a critical aggregation concentration (cac) of 4-5 mmolal SDS, which is below the critical micelle concentration for SDS, both of these polymers bind SDS in a cooperative manner [22,23]. Below this concentration, only HM-EHEC binds SDS non-cooperatively. It has also been shown that the bulk rheological properties of these polymers are greatly enhanced by SDS [8], but it has up to now been unclear to what degree the surface properties of EHEC are influenced by hydrophobic modifications and surfactant interaction. The adsorption characteristics of these systems are measured in this study by means of the axisymmetric drop shape analysis instrument, and the surface dilatational rheology is measured by means of the oscillating bubble method. A Langmuir surface balance is used for the measurements on the spread polymer layers [10].

Experimental

Instruments

The dynamic surface tension and surface rheological measurements were performed with an axisymmetric drop shape analysis instrument developed by this lab. This instrument has been described elsewhere, [5-7], and will only be shortly reviewed here. All experiments were performed at 25 ± 0.5 °C. The instrument consists of a goniometer (Ramé-Hart) fitted with a macro lens and a CCD video camera. The video frames are captured by a DT3155 frame grabber (Data Translation) in a Pentium PC. The drop is controlled by a dispenser and a specially designed oscillation unit consisting of a syringe with an excenter mounted piston that is motor driven. The dispenser is controlled by the PC. The dispenser and oscillation units are mounted in series with stainless steel pipes that are filled with distilled water. The drops and bubbles are extended from the tip of a small PTFE ("Teflon") tube into a cuvette inside a thermostatted and water filled chamber with glass windows. The Teflon tube contains an air pocket toward the water in the steel pipe. The specially written DROPimage computer program can control the frame grabber and the PC's RAM for picture storage, thus the program can capture pictures directly to RAM in real-time and calculate the results later. A maximum capture rate of 25 Hz in CCIR video is possible, and the rate of calculation is less than 0.2 s per picture (processor dependent). The program has many facilities for different measurement strategies, and is also able to keep the drop or bubble volume constant during a long period of time by means of a feedback method. This is very important when measuring adsorption in highly elastic films, where small deviations in volume (i.e. area) may cause significant measurement errors. The results that are calculated are the surface tension, shape factor (β), radius of curvature (R_0) the drop volume, height and width, the surface area, and the contact angle with the horizontal plane. A description of the instrument and the program may be found at http://www.uio.no/~fhansen/dropinst.html.

An automatic Langmuir surface balance (Minitrough, KSV Instruments Ltd. Finland) was used to record the ΠA isotherms. The trough is made from a single block PFTE and has a surface area of approximately 250 cm^2. The surface balance is kept in a glass cabinet placed on a stone table. The subphase temperature is controlled and maintained by water circulation, and was 25 °C throughout the study. The surface pressure of the monolayers is measured by means of a Wilhelmy plate, using a roughed platinum plate.

Chemicals

The ethyl(hydroxyethyl) cellulose (EHEC) and the hydrophobically modified ethyl(hydroxyethyl) cellulose (HM-EHEC) were manufactured by Akzo Nobel AB, Stenungsund, Sweden and purified at the University of Lund. Both the unmodified and the hydrophobically modified polymers are ethyl(hydroxyethyl) cellulose ethers with the same molecular weight (Mw ≈ 100.000 g mol^{-1}). The average degrees of substitution of ethyl and hydroxyethyl groups are 0.6-0.7 and 1.8, respectively. These values correspond to the number of ethyl and hydroxyethyl groups per anhydroglucose unit of the polymer [8]. The HM-EHEC polymer is equivalent to the EHEC polymer sample, but with branched nonylphenol chains grafted to the cellulose backbone. The degree of nonylphenol substitution has been determined to be 1.7 mol %

(approximately 6.5 groups per molecule) relative to the repeating units of the polymer [8].

Sodium dodecyl sulfate (SDS) of analytical grade was purchased from Fluka and used without further purification. Sodium acetate and isopropanol of analytical grade were supplied from Fluka and used without further purification.

Water was de-ionized followed by distillation in an all glass still. The surface tension was measured to 72.3±0.1 mN m^{-1} at 21 °C and the conductance was measured to 1.8 μS cm^{-1}. The water was checked for surface-active impurities by compressing a 250 cm^2 water surface on the Langmuir surface balance. Maximum compression gave surface pressures less than 0.1 mN m^{-1}.

Polymer stock solutions (0.1 wt% (~1 g L^{-1})) were prepared and allowed to equilibrate in a refrigerator for one month before further dilution. From these stock solutions new stock solutions of 0.01 wt% were made and equilibrated in refrigerator for seven days prior to use. The pure polymer and SDS/polymer solutions were made from these solutions by weighing a suitable amount of polymer and SDS and diluting this to the set concentration. This solution was let to equilibrate for at least two hours before use.

Procedure and data analysis

The procedure has been described elsewhere [7] and only a short description will be given. A small amount of air (ca 150 μL) was sucked into the tip of the PTFE tube. The tip was dipped into ca. 10 mL of the polymer/SDS sample solution in the cuvette and fixed to the experimental chamber. After equilibration of a ca 10 μL bubble the computer was programmed to increase the bubble volume to 50 μL with immediate start of the surface tension measurement. The surface tension was measured as function of time for 50.000 s, until near equilibrium conditions were reached. In order to avoid compression effects due to bubble shrinkage the bubble volume was kept constant during the time of the experiment. After 50,000 s and completion of the dynamic surface tension measurement the drop was oscillated around the near equilibrium position with amplitude of ca 2.5 μL, corresponding to a surface area oscillation of approximately 1.5 mm^2. The oscillation was performed over a frequency interval between 0.2 and 2 Hz, and the presented rheological results are averages of six separate frequency sweeps.

The calculation of the surface rheological parameters has been described elsewhere [7], but for completeness, as short description will be given here. For an oscillating bubble, we vary the surface area by changing the bubble volume in a sinusoidal manner with an angular rate ω, and provided that the volume change is small, this results in a corresponding sinusoidal variation in the bubble surface area. This also leads to a corresponding surface tension variation. We can write this

$$\Delta A = A - A_0 = A_a \sin(\omega t) \quad \text{and} \quad \Delta \gamma = \gamma - \gamma_0 = \gamma_a \sin(\omega t + \delta)$$

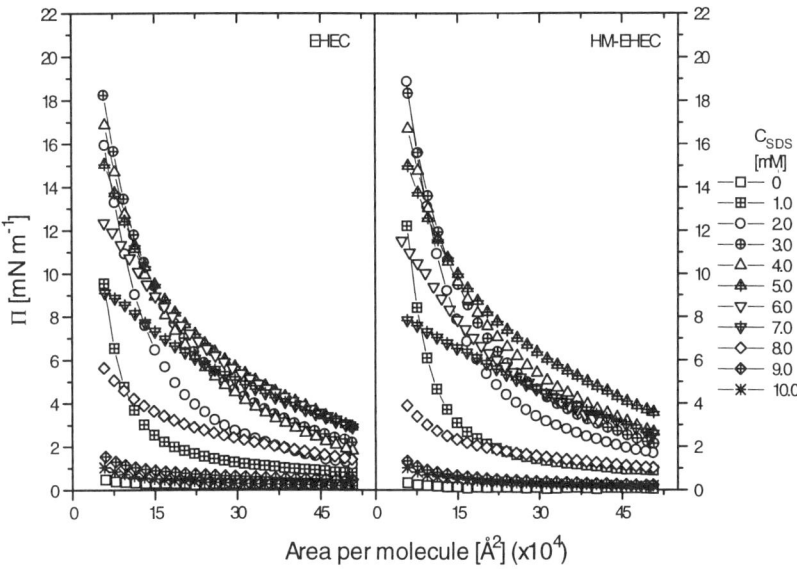

Figure 2. Excess surface pressure/area isotherms of EHEC and HM-EHEC monolayers spread on SDS solution with different concentration. (For the sake of clarity only 11% of the data points are shown.)

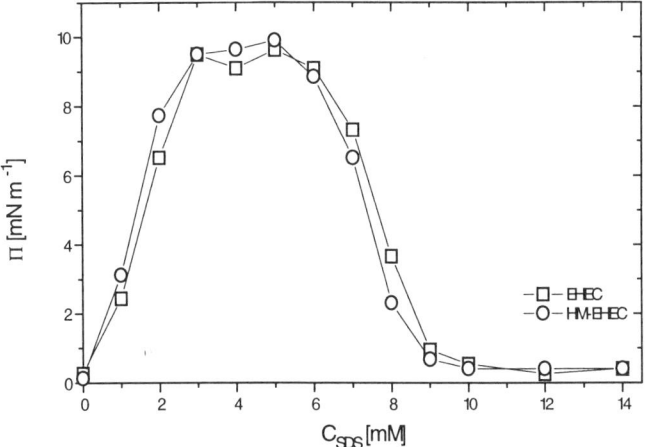

Figure 3. The surface pressure of EHEC and HM-EHEC at a molecular area of 15000 $Å^2$ from figure 2, is plotted as function of the SDS concentration in the subphase. The molecular area corresponds to a surface concentration of ca 0.11 mg m^{-2}.

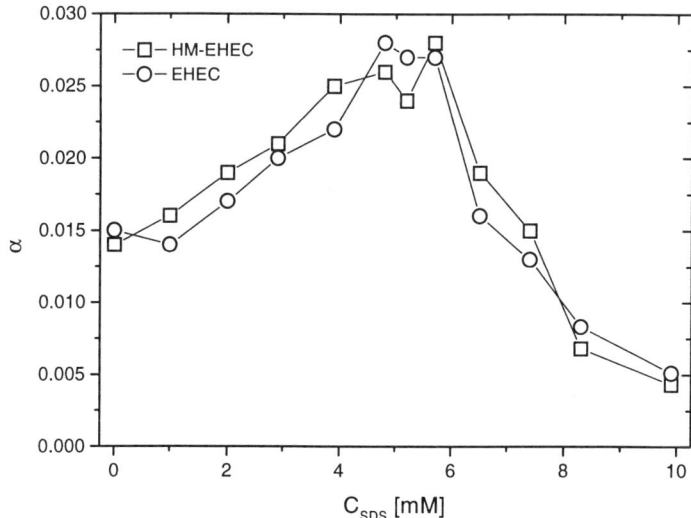

Figure 4. The long time power law exponent, α, presented as function of the SDS concentration.

From Figure 1 it can also be seen that in this regime some of the surface tension curves drop below the curve of 9.9 mM SDS. This may also serve as an illustration of the presence of a polymer/SDS complex that is more surface-active than the separate compounds. Even after 50,000 seconds equilibrium or steady state is not reached, and one can speculate on whether the low SDS concentration systems also after long enough times will drop below the 9.9 mM SDS system. After spreading of 0.8 μg polymer on different SDS solutions, the growth in surface pressure as function of time can be measured with a Langmuir surface balance. These results are reported in Figure 5. The figure clearly shows how the rate of surface pressure increase is affected by different SDS concentrations. Surfactant concentrations from 1 mM up to 5 mM cause an increase in the surface pressure growth rate, and also in the plateau value that is approached after more than 2000 seconds. Increasing the SDS concentration further beyond 5 mM causes a reduction in the surface pressure growth rate, with declining plateau values. HM-EHEC spread on 9 and 10 mM SDS hardly shows any increase in surface pressure relative to the system without SDS. These observations are in line with the previous interpretations of enhanced surface activity due to the presence of a more surface-active complex.

The presence of polymer/surfactant complexes should indeed be revealed through surface rheology measurements. It is expected that a more surface-active complex should give a higher response in the elastic behavior than the case is for a less surface active compound [21]. Figure 6 shows the frequency dependency of the elastic and viscous moduli of adsorbed EHEC/SDS and HM-EHEC/SDS layers and shows what already was mentioned.

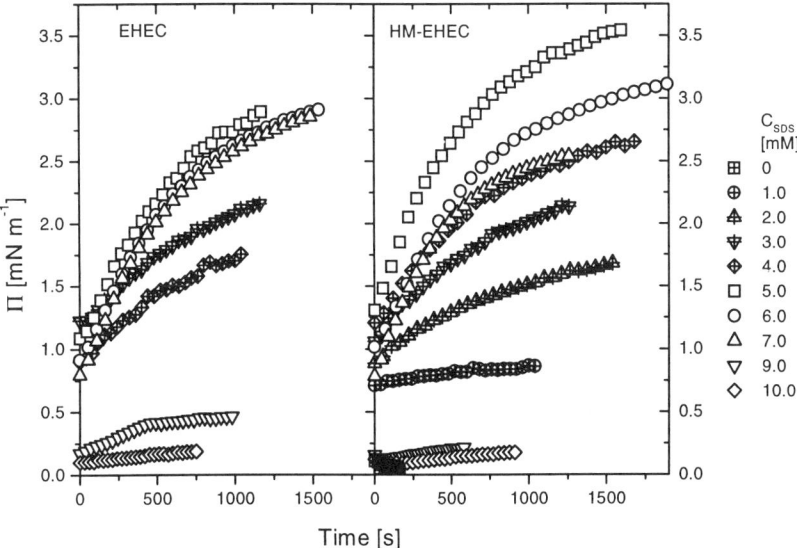

Figure 5. The increase in surface pressure after 0.8μg of EHEC or HM-EHEC has been spread on subphases with different SDS concentration. Surface concentration ca 0.11 mg m^{-2}.

Figure 6 also shows that the elastic moduli, E', are much higher than the corresponding viscous moduli, E'' which are close to zero. This is a consequence of a very low phase angle, which reflects the elastic nature of the interfacial layers and the lack of transport between the bulk and the surface in the time scale of an oscillation period. Within the experimental errors the viscous moduli (±1.6 mN m^{-1} on a 95% confidence limit level [7]) all extrapolate to approximately zero as the frequency vanishes. This reflects the insoluble nature (irreversibly adsorbed) of these layers since a higher value at low frequencies can be interpreted as a result of desorption [12,25]. The figure also shows that the elastic moduli increase with frequency, a behavior that is characteristic for insoluble monolayers/irreversibly adsorbed layers [12]. From the figure one can also see that within this frequency regime there are no major differences between the EHEC/SDS and HM-EHEC systems except that the elastic modulus for the EHEC/SDS systems is slightly higher in the 4-5 mM SDS interval. For both systems the level of the elastic moduli is shifted higher as the SDS concentration increases. This behavior is observed up to a certain SDS concentration. Beyond this SDS concentration the elastic modulus levels off and declines towards lower values again. This behavior is summarized in Figure 7 where the elastic moduli at 0.6 and 1.0 Hz are presented as function of SDS concentration for the two polymer systems.

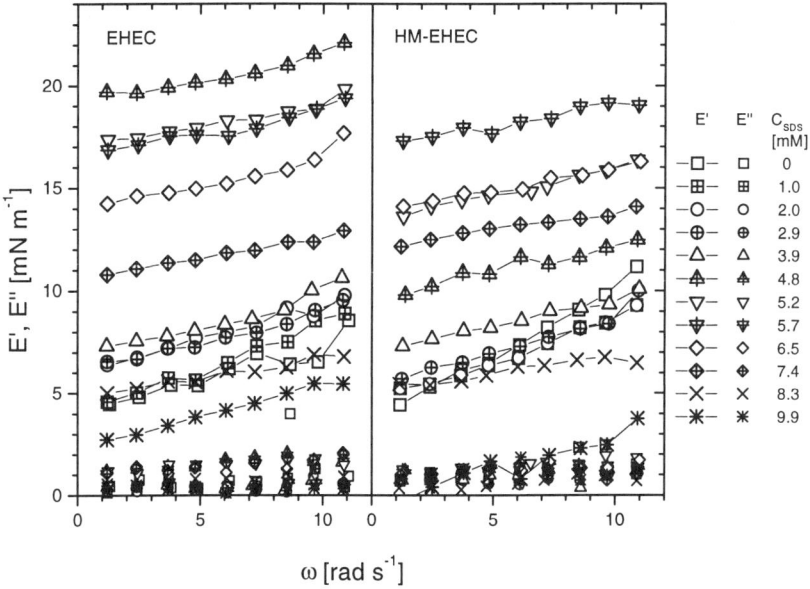

Figure 6. Surface elastic and viscous moduli measured as function of frequency for EHEC/SDS and HM-EHEC/SDS systems.

The increase in elasticity observed in Figure 7 represents in addition to the presence of a more surface active polymer/SDS complex the ability the surface layer has to restore surface tension uniformity upon deformations. This ability is a property of the greatest importance for foaming and emulsification processes [24]. Through this figure it can also be observed that there is a small difference between the EHEC/SDS and HM-EHEC/SDS systems when the location of the optimum composition for enhanced surface elasticity is considered. The EHEC/SDS system reaches the optimum value in surface elasticity at a lower SDS level than for the corresponding HM-EHEC/SDS system. If this effect is significant (and not an artifact) it may be caused by the more hydrophobic HM-EHEC and this requires more SDS in order to obtain the optimal surface-active complex (optimal hydrophobic/hydrophilic balance). The absolute value of the elastic modulus is also higher for EHEC than HM-EHEC, and can probably be explained in the same manner.

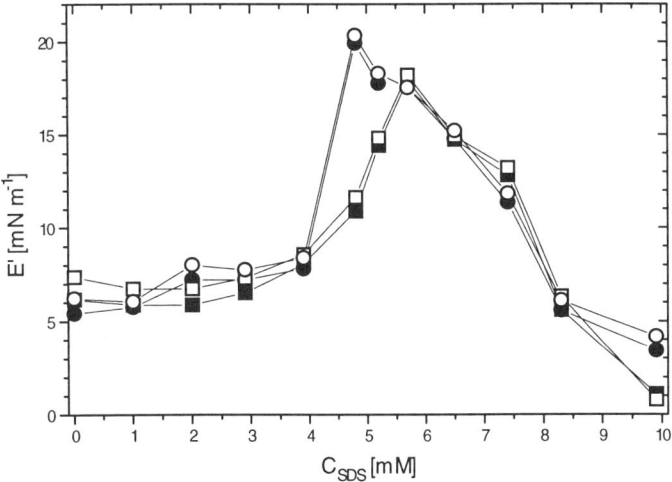

Figure 7. Surface elastic moduli at 0.6 Hz (solid symbols) and 1.0 Hz (open symbols) for EHEC (circles) and HM-EHEC (squares) adsorbed layers as function of SDS concentration.

Conclusion

The results from dynamic and static experiments performed on adsorbed and spread EHEC/SDS and HM-EHEC/SDS layers have shown that the interfacial behavior can be designed from the composition of the system. Enhanced surface properties, such as dynamic surface tension, dilatational elastic modulus and the surface pressure serve to show that special synergism between these two polymers and the surfactant SDS takes place. The formation of what is believed to be a more surface-active polymer/SDS complex can be observed through four independent types of experiments. The SDS concentration that give the optimal composition for the formation of the most surface active polymer/SDS complex is determined to be 5-6 mM and was observed in all four experiments. This concentration coincides quite well with the cac of the parent polymer system (EHEC). Above the cac both polymers bind SDS in a cooperative manner [22], which makes the polymer more soluble and thus less surface active. Below the cac there is definitely a polymer-surfactant interaction that materializes in a surface pressure and elasticity increase. Only small or no differences between the EHEC/SDS and the HM-EHEC/SDS systems could be observed. This is in strong contrast to bulk behavior of these polymer/surfactant systems, where a strong effect of the hydrophobic modification is observed [8]. The non-cooperative binding of SDS to HM-EHEC polymer thus does not seem to result in enhanced surface activity. The rationale for this difference is probably connected to the fact that the hydrophobic modification is relatively small, and probably grafted to the polymer backbone in regions of an already hydrophobic character. Because EHEC is considerably surface active in its unmodified form, this modification is not sufficient to alter the overall surface activity of the polymer.

References

1. *Pharmaceutical Technology*, Yearbook **1998**, *32*.
2. Persson, B., Nilsson, S. and Sunderlöf, L.-O. *Carbohydrate Polymers* **1996**, *29*, 119.
3. Ybert, C. and di Meglio, J.-M. *Langmuir* **1998**, *14*, 471.
4. Nahringbauer, I. *Langmuir* **1997**, *13*, 2242.
5. Hansen, F.K. and Rødsrud, G. *J. Colloid Interface Sci.* **1991**, *141*, 1.
6. Hansen, F.K. *J. Colloid Interface Sci.* **1993**, *160*, 209.
7. Myrvold, R. and Hansen, F.K. *J. Colloid Interface Sci.* **1998**, *207*, 97.
8. Thuresson, K., Nyström, B., Wang, G. and Lindman, B. *Langmuir* **1995**, *11*, 3730.
9. Ställberg, S. and Teorell, T. *Trans. Faraday Soc.* **1939**, *35*, 1413.
10. Myrvold, R., Hansen, F.K., and Balinov,B. *Colloids and Surfaces A* **1996**, *117*, 27.
11. Nahringbauer, I. *Progr. Colloid Polym. Sci.* **1991**, *84*, 200.
12. Dukhin, S.S., Kretzschmar, G. and Miller, R. *Dynamics of Adsorption at Liquid Interfaces, Studies in Interfacial Science*; Elsevier: Amsterdam, 1995.
13. Tsay, R-Y., Lin, S-Y., Lin, L-W. and Chen, S-I. *Langmuir* **1997**, *13*, 3191.
14. Hansen, F. K. and Myrvold, R. *J. Colloid Interface Sci.* **1995**, *176*, 408.
15. Boury, F., Ivanova, T. Z. Panaïotov, I., Prost, J. E., Bois, A. and Richou, J. *J. Colloid Interface Sci.* **1995**, *169*, 380.
16. Jiang, Q. and Chiew, Y. C. *Macromolecules* **1994**, *27*, 32.
17. Vilanove, R. and Rondelez. F., *Phys. Rev. Lett.* **1980**, *45*, 1502.
18. Vilanove, R., Poupinet, D. and Rondelez, F. *Macromolecules* **1988**, *21*, 2880.
19. Kawaguchi, M., Komatsu, S., Matsuzumi, M. and Takahashi, A. *J. Colloid Interface Sci.* **1984**, *102*, 356.
20. Takahashi, A., Yoshida, A. and Kawaguchi, M. *Macromolecules* **1982**, *15*, 1196.
21. Regismond, S. T. A., Gracie, K. D., Winnik, F. M. and Goddard, E. D. *Langmuir* **1997**, *13*, 5558.
22. Thuresson, K., Söderman, O., Hansson, P., Wang, G. *J. Phys. Chem.* **1996**, *100*, 4909.
23. Thuresson, K., Lindman, B. *J. Phys. Chem. B* **1997**, *101*, 6460.
24. Lucassen, J. and van den Tempel, M. *Chem. Eng. Sci.* **1972**, *27*, 1283.
25. Lucassen-Reynders, E. H. *Surfactant Science Series* **1991**, *11*, 173.
26. Myrvold, R., Hansen, F. K., Balinov, B. and Skurtveit, R. *J. Colloid Interface Sci.* **1999**, *215*, in press.
27. Edwards, D. A., Brenner, H. and Wasan, D. T. *Interfacial Transport Processes and Rheology*; Butterworths-Heinemann Publishers, 1991.

Chapter 19

How Much Surfactant Binds to an Associating Polymer? The HMHEC/SDS Case Revisited

Lennart Piculell, Susanne Nilsson, Jesper Sjöström, and Krister Thuresson

Physical Chemistry 1, Department of Chemistry and Chemical Engineering, Lund University, Box 124, S–221 00 Lund, Sweden

The binding of sodium dodecyl sulfate (SDS) to hydroxyethyl cellulose (HEC) or hydrophobically modified HEC (HMHEC) is reexamined. Data are presented on solution viscosities and the SDS self-diffusion for mixed semi-dilute solutions, and on the equilibrium swelling of covalently crosslinked HEC or HMHEC gels immersed in SDS solutions. All techniques reveal a cooperative binding of SDS to unmodified HEC, commencing at a critical aggregation concentration (cac) of 6 mM SDS. In contrast, the binding to HMHEC starts already at very low SDS concentrations, but the binding initially remains low until the free SDS concentration approaches the cac, where a strong cooperative binding sets in. The results are interpreted in terms of a mixed micellization of the surfactant molecules and the HMHEC hydrophobes. A classification of the mixed micellar aggregates in three categories (polymer-dominated, transitional, or surfactant-dominated) is found to be useful. The classical large viscosity increase in mixed solutions occurs in the transitional region, where the number of mixed micelles is nearly unchanged, but their aggregation number increases due to added SDS. The viscosity decrease occurs in the surfactant-dominated regime, where there is a strong increase in the number of mixed micelles.

A hydrophobically associating polymer (HAP) is a water-soluble polymer modified by hydrophobic units. These units, the HAP hydrophobes, may be situated at the ends of the polymer or as side-chains grafted, more or less randomly, onto the backbone. Surfactants typically have striking effects on aqueous HAP solutions. A progressive increase in the surfactant concentration may give rise to a pronounced viscosity maximum in semidilute solutions *(1-10)* and/or a phase separation followed by a redissolution *(11-18)*. Recently, it has also been shown that the equilibrium swelling of covalently crosslinked HAP gels immersed in an aqueous solution can be strongly affected by the surfactant content in the surrounding solution *(19,20)*.

It is now well established that the effects of the added surfactant may be attributed to the formation of *mixed micelles* between the surfactants and the HAP hydrophobes. In fact, pure HAP micelles, formed by the self-association of HAP

© 2000 American Chemical Society

hydrophobes, typically exist also in the absence of added surfactant *(8,15,16,21)*. These micelles can incorporate surfactant molecules already at vanishingly low surfactant concentrations *(8,15,16,21,22)*. However, as was recently pointed out *(16,19,23,24)*, the observation of an immediate binding should not lead to the erroneous conclusion that *all*, or even most of, the added surfactant molecules are incorporated in the mixed micelles. Hence the question formulated in the heading of this chapter.

Figure 1 gives a schematic presentation of the different species present in HAP/surfactant mixtures, i.e.,

- mixed micellar aggregates (with a variable composition),
- "free" surfactant monomers, and
- pure surfactant micelles (at a sufficiently large excess of surfactant).

The relation between the concentration of free surfactant monomers, c_f, and the mixed micellar stoichiometry, expressed as the ratio c_b/c_h of bound surfactants to HAP hydrophobes, is given by the surfactant binding isotherm. Recently *(16,19,23,24)*, it was pointed out that the overall features of this isotherm should be similar to those for the binding of a surfactant to a solution of another micellar surfactant. A highly idealized, but nevertheless instructive, reference case is thus given by the pseudo-phase separation description of surfactant micellization *(25)*. In the simplest case of ideal mixing within the micellar "phase" the corresponding binding isotherm is given by the relation

$$c_f = x_b cmc = [c_b/(c_b+c_h)] \cdot cmc \tag{1}$$

where x_b is the molar fraction of "bound" surfactant in the mixed micelle, and cmc is the critical micelle concentration of the pure surfactant. The ideal binding isotherm is reproduced in Figure 1. The real binding isotherm for a surfactant to an HAP shows the same features *(22,23)*: there is a gradual binding at low c_f, and a very cooperative binding close to the cmc, where free micelles would form in the absence of the HAP.

There is also a more intuitive way of interpreting eq 1, by realizing that it is analogous to the familiar Raoult's law. The latter gives the partial pressure of a component A over an ideal mixed liquid phase:

$$P_A = x_A P_A^\circ \tag{2}$$

Here x_A is the molar fraction of A in the liquid, and P_A° is the vapor pressure of A over pure A. In this analogy, the micelle is viewed as a mixed liquid phase in equilibrium with a surrounding "gas phase" of monomeric surfactants. A surfactant with a high cmc may thus be thought of as a "volatile" surfactant; therefore, the free concentration of such a surfactant in equilibrium with a mixed micelle is generally comparatively high. We note that, like Raoult's law, eq 1 becomes increasingly accurate for non-ideal mixtures as x_b approaches unity.

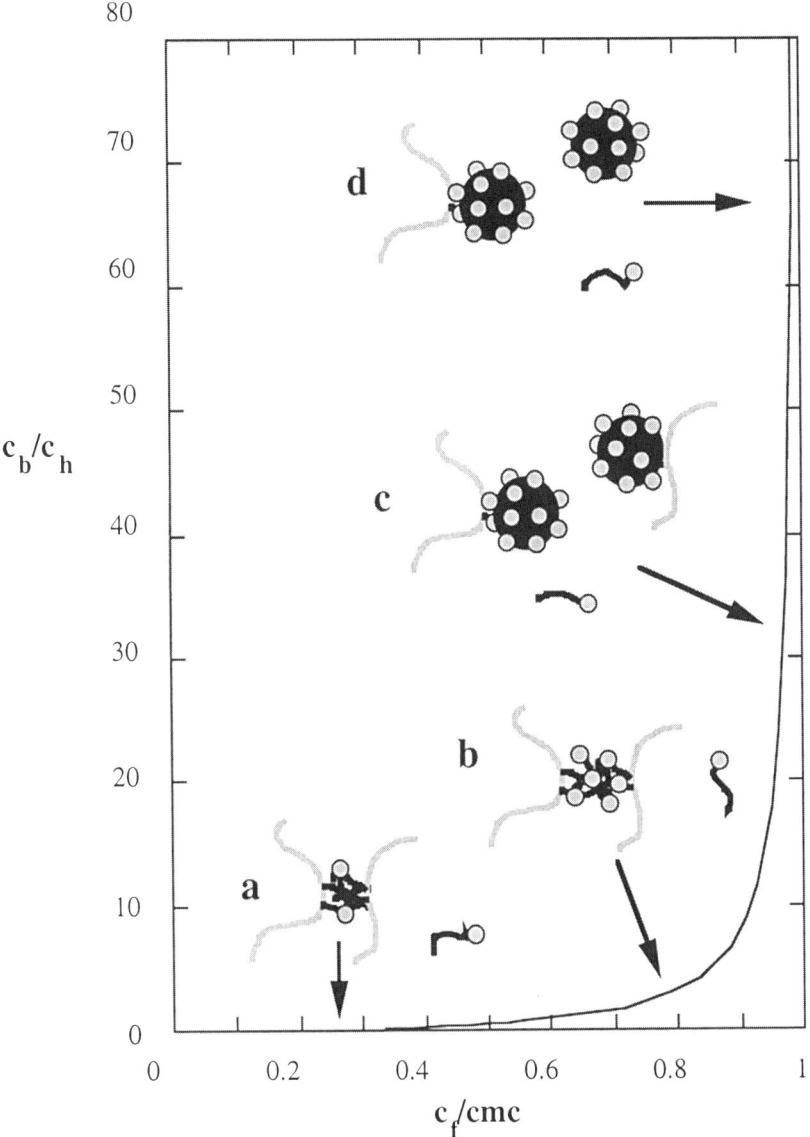

Figure 1. Schematic binding isotherm (based on the ideal mixing model) and mixed micellar compositions (cartoons) for mixtures of an HAP with a surfactant. Free (monomeric) surfactant is present at all finite surfactant concentrations. The aggregates present are a) HAP-dominated mixed micelles, b) transitional mixed micelles, c) surfactant-dominated mixed micelles, and d) a mixture of surfactant-dominated and pure surfactant micelles. Hydrophilic regions of the components are in grey, hydrophobic in black.

At this point, we should consider the consequences of the fact that a mixed HAP/surfactant micelle is inherently non-ideal. For instance, a possible blockiness in the substitution pattern of the HAP could prevent the formation of mixed micelles containing only one HAP hydrophobe. Indeed, for finite mixed micelles, we must, in the limit $x_b \rightarrow 1$, have a mixture of pure surfactant micelles and mixed micelles with a variable, but finite hydrophobe content. Nevertheless, the mixed micellar analogy, and eq 1, is very valuable for qualitative or even semi-quantitative considerations. Thus, when the free surfactant concentration is approaching the cmc, then the mixed micelle should be surfactant-dominated (*cf.* Figure 1), and *vice versa*. Moreover, if c_h « cmc (as is often the case), it is clear from Figure 1 that c_f will dominate over c_b in the early part of the binding isotherm; i.e., only a minor fraction of the surfactant is actually bound to the mixed micelles.

Useful as it is, however, the pseudo-phase separation model says nothing about the aggregation numbers of the mixed micellar aggregates. For an HAP/surfactant micelle, the average aggregation number, $N_{tot} = N_s + N_h$ (N_s and N_h are the average numbers of surfactant molecules and hydrophobe chains, respectively, in the mixed micelle), is generally expected to vary with the micellar composition, as indicated in the cartoons in Figure 1. For ionic surfactants, N_{tot} also varies with the concentration of added salt. The aggregation number has important consequences for the mixed micellar junctions connecting HAP, since the functionality of the junctions should depend on $N_h = c_h/c_{mic}$, where c_{mic} is the total concentration of micelles *(5)*. Moreover, at sufficiently large concentrations of added surfactant, c_{mic} must eventually become larger than c_h. At this point, some of the micelles must necessarily be totally free from HAP hydrophobes.

In the present chapter, we wish to illustrate the above points by revisiting the classical *(1-3,11,15,21)* mixture of hydrophobically modified hydroxyethyl cellulose (HMHEC) with sodium dodecyl sulfate (SDS). HMHEC is a graft-modified polymer *(26)*, but the qualitative conclusions on surfactant binding should be valid also for end-modified HAP architectures. In our laboratory, we have recently studied the HMHEC/SDS system by viscometry and time-resolved fluorescence quenching methods *(5)*. Here we complement these data with new surfactant self-diffusion measurements, which enable us to obtain the surfactant binding isotherm. By combining the previous and the new data we can correlate the observed macroscopic viscosity of HMHEC/SDS mixtures with detailed microscopic information on the stoichiometries and aggregation numbers of the mixed micellar aggregates. We, furthermore, make comparisons with solutions of non-modified HEC, and find that also in this case, the binding of SDS is quite significant. Although evidence of an SDS-HEC association has been presented previously *(11,15,21)*, the interaction has escaped notice in a number of other investigations *(2,3,13,17,18,27,28)*.

In addition to the solution data, we present results on the equilibrium swelling, in surfactant solutions, of gels of chemically cross-linked HEC or HMHEC. These experiments demonstrate additional (to viscosity and self-diffusion) consequences of surfactant binding to an HAP and, as we will show, their interpretation is intimately related to the generic features of the surfactant-HAP binding isotherm.

Experimental

Materials

HEC and HMHEC with the commercial names Natrosol 250 GR and Natrosol Plus grade 330, respectively, were obtained from Aqualon. According to the manufacturer, both samples have a molecular mass of *ca.* 250,000. The degree of substitution of hydroxyethyl groups per repeating anhydroglucose unit of the polymer is 2.5 for HEC and 3.3 for HMHEC. In addition, HMHEC contains grafted C_{16} alkyl hydrophobes amounting to 1.2% (w/w) of the (total) dry sample mass according to a previous investigation *(5)*. This corresponds to 0.54 mM alkyl chains in a 10 g/l aqueous polysaccharide solution. With the hydroxyethyl substitution degree as given by the manufacturer, 1.7 mol% of the anhydroglucose units in HMHEC contain alkyl chains. Sodium dodecyl sulfate (SDS), (specially pure) from BDH was used as supplied. NaOH (analytical grade from Eka Nobel) and divinyl sulfone (DVS) from Sigma were used in the crosslinking of the polysaccharides. D_2O from Dr. Glaser AG (99.8% pure) was used as a solvent for the NMR measurements.

Sample Preparation

The HEC and HMHEC chemicals used for solution studies were purified as follows. HMHEC was first extensively washed with acetone to remove unreacted alkyl chains and then dried. HEC or acetone washed HMHEC was dissolved in water to a concentration of 10 g/l. The HMHEC solution was centrifuged at 10000 g to remove particulate impurities such as unreacted cellulose. Both samples were dialyzed against Millipore water in a Filtron Ultrasette device until the conductivity of the expelled water was < 2 µS/cm. After freeze drying, the polymers were stored in desiccators. Solution samples at a polymer concentration of 10 g/l were prepared from stock solutions by weight. Prior to measurements, the samples were mixed by a magnetic stirrer for at least 1 day. Solutions for NMR measurements were made in D_2O. The HEC samples investigated by viscometry were the same as those investigated by NMR.

Viscometry

Viscosity measurements on HMHEC solutions were performed at 25°C by oscillatory measurements on a Carri-Med controlled stress CSL 100 rheometer, as reported previously *(5)*. The complex viscosity, η*, was calculated through

$$\eta^* = [(G'^2 + G''^2)/(2\pi f)^2]^{1/2} \tag{3}$$

where G' is the storage modulus, G" the loss modulus and f the frequency. Viscosities of HEC solutions were determined in the constant shear mode at 25°C with a Physica UDS 200 rheometer equipped with a 1° cone and plate geometry (7 cm diameter). The shear rate range was 0.01 - 2000 s^{-1}, and the reported viscosity values, η, correspond to the Newtonian limit.

NMR Measurements

The NMR measurements were performed on a Bruker DMX200 spectrometer equipped with a Bruker field gradient probe. All experiments were carried out at 24.1 °C in 5 mm NMR tubes. Before measurement the samples were equilibrated in the probe for at least 10 minutes. The self-diffusion coefficients of SDS were measured by the pulsed field gradient (PFG) method with an ordinary spin-echo (90_x-180_y) sequence *(29,30)*. The gradient pulse length (δ) was kept constant at 1.4 ms while the gradient strength (g) was gradually increased to a maximum value between 0.7 and 4.8 Tm^{-1}, depending on the value of the diffusion coefficient measured. In all cases the surfactant signal intensity showed a single exponential decay when plotted against the variable $(\gamma g \delta)^2 (\Delta - \delta/3)$, in accordance with the Stejskal-Tanner relation

$$I = I_0 \exp[-(\gamma g \delta)^2 (\Delta - \delta/3) D] \tag{4}$$

Here γ is the proton magnetogyric ratio, Δ the time (here 70 ms) between the leading edges of the gradient pulses, and D the self-diffusion coefficient. The data were evaluated by a Levenberg-Marquat fitting procedure used on the Matlab package on Macintosh computers.

Gel Swelling Experiments

Chemical gels were made by crosslinking HEC or HMHEC (without prior purification) with DVS in 20 mM alkaline solution *(20,28,31)*. The concentration of polysaccharide at synthesis, c_0, was 10 or 20 g/l and the amount of DVS was varied in the range 0.1 - 0.8 μl/g polysaccharide. DVS was added during stirring and the crosslinking reaction was allowed to proceed for about 24 hours in a water bath at 50°C in glass tubes with an inner diameter D_0 = 1.4 mm. The gels were cut into approximately 1.4 mm long rods. Residual chemicals and low-molecular impurities were leached out by immersing the gel rods in a large excess of deionized water followed by Millipore water during 5 days in total.

The washed gel rods were immersed in vials each containing 1 gel rod and 5 ml aqueous surfactant solution. Owing to the much larger volume of the swelling medium compared to the volume of the gel rods, the decrease of the surfactant concentration in the swelling medium due to absorption into the gel was always negligible. The gel rods were allowed to equilibrate for at least 3 days at room temperature to reach the equilibrium degree of swelling. The HMHEC gel with the highest degree of swelling (*cf.* below) was measured also after 30 days, and showed no significant additional swelling.

The equilibrium swelling is given as V/m_0, where V is the equilibrium volume of a gel piece prepared by crosslinking a mass m_0 of polysaccharide. V/m_0 was calculated as $(D/D_0)^3/c_0$, where D is the measured gel diameter at equilibrium swelling. Both the diameters of the gels, D, and the inner diameters of the glass tubes, D_0, were measured in the vials by a video camera calibrated with a 0.1 mm scale with the help of an image computer program. Reproducibility measurements on a large number of gel rods at different positions in the vials gave a typical variation in D of ca. 5%, corresponding to a 15% uncertainty in the gel volume.

Results and Discussion

Viscosity and Self-Diffusion

Figure 2 illustrates the classical variation of the viscosity of semi-dilute solutions of HMHEC at different total concentrations of added anionic surfactant *(1-3,5,21)*: Addition of surfactant leads, *via* a pronounced viscosity increase, to a maximum followed by a decrease and, finally, a levelling off at values considerably lower than that for HMHEC alone. In the same figure, we also show the concentration of micellar aggregates in the system, measured by the time-resolved fluorescence quenching technique (*cf.* the original reference *(5)* for experimental details). Note that pure HMHEC micelles, with $N_h \approx 5$ (*cf.* below), exist even before the surfactant is added. As expected, the concentration of micelles increases with increasing SDS concentration, but the increase is slow at low surfactant concentrations. In addition to the HMHEC data, which are all taken from a recently published study in our laboratory *(5)*, Figure 2 shows new data on the viscosity of non-modified HEC in SDS solutions. These data demonstrate that HEC solutions also display a significant - but small - variation of the viscosity with the SDS concentration. This effect is most clearly seen in the figure inset. In addition to the huge quantitative differences in the effects of SDS on the two polymers, there is also a qualitative difference: The viscosity of HEC is only affected by SDS at surfactant concentrations above ca. 6 mM.

If Figure 2 can be viewed as an illustration of the effect of added surfactant on the polymers, effects of the polymers on the surfactant properties are shown in Figure 3. Here the SDS self-diffusion, as a function of the SDS concentration, is compared for SDS dissolved in pure D_2O and in 10 g/l solutions of HEC or HMHEC in D_2O, respectively. The result for the surfactant in pure heavy water is again classical *(30,32)*: The break point at 8 mM is due to the formation of micelles, that have a much slower diffusion constant than the monomeric surfactant.

Both HMHEC and non-modified HEC slow down the diffusion of SDS, giving evidence of a polymer-surfactant association. The effect of HEC on the SDS diffusion is insignificant at low surfactant concentrations, implying that monomeric SDS is unaffected by HEC. Above ca. 6 mM of SDS, however, the SDS diffusion is substantially reduced in the HEC solution, compared to aqueous SDS alone. Note that the onset of the effect occurs at the same total SDS concentration in the self-diffusion experiments as in the viscosity data. Clearly, the cmc is lowered in presence of the polymer. A similar slight reduction of the cmc has previously been noted by fluorescence for SDS in the presence of HEC *(15,21)*. Clear evidence of an HEC-SDS association at SDS concentrations close to 6 mM was also presented in the early

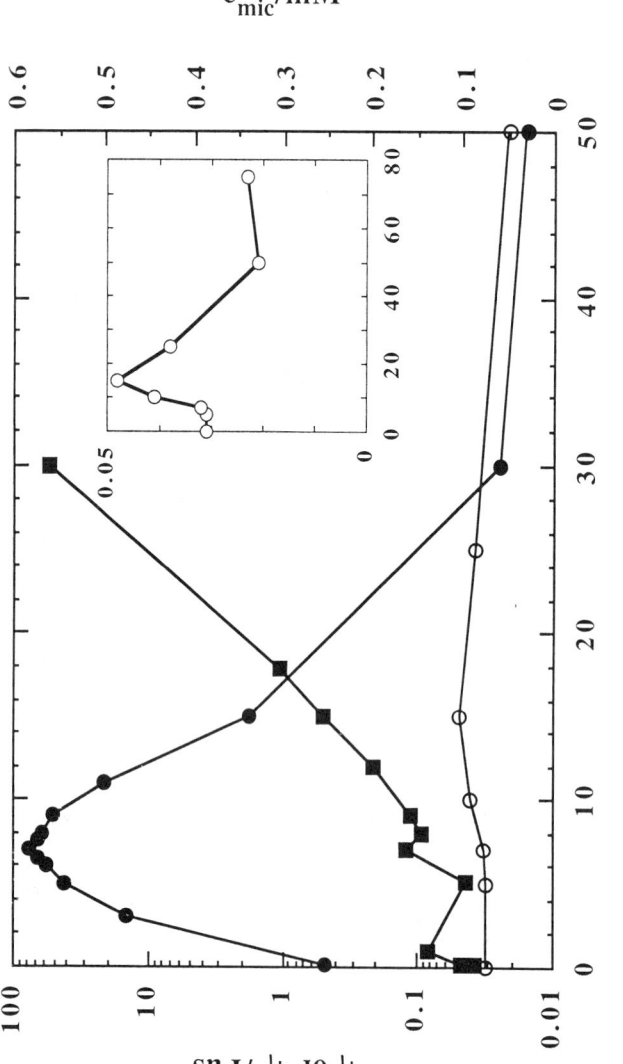

Figure 2. Viscosities of HMHEC (●) or HEC (○) solutions (lefthand scale), and concentration of micellar aggregates in HMHEC solutions (■, righthand scale) at varying total concentrations of SDS. All data refer to 25°C and a polysaccharide concentration of 10 g/l. The inset shows the viscosity data for HEC on a linear scale. All data for HMHEC are taken from ref. (5).

viscometric study on dilute polysaccharide solutions by Goddard and Hannan *(11)*. Many other "slightly hydrophobic" *(24)* water-soluble polymers have a similar effect on ionic surfactants *(33)*. The lower critical micellization concentration in a polymer solution is commonly referred to as the critical aggregation concentration, cac, of the surfactant in the presence of the polymer.

Turning to HMHEC, the largest difference between the HMHEC and the HEC results is that there is a strong reduction of the SDS diffusion in HMHEC even at the lowest SDS concentrations. At sufficiently high SDS concentrations, however, the data for HMHEC and HEC seem to merge.

The observed diffusion coefficient of a surfactant may be interpreted as a population weighted average over the diffusion coefficients of free and bound (micellized) surfactant molecules *(30)*:

$$D_{obs} = p_f D_f + p_b D_b \tag{5a}$$

Here p_f and p_b denote the fractions of free and bound surfactant molecules, respectively. For polymer-bound micelles, henceforth called mixed micelles, D_b is usually assumed to be given by the diffusion of the polymer. Typically, the polymer diffusion is sufficiently slow in this type of system (this was confirmed in our case by measurements on HMHEC) so that D_b may be neglected *(22)*. This leads to the simple expression

$$D_{obs} = p_f D_f = (c_f/c_{tot}) D_f \tag{5b}$$

where c_{tot} is the total surfactant concentration. This approximation should hold as long as no pure SDS micelles are formed; *cf.* Figure 1. For a simple polymer-induced micellization, one expects that c_f = cac at $c_{tot} \geq$ cac. Then, according eq 5b, D_{obs} should be proportional c_{tot}^{-1} above the cac. The straight line in Figure 3 shows a fit of this relation to the HEC data. Indeed, the fit is good, at least at the lower surfactant concentrations. From the same analysis, cac is obtained as the concentration where the fitted line intercepts the constant value of $D_{obs} = D_f$ at low surfactant concentrations. We thus obtain cac = 6.0 mM for SDS in HEC solutions.

Surfactant Binding

With the help of eq 5b and a value of $D_f = 4.21 \cdot 10^{-10}$ m^{2}s^{-1} (the average value measured for SDS in 10 g/l HEC at the two concentrations below cac) we calculated the bound and free surfactant concentrations in the polymer solutions as functions of c_{tot}. The results are shown in Figure 4. As expected, the main difference between HEC and HMHEC is seen at low SDS concentrations: For HMHEC, there is a significant surfactant binding in mixed micelles also below 6 mM. Note, however, that most of the added surfactant is actually in the monomeric state until the total surfactant concentration is raised above 10 mM. As discussed above, this is a consequence both of the nature of the surfactant binding and of the low concentration of HMHEC hydrophobes (c_h = 0.54 mM) in the system. A comparison of the results

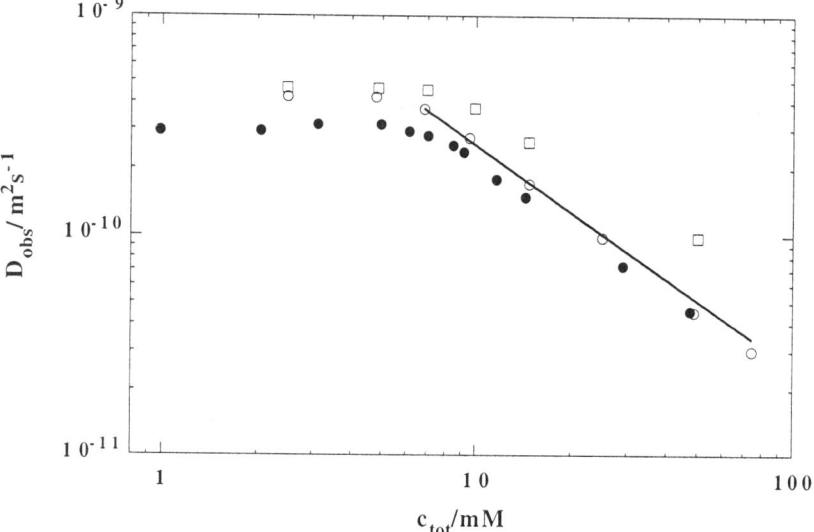

Figure 3. Self diffusion coefficients for SDS at 24.1°C in pure D2O (□) and in 10 g/l solutions of HEC (O) or HMHEC (●) in D_2O at varying total concentrations of SDS. The solid line is a fit of the HEC data for c_{tot} > 6 mM to eq 5b.

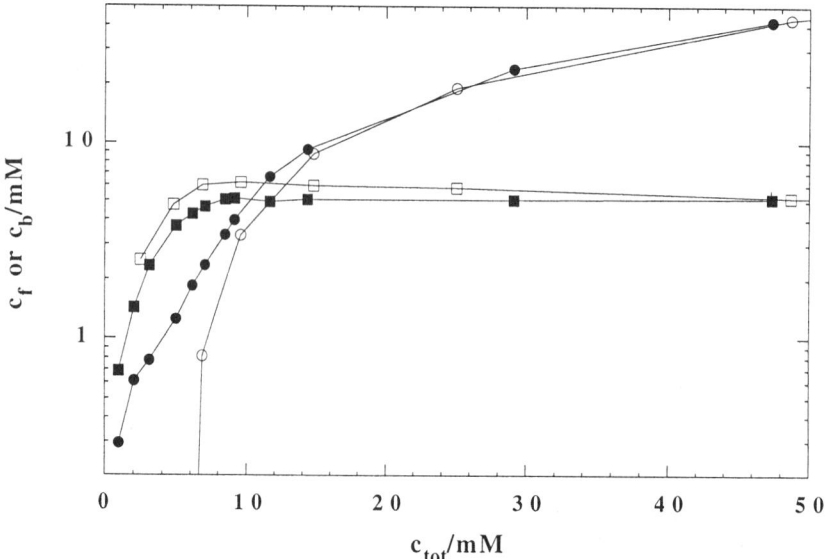

Figure 4. Concentrations of free (squares) and bound (circles) SDS in 10 g/l solutions of HEC (unfilled symbols) or HMHEC (filled symbols) in D_2O at varying total concentrations of SDS, obtained from the data in Figure 3 as described in the text.

from Figures 4 and 2 shows, for instance, that only 1/3 of the added SDS is bound in the mixed micelles at the viscosity maximum of the SDS/HMHEC mixtures, which occurs at c_{tot} = 7 mM.

In Figure 5, we have used the data from Figure 4 to construct a binding isotherm for SDS to HMHEC, assuming that the fraction of non-micellized HMHEC hydrophobes is negligible. For comparison, we have also plotted the corresponding ideal binding isotherm. In constructing the latter isotherm, we have assumed that cmc = cac for the surfactant in the solution of non-modified HEC. This is the relevant cmc in the present case, since it is the concentration where micelles would start to form in the same system, but in the absence of the HAP hydrophobes. The general trends of the two binding isotherms are clearly similar, as we anticipated. However, the experimental binding is always significantly larger than that predicted from the simple model, both at low and (especially) at high surfactant concentrations.

The positive deviation at low c_f is expected for an ionic surfactant. Since there is no electrostatic repulsion between a pure HAP micelle and an SDS molecule, the binding should initially be much stronger than the model predicton based on the cmc of SDS. From this we would expect a merging of the curves at higher c_f, as the mixed micelles become progressively more charged and surfactant-rich. Such a tendency is actually seen as c_f is increased up to ca. 5 mM.

At c_f > 5 mM there is a pronounced deviation between the experimental and the ideal binding isotherms. Two effects could contribute to this discrepancy. The first is the general effect that, for electrostatic reasons, c_f actually decreases with increasing surfactant concentration for an ionic surfactant (34). Experimental data (35) as well as theoretical calculations (34) for SDS in pure water indicate that the decrease is substantial already at $c_{tot} \approx 2 \times$ cmc. This means that cmc is not the true limiting value for c_f of an ionic surfactant at high degrees of binding to a polymer. Indeed, the data on c_f versus c_{tot} for HEC (Figure 4) displays a significantly decreasing trend, and at high surfactant concentrations they approach the value ($c_f \approx$ 5 mM) found for HMHEC. The other possible effect is that the value of cac for the HMHEC sample could be lower than 6 mM, which is the value we have assumed in Figure 5. A possible reason for a lower cac could be that the HMHEC sample has a higher degree of hydroxyethyl substitution than the HEC sample (cf. the experimental section).

Relation Between Microscopic Structure and Viscosity

We can now combine the data from all techniques to show how the microscopic (average aggregation numbers) and macroscopic (viscosity) properties of the SDS/HMHEC system vary with the hydrophobe stoichiometry, c_b/c_h, of the mixed micelles. This is done in Figure 6. We believe that the hydrophobe stoichiometry is the relevant variable for the viscosity in SDS/HMHEC mixtures, in view of the comparatively minor viscosifying effect of SDS on non-modified HEC. The microscopic parameters describing the micelles in Figure 6 are the average numbers of surfactant molecules, N_s, and HMHEC hydrophobes, N_h, in a mixed micelle. These numbers were obtained from c_{mic} data combined with c_b and c_h, respectively.

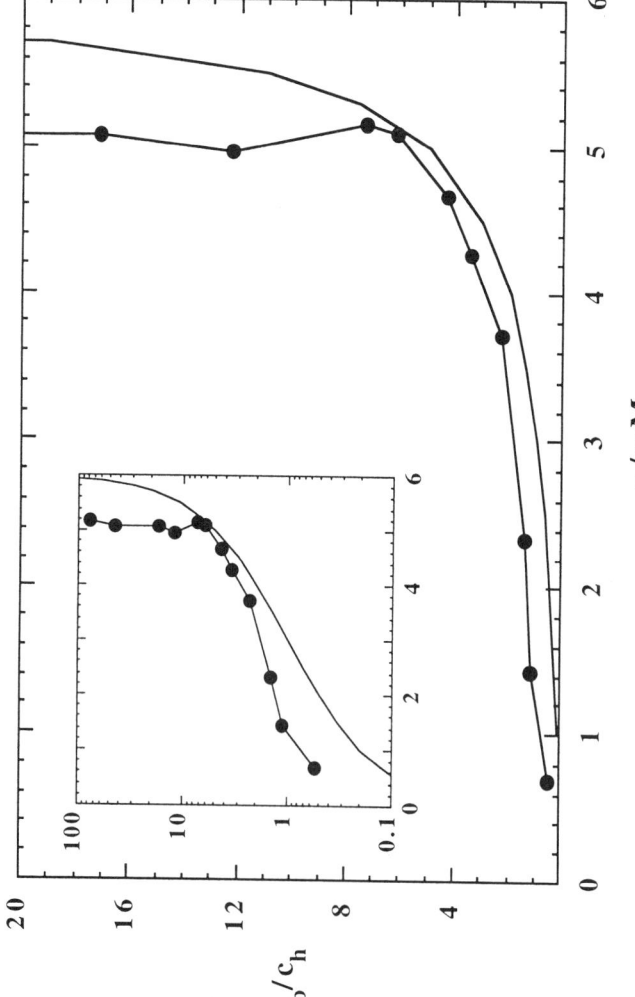

Figure 5. Ratio of bound SDS to HMHEC hydrophobes in the mixed micelles as a function of the free SDS concentration in 10 g/l solutions of HMHEC, obtained from the data in Figure 3 as described in the text. The inset shows the micellar compositions on a logarithmic scale.

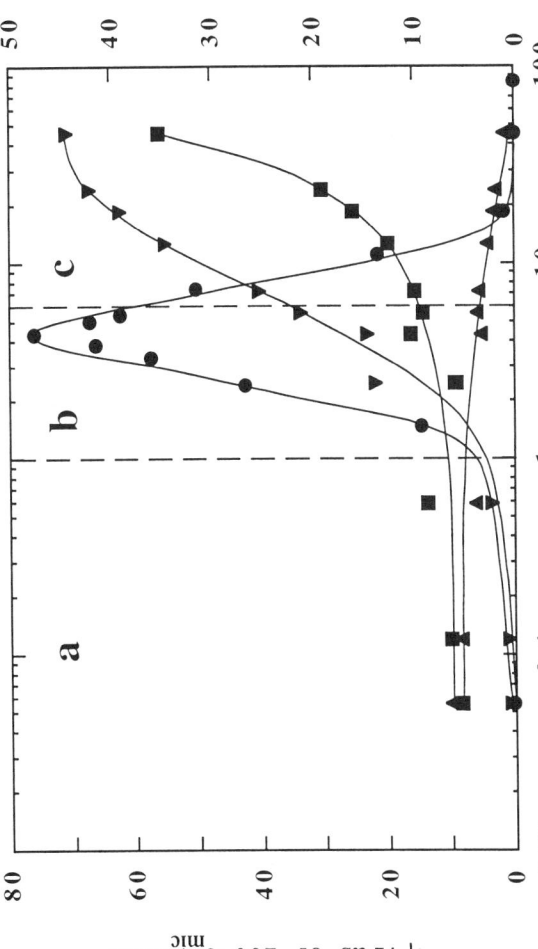

Figure 6. Viscosity (●, left scale), micelle concentration (■, left scale), and numbers of SDS molecules (▼, right scale) or HMHEC hydrophobes (▲, right scale) per micelle in aqueous SDS solutions containing 10 g/l HMHEC plotted against the mixed micellar stoichiometry. Solid lines are guides to the eye only. Regions separated by dashed lines indicate HAP dominated (a), transitional (b), and surfactant dominated mixed micelles (c).

Thus, in calculating N_h, we assume as above that essentially all HMHEC hydrophobes are in the micelles.

Referring to Figure 1, we will discuss the data in Figure 6 in terms of the nature of the mixed micelles, and we distinguish between the *HAP-dominated, transitional,* and *surfactant-dominated* composition regimes. There are no sharp boundaries between these regimes. Here we have, somewhat arbitrarily, chosen the average total aggregation number of the mixed micelle as our criterion. The HAP- or surfactant-dominated regimes are obviously at the left and right sides of Figure 6, respectively, and the transitional regime is considered to be entered when N_{tot} deviates by more than a factor of 2 from its value at either extreme of the mixing region. As seen in Figure 6, the transitional regime then corresponds to a region where the mixed micellar stoichiometry lies in the interval $1 < c_b/c_h < 6$.

In the HAP-dominated regime, the concentration of mixed micellar aggregates and, consequently, N_h stay roughly constant. We note that the (average) number of HMHEC hydrophobes in an aggregate is quite small; *ca.* 5. In this region, the bound surfactant molecules are essentially just added to the aggregates already present, the micelles remain small, and the system properties do not change very much.

A more pronounced micellar growth, by added surfactant, is the main microscopic characteristic of the transition region. Here N_s shoots up, while the number of micelles and N_h still remain nearly constant. Macroscopically, this change in composition of the mixed micelles has dramatic consequences: This is the region where the entire (huge) viscosity increase occurs. As previously *(5)*, we interpret the origin of this effect mainly in terms of an increase in the lifetimes of the mixed micellar junctions, since the cross-link functionality of the junctions (given by N_h) hardly changes. Our measurements indicate that the viscosity maximum occurs at $c_b/c_h \approx 4$. This stoichiometry at the maximum is quite close to the value 3 that we previously estimated *(19)*, rather more indirectly, from experimental data *(3)* for HMHEC in mixtures with a range of different surfactants. We should also note that the present value is based on a more accurate knowledge of c_h than was available for the calculations in ref. *(19)*.

In the surfactant-dominated regime, finally, the increase in the aggregation number slows down and eventually levels off at a value similar to (but slightly lower than) that of a pure SDS micelle. In contrast to the situation at lower surfactant contents, the increased concentration of bound surfactant here gives rise to not only an increase in N_s but also an increasing number of micelles. The latter results in a decreasing connectivity, which shows up macroscopically as a decrease in viscosity. The viscosity levels off when N_h has decreased below *ca.* 2, which seems quite reasonable. At this point, N_s also levels off, indicating that further addition of surfactant leads almost exclusively to the formation of more micelles, rather than a micellar growth. This marks the onset of a fourth regime where micelles free from HAP hydrophobes start to form. We may note, however, that the first "hydrophobe-free" micelles that form should not be pure SDS micelles, but rather mixed micelles incorporating hydrophobe-free stretches of the HEC chain; *cf.* the discussion above. This may be the reason that N_s here has a lower value than 60-70, as found for pure SDS micelles *(36-38)*.

Gel Swelling Experiments

Owing to their surfactant binding properties, water-swollen chemical gels based on cross-linked HAP might be useful in absorbent, drug delivery or water purification applications. Typically, a surfactant-binding gel responds to the surfactant binding by changing its equilibrium volume (swelling or shrinking). The gel swelling experiment is thus appealing as a conceptually simple method to study polymer-surfactant interactions, and this is the aspect that we will emphasize here.

Figure 7 shows the equilibrium swelling of gels of covalently crosslinked HEC or HMHEC immersed in solutions of SDS at varying concentrations. Two different HMHEC gels were investigated, differing in the crosslink density (as given by the ratio of crosslinker to polysaccharide) and in the polysaccharide concentration used in the crosslinking reaction. The abscissa gives the concentration of surfactant in the external swelling medium, c_{ex}, and the ordinate gives the equilibrium volume of a gel divided by its initial mass content of polysaccharide at synthesis. Providing that negligible amounts of possible unreacted polysaccharide leached out in the gel washing step (*cf.* the experimental section), the quantity V/m_0 thus corresponds to the inverse of the weight/volume concentration of polysaccharide in a gel at equilibrium swelling. Note that the polysaccharide concentration range in the swollen gels (6 - 17 g/l) is quite comparable to the concentration (10 g/l) used in the solution studies above.

The surfactant response is strikingly similar for the HMHEC and HEC gels, and it is also similar to that observed previously for other gels from slightly hydrophobic polymers *(19,28,39,40)*: Around the cac of the surfactant there is a sharp volume increase, followed by a slow decrease at c_{ex} slightly above the cmc. The gel swelling experiment thus confirms the association between HEC and SDS. Interestingly, the small difference between the cmc and the cac for SDS binding to HEC leads to a quite sharp volume increase in a narrow SDS concentration interval; gels based on polymers giving a lower cac show a correspondingly more gradual transition *(19,28,39,40)*. As has been shown previously *(20,28,40)*, it is possible to shift the swelling maximum by adding salt, which leads to shifts in both cmc and cac.

In a previous study in our laboratory, we failed to detect significant volume changes for a HEC gel when immersed in SDS solutions *(28)*. However, the previously studied gel had a higher degree of crosslinking than the present one, and this could explain the discrepancy. The results for the two different HMHEC gels in Figure 7 clearly demonstrate that the degree of crosslinking has an effect on the extent of the gel swelling (but not on the characteristic SDS concentrations at the swelling extrema). We tried to make HEC gels as lightly crosslinked as the least crosslinked HMHEC gel in Figure 7, but those HEC gels turned out to be too fragile to handle.

The results for the hydrophobically modified gels show that it is only in the highly cooperative part of the binding isotherm, corresponding to surfactant-dominated mixed micelles, that the surfactant binding gives rise to a gel swelling. Similar results were obtained previously for gels of hydrophobically modified EHEC *(19)*. It is also notable that the sharp increase in swelling starts at c_{ex} = 6 mM rather than at 5 mM, as might have been expected on the basis of the binding isotherm in Figure 5. The reason for this discrepancy is unclear, but it is possible that the monomer concentration is actually lower inside the gel, due to Donnan effects.

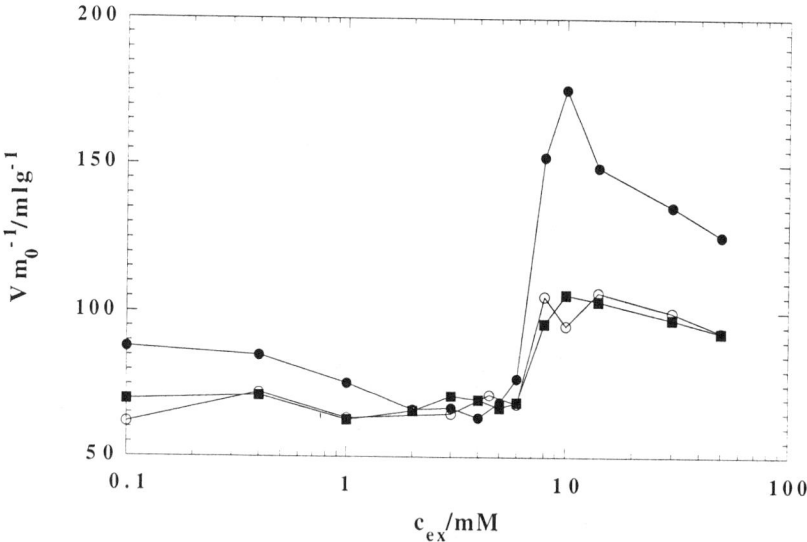

Figure 7. Equilibrium volume per intitial mass of polysaccharide in crosslinked gels of HMHEC or HEC as functions of the SDS concentration in the external aqueous SDS solution. The crosslinking conditions were 20 g/l HMHEC with 0.1 ml DVS/mg polysaccharide (●), 10 g/l HMHEC with 0.8 ml DVS/mg polysaccharide ■), and 20 g/l HEC with 0.3 ml DVS/mg polysaccharide (○).

The absence of a swelling of the hydrophobically modified gels at low c_{ex} might seem to contradict the clear evidence (from both viscometry and NMR) of a continuous surfactant binding to HMHEC at free surfactant concentrations far below the cac. However, the extent of ionic swelling is expected to be low in this part of the binding isotherm, since the degree of binding is low. (*Cf.* the dominance of the free surfactant fraction at low surfactant concentrations in solutions with a similar polysaccharide content; Figure 4.) Moreover, for the most lightly crosslinked HMHEC gel a small, but significant, *shrinking* occurs with increasing c_{ex} below 6 mM, indicating that there is indeed an interaction. This points to the presence of mechanism(s) which counteract the expected ionic swelling at low degrees of surfactant binding, such as a more efficient mixed micellar crosslinking.

The swelling of HAP gels in ionic surfactant solutions thus reflects an interplay between hydrophobic crosslinking, ionic swelling, and electrostatic screening. This interplay was also clearly demonstrated in recent experiments *(20)* on gels made from cat-HMHEC, a HEC modified by cationic hydrophobes. The cat-HMHEC gels are salt-sensitive, and collapse already at low levels of monovalent salt (*ca.* 15 mM). A similar collapse of cat-HMHEC gels with increasing c_{ex} was obtained for added cationic surfactant, if the surfactant had a cmc above 15 mM, and if c_{ex} was kept below cmc. This was explained *(20)* by the same mechanism as we invoke here, i.e., far below the surfactant cmc, most of the added surfactant ended up as a screening electrolyte (contributing to the deswelling) and only a small fraction was bound in the mixed micelles. However, when the surfactant cmc was approached, a sharp reswelling occurred, similar to that shown in Figure 7, as a result of the cooperative binding.

Conclusions

We have here presented new data on the interactions of HEC and hydrophobically modified HEC with SDS, in solutions and in chemically crosslinked gels. From the data on the self-diffusion of SDS in solutions, binding isotherms could be constructed. These binding data show, rather surprisingly, that there is a substantial binding of SDS also to non-modified HEC. The binding isotherm for SDS to HMHEC shows the expected features, but there are clear indications that the cooperative binding occurs as the cac of SDS in HEC, rather than the bulk cmc, is approached.

Careful measurements reveal that the interaction between HEC and SDS is evident also in the viscosity of semidilute HEC solutions. The viscosity effect is quite small, however, for reasons that are unclear at this stage. The explanation may be sought in either the structure or the dynamics of the mixtures. Thus, there could be a low degree of mixed micellar crosslinking (i.e., most micelles bind to only one HEC polymer), but there could also be a rapid exchange between micelle-bound and bare parts of the HEC chain. In any case, the binding to the unmodified parent HEC is not expected to contribute much to the strong viscosity enhancement found for HMHEC with SDS. It is notable that the strongest viscosity enhancement for the modified polymer occurs in a surfactant concentration region where there is no binding to non-modified HEC.

An analysis combining viscosity, fluorescence and NMR self-diffusion data indicates that the variation in viscosity reflects the microscopic nature of the mixed micelle, which may be classified as HAP-dominated, transitional, or surfactant-dominated. The large viscosity increase occurs in the transitional regime, where the mixed micelles grow substantially by added surfactant, but the number of junctions remains rather constant. At the viscosity maximum, the average mixed micelle is still fairly small, containing on the average 15 SDS molecules and 4 HEC hydrophobes. The largest viscosity decrease occurs in the surfactant-dominated regime, where the aggregation number approaches that of pure SDS micelles, and the connectivity is gradually lost.

The binding of SDS both to HEC and to HMHEC was confirmed by gel swelling experiments. For HMHEC gels, the early binding of the surfactant (below the cac), resulted in a small shrinking, rather than a swelling, of the gel.

Acknowledgements

This project was funded by the Centre of Competence for Amphiphilic Polymers from Renewable Resources at Lund University (CAP) and by the Swedish Research Council for Engineering Sciences (TFR).

References

1. Gelman, R. *TAPPI Proceedings of the International Dissolving Pulps Conference,* Geneva, 1987, p 159.
2. Dualeh, A. J.; Steiner, C. A. *Macromolecules* **1990**, *23*, 251.
3. Tanaka, R.; Meadows, J.; Williams, P. A.; Phillips, G. O. *Macromolecules* **1992**, *25*, 1304.
4. Biggs, S.; Selb, J.; Candau, F. *Langmuir* **1992**, *8,* 838.
5. Nilsson, S.; Thuresson, K.; Hansson, P.; Lindman, B. *J. Phys. Chem. B* **1998**, *102,* 7099.
6. Iliopoulos, I.; Wang, T. K.; Audebert, R. *Langmuir* **1991**, *7*, 617.
7. Sarrazin-Cartalas, A.; Iliopoulos, I.; Audebert, R.; Olsson, U. *Langmuir* **1994**, *10*, 1421.
8. Magny, B.; Iliopoulos, I.; Zana, R.; Audebert, R. *Langmuir* **1994**, *10*, 3180.
9. Annable, T.; Buscall, R.; Ettelaie, R.; Shepherd, P.; Whittlestone, D. *Langmuir* **1994**, *10*, 1060.
10. Persson, K.; Wang, G.; Olofsson, G. *J. Chem. Soc. Faraday Trans.* **1994**, *90*, 3555.
11. Goddard, E. D.; Hannan, R. B. in *Micellization, Solubilization, and Microemulsions;* Mittal, K. L., Ed.; Plenum: New York, 1977; Vol. 2, p 835.
12. Goddard, E. D.; Leung, P. S. *Langmuir,* **1992**, *8*, 1499.
13. Goddard, E. D. *J. Colloid Interface Sci.* **1992**, *152*, 578.
14. Goddard, E. D.; Leung, P. S. Colloids Surfaces, **1992**, *65*, 211.
15. Sivadasan, K.; Somasundaran, P.; *Colloids Surfaces* **1990**, *49*, 229.
16. Guillemet, F.; Piculell, L. *J. Phys. Chem.* **1995**, *99*, 9201.
17. Kästner, U.; Hoffman, H.; Dönges, R.; Ehrler, R. *Colloids Surfaces A* **1996**, *112*, 209.
18. Hoffman, H.; Kästner, U.; Dönges, R.; Ehrler, R. *Polymer Gels Networks* **1996**, *4*, 509.

19. Piculell, L.; Thuresson, K.; Ericsson, O.; *Faraday Discuss.* **1995**, *101*, 307.
20. Rosén, O.; Sjöström, J.; Piculell, L. *Langmuir* **1998**, *14*, 5795.
21. Tanaka, R.; Meadows, J.; Phillips, G. O.; Williams, P. A. *Carbohydr. Pol.* **1990**, *12*, 443.
22. Thuresson, K.; Söderman, O.; Hansson, P.; Wang, G. *J. Phys. Chem.* **1996**, *100*, 4909.
23. Piculell, L.; Guillemet, F.; Thuresson, K.; Shubin, V.; Ericsson, O. *Adv. Colloid. Int. Sci.*, **1996**, *63*, 1.
24. Linse, P.; Piculell, L.; Hansson, P. in *Polymer-Surfactant Systems;* Kwak, J. C. T., Ed.; Marcel Dekker: New York, 1998, p 193.
25. *Mixed Surfactant Systems*; Ogino, K.; Abe, M., Eds.; Marcel Dekker: New York, 1993.
26. Landoll, L.M. *J. Polym.erSci., Polymer Chem.* **1982**, *20*, 443.
27. Ohbu, K.; Hiraishi, O.; Kashiwa, I. *J. Am. Oil Chem. Soc.* **1982**, *59*, 108.
28. Rosén, O.; Piculell, L. *Polymer Gels Networks* **1997**, *5*, 185.
29. Stilbs, P. *Prog. Nucl. Magn. Reson. Spectrosc.* **1987**, *19*, 1.
30. Söderman, O.; Stilbs, P. *Prog. Nucl. Magn. Reson. Spectrosc.* **1994**, *27*, 445.
31. Kabra, B. G.; Gehrke, H. S.; Spontak, R. J. *Macromolecules* **1998**, *31*, 2166.
32. Lindman, B.; Wennerström H. *Top. Curr. Chem.* **1980**, *87*, 1.
33. Goddard, E. D. *Colloids Surf.* **1986**, *19*, 255.
34. Gunnarsson, G.; Jönsson, B.; Wennerström H. *J. Phys. Chem.* **1980**, *84*, 3114.
35. Cutler, S. G.; Meares, P.; Hall, D. G. *J. Chem. Soc. Faraday Trans. 1* **1978**, *74*, 1758.
36. Almgren, M.; Löfroth, J.-E. *J. Coll. Int. Sci.* **1981**, *81*, 486.
37. Hayter, J. B.; Penfold, J. *Coll. Polym. Sci.* **1983**, *261*, 1022.
38. Cabane, B.; Duplessix, R.; Zemb, T. *J. Phys.* **1985**, *46*, 2161.
39. Piculell, L.; Hourdet, D.; Iliopoulos, I. *Langmuir* **1993**, *9*, 3324.
40. Rosén, O.; Piculell, L.; Hourdet, D. *Langmuir* **1998**, *14*, 777.

HASE Thickeners

Chapter 20

Determination of the Thickening Mechanism of a Hydrophobically Modified Alkali Soluble Emulsion Using Dynamic Viscosity Measurements

C. M. Miller, K. R. Olesen, and G. D. Shay[1]

UCAR Emulsion Systems, Union Carbide Corporation, Cary, NC 27511

The viscosity of a HASE latex was measured using a reactor calorimeter by developing a relationship between a measured variable closely related to agitator torque and viscosity. A particular advantage of this technique was its ability to continuously monitor the solution viscosity as the pH was adjusted by the addition of sodium hydroxide. Thus, the technique permitted continuous monitoring of the dynamics of the neutralization process, and particularly the equilibration behavior as a function of degree of neutralization.

The results of this study show that the neutralization behavior of a HASE thickener is similar to that reported previously for ASE thickeners. Specifically, the viscosity of any given thickener is a function of degree of neutralization and concentration of the thickener. Furthermore, at sufficiently high concentrations a pronounced viscosity spike is observed at degrees of neutralization between 45 and 55%. Rheology and light scattering measurements suggest that the cause of this peak is predominantly due to the large increase in hydrodynamic volume of the latex particles in their highly water-swollen state immediately prior to their dissolution. Dynamic equilibration experiments revealed that the equilibration rate of a HASE thickener is a strong function of degree of neutralization, with the longest equilibration times occurring between 25 and 44% neutralization.

[1]Corresponding author.

Introduction

Alkali-swellable and alkali-soluble thickeners (AST's) are carboxyl functional polymers produced by free radical polymerization of ethylenically unsaturated monomers (1). These polymers are substantially insoluble in water at a low pH, however, at higher degrees of ionization (higher pH) they become swellable or soluble in water and thus exhibit thickening behavior. As a result of their pH dependent solubility in water, AST polymers can be prepared either as polymer solutions at high pH or as latexes at low pH. Among these, the latex AST's are the most industrially important due to their much lower viscosity in the latex state.

In their ionic form, AST latexes are generally broken down into two classes of materials, either alkali-soluble (or swellable) emulsions (ASE's) or hydrophobically modified alkali-soluble (or swellable) emulsions (HASE's). The difference between these two classes of latexes pertains to the nature of the polymer backbone, which in turn impacts the thickening mechanism. For ASE's, the polymer backbone is generally comprised of acrylic monomers and carboxylic acid monomers, and thus at high pH these materials swell or dissolve in water to thicken by an intermolecular entanglement mechanism, similar to that observed for conventional polymers in organic medium. Electrostatic repulsion of the carboxylate anions in the ASE promotes molecular coil expansion increasing the hydrodynamic volume and chain entanglement. For HASE's, the polymer backbone is essentially the same, however in this case the acrylic polymer backbone is modified with hydrophobically terminated ethoxylated macromonomers which provide a secondary thickening mechanism. For these polymers the mechanism of thickening by intermolecular entanglement is retained, as described above, however, additional thickening is derived from a micelle-like association of hydrophobic moieties along the polymer backbone. This can impart unique rheological properties to HASE thickeners compared to the ASE thickeners.

Regardless of the thickening mechanism, ASE and HASE latexes find great utility in a variety of practical applications including architectural coatings, carpet backing, printing inks, paper coatings, and adhesives. In these applications, copolymers and terpolymers containing predominately ethyl acrylate and methacrylic acid are most frequently used. As a consequence, a large number of papers and patents have been published describing the synthesis and rheological properties of AST's. In addition to the synthesis and rheology of AST's, the mechanism for transition from a water insoluble latex to a water soluble polymer has also received a good deal of attention in the literature, however, in this case most work has been done with ASE's and relatively little work has been published on HASE Thickeners. The remainder of this introduction reviews some of the work on the mechanism for the ASE latex to polymer-solution transition.

Fordyce et al. (2) reported a marked viscosity increase upon neutralization of alkali-soluble methacrylic acid co alkyl acrylate emulsion copolymers (ASEs), where the properties of the alkali responsive thickeners examined depended on the ratio of monomer used, the molecular weight, and the extent of crosslinking present. Fordyce et al. also observed viscosity maxima at degrees of neutralization less than 100%, i.e., where the fraction (α) of carboxyl groups neutralized was in the range of $\alpha = 0.10-0.80$. They concluded that at the viscosity peak, a true solution does not exist, but instead, a highly swollen insoluble state contributes to the high solution viscosity the magnitude of which was dependent on polymer concentration. Finally, in this work it was also reported that the solubilization of the ASE thickeners occurred at a high rate and was controlled almost entirely by the rate at which the alkali could be uniformly distributed through the emulsion, although it is unclear how this conclusion was arrived at.

Yudelson and Mack (3) also observed extraordinary viscosity maxima at degrees of ionization substantially less than 100% for certain acrylic acid co alkyl acrylate copolymers which had been prepared by solution polymerization in nonaqueous media. The pH at which the maxima occurred for the aqueous dispersion of dried polymer was referred to as the "gelation pH". Because the gelation behavior was only observed at concentrations above some critical value, they concluded that the gelation phenomenon must be due to intermolecular interactions. Attractive forces due to hydrogen bonding of carboxyl groups with ester groups were in balance with repulsive forces due to coulombic repulsion of the carboxyl anions.

In a comprehensive examination of latex to solution transition, Verbrugge (4,5) identified three general profiles of viscosity as a function of neutralization degree for a large number of methacrylic acid containing copolymer emulsions. The profile shape was attributed to a combined variety of factors foremost of which was the %MAA in the copolymer, followed by the relative copolymer hydrophilicity, and lastly the copolymer glass transition temperature). He concluded that a single mechanism involving varying degrees of particle swelling along with solubilization when the copolymer hydrophilicity is high enough explains the thickening transition for all acid containing latices. Once again, a manifestation of this investigation was the occasional observance of a viscosity spike at intermediate degrees of neutralization similar to what Murio and Yudelson had reported. A viscosity spike was one of the three basic shapes designated. Absent from Verbrugge's work was the effect of ASE concentration noted by Fordyce and Yudelson. Finally, although no data was shown in his papers, Verbrugge stated that the viscosity of freshly prepared ASE solutions decreased with time at all pH's, and required at least one day to fully equilibrate.

More recently, Quadrat et.al (6) examined the swelling and dissolution behavior of ethyl acrylate/methacrylic acid ASE copolymers during neutralization. For these copolymers they reported that acid contents less than 20% only swell, between 20-40% they may swell but decompose to smaller units of supermolecular aggregates, and at 40% and above, they may initially swell but are molecularly dissolved. Here again, concentration dependent viscosity maxima were observed both above and well below 100% neutralization depending on the MAA concentration. With increasing acid content, the height of the maxima decreased and was shifted to lower degrees of neutralization. Another profound observation was the effect of flow rate on the viscosity spikes. As the velocity gradient increased, the viscosity maxima decreased and the peaks eventually disappeared altogether beyond some maximum flow. The explanation was decomposition of swollen disperse particles to a system which predominantly contains supermolecular aggregates and only a small proportion of dissolved macromolecules.

Several investigators (7-13) have used potentiometric and conductometric titration techniques to determining the effects of carboxyl group distribution within the latex particles on alkali-swelling behavior of carboxyl functional copolymers. The effect of acid distribution on dissolution behavior was found to be a function of acid type, comonomer type, monomer ratios, copolymer Tg, polymerizaton procedure and the method of monomer addition.

In this paper, the viscosity of a model HASE polymer exhibiting the viscosity spike previously observed in some ASE's was measured as a function of extent of neutralization in order to gain insight into the mechanism for transition from a low viscosity latex to a relatively high viscosity polymer solution. The types of measurements described in this study were primarily dynamic in nature, thus a particularly unique aspect of this paper is the viscosity equilibration of the model HASE under varying conditions within a reaction calorimeter. It is apparent from the literature survey above that this is an area that has received relatively scant previous consideration.

Experimental

HASE Thickener Selected for Study

The HASE thickener prepared for this study is referred to as HASE-EO40NP and has the general structure shown in Figure 1 as described elsewhere (*14,15*). The specific ratio of monomers for HASE-EO40NP is W = 60, X = 35, and Y = 5 as a weight percent of the polymer. The polymer was prepared in a 3 liter jacketed glass reactor as a latex at 25% solids by standard emulsion polymerization techniques using anionic surfactant and persulfate initiation. The polymer was characterized by GPC (THF solvent, polystyrene standards) to have a number average molecular weight of 25,000 and a polydispersity index of 2.1. The contrasting hydrophobicity of this polymer in acid form relative to it's very hydrophilic Na+ salt as determined by water vapor sorption has been previously reported (*16*).

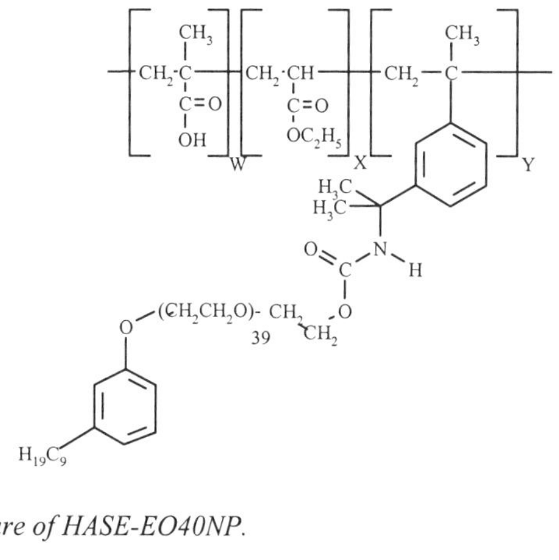

Figure 1. Structure of HASE-EO40NP.

Dynamic Viscosity Measurements

A Mettler RC1 calorimeter (reactor type AP01) was used to continuously monitor the viscosity of the HASE thickener during neutralization. The reactor was equipped with an anchor impeller, pH probe, thermocouple, calibration element, and two feed lines. The feed lines were positioned above and below the pH probe to allow for uniform mixing of the sodium hydroxide solution during the neutralizations. During the experiments, the agitator was run at 75 RPM and the temperature was controlled at 25±0.5°C. The rate of agitation was selected by trial and error to provide the optimum balance of uniform mixing and laminar flow over the broadest range of experimental conditions.

The Mettler RC1 provides a convenient means for estimating the viscosity of a solution through the storage of a variable called "R_{int}" which is the proportional part of the Proportional Integral stirrer controller and is directly related to the required power to achieve a set rate of agitation. While this variable does not directly correspond to the viscosity of the solution being agitated, it is indirectly related to the torque on the stirrer motor, which in turn is directly related to the viscosity of the solution. In fact, it has been shown that under most conditions the R_{int} parameter is directly proportional to the torque on the agitator (17).

The RC1 was calibrated using silicon fluids with viscosities of 9.9, 97.5, 985, and 12640 cP (from Brookfield). Experiments were performed where equivalent volumes of these materials were added to the RC1, and the R_{int} value was obtained at 75 RPM and 25°C. For each standard, R_{int} was determined at two different volumes, 1.15 and 1.25 liters. These volumes correspond to the minimum and maximum volumes present at any time during the neutralization experiments. Using this data, an empirical relationship relating R_{int} to viscosity and reactor volume was developed.

External Measurements

In certain cases samples of the partially neutralized HASE Thickener were removed from the calorimeter and characterized for rheology, particle size, and turbidity. The rheology of these samples was determined using a Bohlin rheometer operated at 25°C using a cup and bob sample cell. The particle size of the samples was determined using a Microtrac UPA particle sizer. This instrument operates under the principles of dynamic light scattering. The turbidity of the samples was determined using an ANALITE portable nephelometer (Model 156, McVan Instruments).

Results and Discussion

Effect of Concentration of HASE Thickener

Experiments were performed with different concentrations of HASE-EO40NP in water. The concentrations investigated were 5, 7.5, and 10% at the start of the neutralization. To each of these latexes, 25% sodium hydroxide solution was added such that the neutralization rate and final degree of neutralization was equivalent. The viscosities of the HASE solutions were then plotted as a function of degree of neutralization and pH. To obtain the degree of neutralization, the equivalence point was determined from a plot of pH vs. grams of NaOH added. This equivalence point agreed fairly well with that calculated using a mass balance. In these experiments, the HASE latex was neutralized with an excess of NaOH, and therefore degrees of neutralization greater than 100% are reported, although it is recognized that it is physically impossible to neutralize beyond 100%.

Figure 2 shows the viscosity of the differently concentrated HASE-EO40NP solutions as a function of extent of neutralization. The data in this figure shows three different types of behaviors. For 10% HASE-EO40NP, a very large increase in the viscosity is observed, followed by a peak and a decrease to a constant value. For 7.5% HASE-EO40NP, the same behavior is observed, but the peak viscosity is much closer to the final plateau viscosity. For both of these latexes the peak viscosity is obtained at about 50% neutralization, or alternatively at a pH of about 6.5. Finally, for 5% HASE-EO40NP a gradual increase in viscosity is observed beginning at about 30% neutralization and ending with a plateau viscosity at about 80% neutralization.

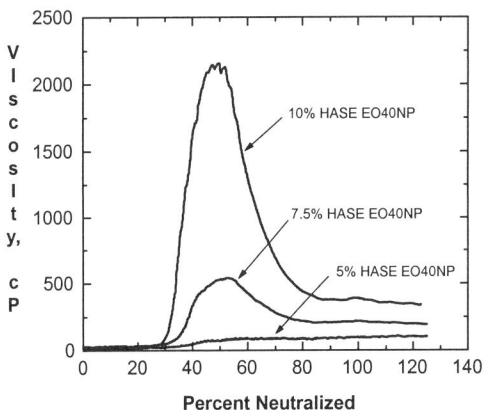

Figure 2. Viscosity plotted as a function of percent neutralization for 5, 7.5, and 10% HASE-EO40NP solutions in water. The rate of neutralization was 50 grams 25% NaOH/hour.

Although all of the references sited in the introduction (*2-13*) deal specifically with neutralization of ASE thickeners whereas this work concerns the neutralization of a HASE thickener, the curves in Figures 2 bare striking resemblance to many of those reported previously. Therefore, it seems likely that the mechanism for the latex to polymer-solution transition for HASE thickeners is similar or identical to that for ASE thickeners, and interpretation of the curves in Figures 2 can be aided by consideration of the previous work.

Effect of Rate of Neutralization

The preceding section showed the effect of degree of neutralization on viscosity at a constant rate of neutralization. It is important to note that in the preceding study the HASE thickener is being continuously neutralized and therefore the value of the measured viscosity at any given time is the "instantaneous viscosity" which may be quite different than the fully equilibrated viscosity. In other words the preceding study, like all others performed using this technique, is dynamic in nature. Two different approaches were used to investigate the nature of the equilibration dynamics of HASE-EO40NP. In this section the first of these two approaches is discussed, neutralization of 10% HASE-EO40NP by feeding 25% NaOH at different feed rates. In order to accomplish this, 100 grams of 25% NaOH solution was fed over 0.5, 2, 4, 8, and 24 hours, corresponding to neutralization rates of 200, 50, 25, 12.5, 4.16 grams 25% NaOH solution per hour.

Figure 3 shows the viscosity of 10% HASE-EO40NP plotted against the amount of base added for the five different feed rates described above. This figure begins to show the importance of the dynamics of the neutralization process, and the utility of the Mettler RC1 to provide insight into the equilibration kinetics. Figure 3 shows that for degrees of neutralization less than 50% (corresponding to the viscosity peak) as the neutralization rate is decreased the measured viscosity is increased. This suggests that the rate of neutralization is important, and the system is not at equilibrium. On the other hand, Figure 3 also shows that for degrees of neutralization greater than 50% the effect of feed rate on viscosity is less pronounced. These trends can be seen more clearly in Figure 4, which shows the viscosity as a function of feed rate at 37.5%, 50%, and 62.5% neutralization (corresponding to 30, 40, and 50 grams 25% NaOH fed, respectively). By extrapolation, this figure can also be used to estimate the equilibrated viscosity which corresponds to the zero feed rate (fully equilibrated) viscosity. Figure 4 shows that at 37.5% neutralization the viscosity increases with

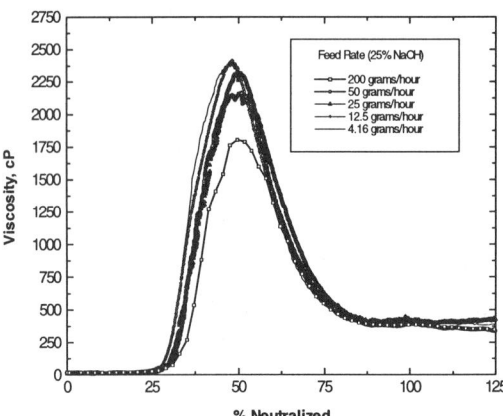

Figure 3. Effect of rate of neutralization on the viscosity as a function of degree of neutralization for 10% HASE-EO40NP aqueous solutions.

decreasing feed rate for all of the rates investigated, with the viscosity rising rapidly when the feed rate approaches zero. This demonstrates that even for the slowest feed rate investigated (4.16 grams NaOH per hour) the HASE polymer does not have time to fully equilibrate. At 50% neutralization this behavior is also apparent, however, at this extent of neutralization the viscosity appears to become constant when the feed rate is less than 25 grams/hour. In other words, for feed rates less than 25 grams per hour, the measured viscosity is independent of feed rate and thus should be the same as the fully equilibrated viscosity which would be measured at a feed rate approaching zero. Finally, at 62.5% neutralization, the viscosity is almost independent of the feed rate. This suggests that at higher degrees of neutralization the equilibration is essentially instantaneous and the measured viscosity is close to the fully equilibrated viscosity.

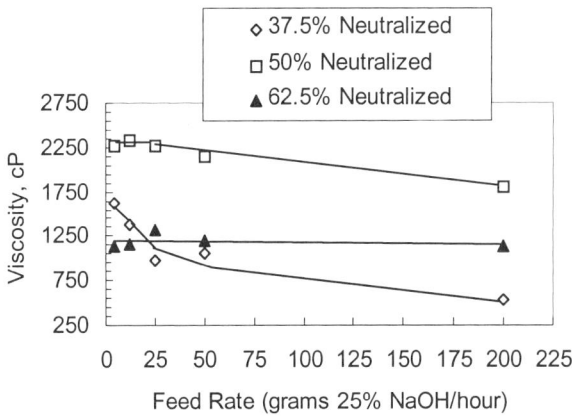

Figure 4. Measured "instantaneous" viscosity versus feed rate for 10% HASE-EO40NP solutions.

The different equilibration behavior at different degrees of neutralization is probably due to the mechanism for thickening at the different extents of neutralization. Specifically, at 37.5% neutralization HASE-EO40NP is opaque and therefore is most likely comprised of water-swollen latex particles. At 67.5% neutralization, HASE-EO40NP is completely transparent and therefore is most likely a polymer solution. In contrast to these degrees of neutralization, at 50% neutralization the physical state of HASE-EO40NP is not as obvious. This is because while HASE-EO40NP turns transparent between 40 and 45% neutralization, the viscosity continues to increase until a maximum at about 50% neutralization. Therefore, at 50% neutralization it is not clear whether the HASE thickener has begun to dissolve or is simply behaving as a highly-swollen latex.

Viscosity Relaxation of HASE Thickener at Different Degrees of Neutralization

The above results showed that the neutralization rate had a pronounced effect on the instantaneous viscosity of HASE-EO40NP, and this effect differed depending upon the extent of neutralization. In this section an alternative approach for studying this behavior is described whereby HASE-EO40NP is neutralized at a constant rate (25 grams/hour) until a desired amount of base is added, after which the feed is stopped and the latex is allowed to equilibrate. The viscosity response during this equilibration is then followed, and the percent change in viscosity and equilibration time are determined. This behavior is referred to as the "viscosity relaxation" of HASE-EO40NP, since the non-equilibrated material is relaxing to its equilibrated state. After equilibration the partially neutralized HASE-EO40NP was removed from the reactor and analyzed for rheology, particle size, and turbidity. Combined, these different measurements were used to infer the mechanism for thickening of HASE-EO40NP.

Figure 5 shows the relaxation behavior for HASE-EO40NP neutralized to 25, 31.25, 37.5, 43.75, 56.25 and 125% neutralization. For all of these experiments 25% NaOH solution was fed at a rate of 25 grams/hour until the desired amount of base was added to the latex, after which the feed was stopped and the HASE was allowed to equilibrate. Figure 6 shows that in each case after the feed was stopped the viscosity changed during an equilibration period, after which it remained constant. The percent change in viscosity and time for equilibration are more clearly illustrated in Figure 6. This figure clearly shows that at all degrees of neutralization the HASE thickener was not instantaneously equilibrated, however, the time and amount of viscosity change upon equilibration was highly dependent upon the actual degree of neutralization. Specifically, between 25 and 37% neutralization HASE-EO40NP requires the greatest amount of time to equilibrate and undergoes the greatest change in viscosity upon equilibration. For the discrete data points measured, at 37.5% neutralization, HASE-EO40NP required greater than 45 minutes to equilibrate and underwent a 175% increase in viscosity during the equilibration. The data in figures 5 and 6 complement the previous data shown in figures 3 and 4 and suggest that the equilibration of a HASE thickener is highly dependent upon its physical state in solution. It appears that at lower degrees of neutralization a significant amount of time is required to achieve a fully equilibrated state, and this behavior is not seen at higher degrees of neutralization. Based on figures 5 and 6 it can further be stated that the longest equilibration times are observed during a very narrow region of degrees of neutralization, apparently when the HASE thickener begins to rapidly swell with water. Further proof of this can be provided by examining Figure 7 which shows the measured particle size and turbidity of the fully equilibrated samples removed from the reactor. This figure shows that the particle size increases rapidly and the turbidity decreases rapidly up to about 37.5% neutralization, which corresponds to the maximum equilibration time. Beyond this degree of neutralization, the viscosity continues to increase and the turbidity continues to decrease, but the equilibration time decreases rapidly. It should be noted that although no particle size could be measured for degrees of neutralization greater than 37.5%, this may be due to limitations in the particle size instrument rather than a true reflection of the physical state of the HASE thickener.

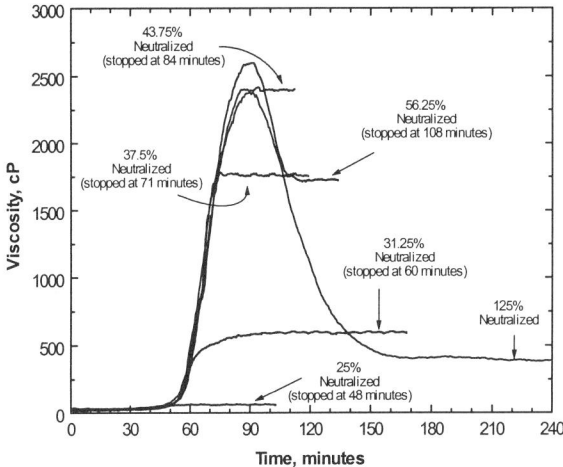

Figure 5. Viscosity Relaxation (Equilibration) for 10% HASE-EO40NP aqueous solutions neutralized to different extents at a feed rate of 25 grams 25% NaOH/hour.

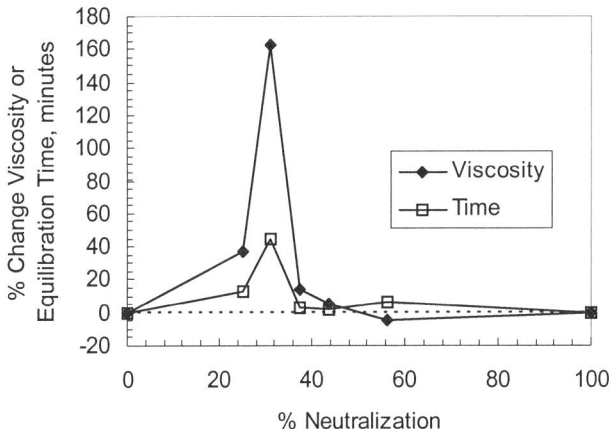

Figure 6. Percent change in viscosity and equilibration time for 10% HASE-EO40NP aqueous solutions neutralized to different extents at a rate of 25 grams 25% NaOH/hour.

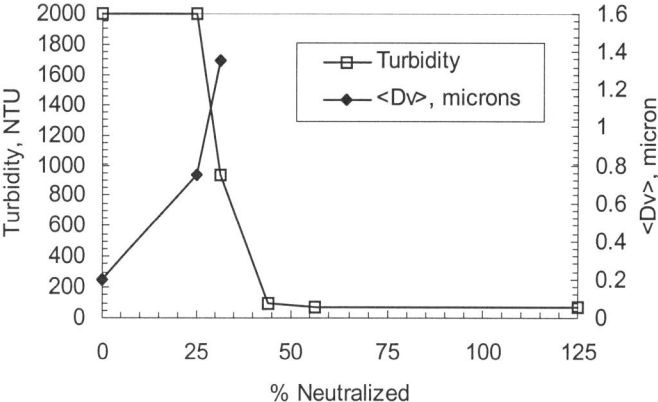

Figure 7. Turbidity and particle size of fully equilibrated samples of 10% HASE-EO40NP aqueous solutions at different extents of neutralization.

The data presented in Figure 7 show an important effect of degree of neutralization on the thickening mechanism, and in turn on the rate of equilibration. However, thus far the data are not able to unambiguously show the physical state of the thickener as a function of extent of neutralization, which is important for inferring what the actual mechanism for thickening is. To this end, further information was obtained by analyzing each of the samples discussed in Figures 5 through 7 above for their rheology using a Bohlin Rheometer. Figure 8 shows the rheology of HASE-EO40NP at each of the degrees of neutralization discussed above, 25, 31.25, 37.5, 43.75, 56.25 and 125% neutralization. Since HASE-EO40NP exhibits shear thinning behavior in its latex form and nearly Newtonian behavior as a polymer solution, the Bohlin Rheometer enables one to determine at what point the thickener begins to resemble a polymer solution in viscoelastic properties. Figure 8 shows that this point is clearly at degrees of neutralization greater than those corresponding with the peak viscosity. This suggests that up to and around the peak viscosity, the system is comprised of discrete, highly water-swollen latex particles, whereas after the peak viscosity the system is comprised of more or less homogeneous polymer solution. Thus, we conclude that a major cause of the peak viscosity is simply hydrodynamic crowding of the swollen latex particles. This is somewhat different than the mechanism proposed by Yudelson and Mack (*3*) and later elaborated upon by Nishida et al. (*12*) who proposed that the cause of the peak viscosity was more due to hydrogen bonding forces than hydrodynamic volume effects.

Validation of Experimental Technique

The viscosity data presented in this paper were all obtained using a rather unconventional technique, namely, calibration of a reactor calorimeter not specifically designed for viscosity measurements. While it is obvious that a great deal of information was obtained from this instrument, the precision of this technique has not yet been established which must bring into question some of the results discussed above. To address this, the Bohlin rheometer data shown in Figure 8 was compared with the corresponding viscosities determined using the calorimeter. Since the samples were not Newtonian in shear behavior, a linear least squares regression was used to estimate the average shear rate in the calorimeter. The results of this analysis revealed that the average shear rate in the calorimeter was approximately 29 s^{-1}, and the Mettler and Bohlin measured viscosities were within 5% of each other. Thus, it appears that when properly calibrated the Mettler RC1 can be accurately used to estimate the viscosity of a solution.

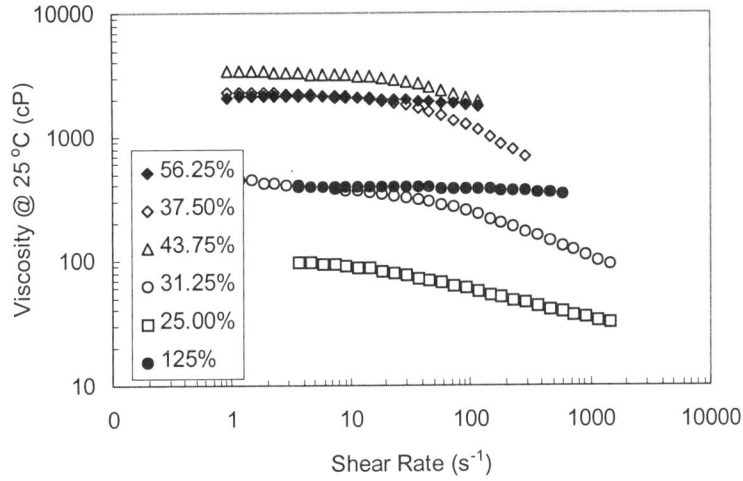

Figure 8. Viscosity as a function of shear rate for fully equilibrated samples of 10% HASE-EO40NP aqueous solutions removed from the Mettler RC1 at different extents of neutralization as indicated in the legend.

Conclusions

Viscosity measurements performed using an automated reactor calorimeter allowed for precise monitoring of the viscosity of a HASE thickener as a function of degree and rate of neutralization. Using this technique, the mechanism for the dissolution of a HASE latex was inferred, and the importance of the equilibration as a function of degree of neutralization was established. It was found that the neutralization behavior of a HASE thickener is similar to that reported previously for ASE thickeners. Specifically, the viscosity of any given thickener is a function of degree of neutralization and concentration of the thickener. Furthermore, at sufficiently high concentrations a pronounced viscosity spike is observed at degrees of neutralization between 45 and 55%. Rheology measurements suggest the HASE thickener to be comprised of discrete polymer particles at this viscosity peak, thus the cause of this peak appears to be predominantly due to the large increase in hydrodynamic volume of the latex particles in their highly water-swollen state immediately prior to their dissolution. Dynamic measurements revealed that the rate of equilibration of a HASE thickener is a strong function of degree of neutralization, with the longest equilibration times occurring between 25 and 44% neutralization. External turbidity, particle size, and rheology measurements showed that during this narrow range of degrees of neutralization the HASE latex rapidly swells with water and accordingly rapidly increases in diameter. Thus it appears that the long equilibration times between 25 and 44% neutralization are caused by a resistance to swelling, either due to slow diffusion of alkali through the outer layers of hydrated latex, or due to viscoelastic resistance of the polymer chains to unentangle from their initial latex form.

References

1. Shay, G. D. In *Polymers in Aqueous Media;* Glass, J. Edward., Ed.; ACS Advances in Chemistry Series No. 223; American Chemical Society: Washington, D.C., 1989; pp. 457-492.
2. Fordyce, D. B.; Dupre', J.; Toy, W. *Offic. Dig.* **1959**, *31*, p. 284.
3. Yudelson, J. S.; Mack, R. E. *J. of Pol. Sci.: Part A* **1964**, *2*, 4683-4695.
4. C.J. Verbrugge *J. Appl. Polym. Sci.* **1970**, *14*, 897-909.
5. C.J. Verbrugge, *J. Appl. Polym. Sci.* **1970**, *14*, p. 911-928.
6. Quadrat, O.; Mrkvickova, L.; Jasna, E.; Snuparek, J. *Colloid & Polym. Sci.,* **1990**, *268*, 493-499.
7. Muroi, S.; Hosoi, K.; Ishikawa, T. *J. Appl. Polym. Sci.* **1963**, *11*, p. 1967.
8. Muroi, S. *J. Appl. Polym. Sci.* **1966**, *10*, 713-729.
9. Hoy, K. L. *J. Coat. Tech.* **1979**, *51, No. 651*, p. 27.
10. Bassett, D. R.; Hoy, K. L. In *Polymer Colloids*; Fitch, R. M., Ed.; Plenum Press: New York, NY, 1980; Vol. II pp. 1-25.
11. Nishida, S., Ph.D. thesis, Lehigh University, Bethlehem, PA, 1980.
12. Nishida, S.; Hosoi, K.; Klein, A.; and Vanderhof, J. W. In *Emulsion Polymerization;* ACS Symposium Series No. 165; American Chemical Society: Washington, D.C., 1980; pp. 1291-1314.
13. Loncar, F. V.; El-Asser, M. S.; Vanderhoff, J. W. *Polym. Mater. Sci. Eng.* **1985**, *52*, 299-303.
14. Shay, G. D.; Rich, A.F. *Journal of Coatings Technology* **1986**, *58, No. 732*, 43-53.
15. Shay, G. D.; Bassett, D. R.; Rex, J. D. *JOCCA, Surface Coatings International* **1993**, *76, No. 11*, 446-453.
16. Shay, G. D.; Olesen, K. R.; Stallings, J. L. *Journal of Coatings Technology*, **1996**, *68, No. 854*, 51-63.
17. Groth, U, Presentation Notes, *Rint - Viscosity - Torque*; 4th RC User Forum, October 7-10th, Montreux (CH), 1990.

Chapter 21

The Network Strength and Junction Density of a Model HASE Polymer in Non-Ionic Surfactant Solutions

W. P. Seng[1], K. C. Tam[1,4], R. D. Jenkins[2], and D. R. Bassett[3]

[1]School of Mechanical and Production Engineering,
Nanyang Technological University, Nanyang Avenue, Singapore 639798
[2]Technical Center, Union Carbide Asia Pacific Inc.,
16 Science Park Drive, The Pasteur, Singapore 118227
[3]UCAR Emulsions Systems, Research and Development,
Union Carbide Corporation, 410 Gregson Drive, Cary, NC 27511

Rheological experiments had been carried out on 1wt.% aqueous solutions of hydrophobically modified alkali-soluble emulsion (HASE) at pH around 9 in the presence of various amounts of non-ionic polyoxyethylene ether type surfactant ($C_{18}EO_{100}$). The steady and dynamic results revealed that the low shear viscosity and dynamic moduli initially increased, reaching a maximum, which then decreased at higher surfactant concentrations. The maximum low shear viscosity was observed at a surfactant concentration 50 times the *cmc*. In addition, other rheological techniques such as temperature effect and superposition of oscillation on steady shear were performed. These experiments confirmed that the strengthening of network at low $C_{18}EO_{100}$ concentration was predominantly driven by an increase in the number, strength of the intermolecular hydrophobic junctions and an expanded polymer conformation due to electrostatic repulsive forces. At higher surfactant concentration, the disruption of network was attributed to a decrease in the number and strength of these junctions.

---

[4]Corresponding author (fax: 65-791-1859).

Hydrophobically modified alkali-soluble emulsion (HASE) is a class of associative thickener that are both environmentally and industrially accepted due to its water-soluble nature and ability to enhance the thickening efficiency. Numerous researchers have reported experimental studies on the effects of adding non-ionic surfactants into hydrophobically-modified polyelectrolyte (1-5). Some observed that the maximum viscosity occurred at surfactant concentration several orders higher than *cmc*. However, the reasons as to why the optimal rheological properties are observed at concentration way above *cmc* and the interacting mechanisms behind this are still not well understood. This chapter aims to discuss the mechanisms responsible for the enhancement of the low shear viscosity of polymer in the presence of a non-ionic surfactant ($C_{18}EO_{100}$). It is generally believed that the presence of surfactant results in two dominant mechanisms, i.e. an increase in the number of inter-molecular hydrophobic junctions and an increase in average lifetime of these junctions, which increases the strength of these junctions. In this study, different techniques such as steady and dynamic shear; temperature effect; and superposition of oscillation on steady shear were used to provide evidence to support the above hypothesis.

Experimental

Materials

HASE is a copolymer that consists of methacrylic acid (MA), ethyl acrylate (EA) and small fractions of associative macromonomers (AM) distributed randomly along the backbone. This polymer exists as latexes with a white, milky appearance at low pH. Upon neutralization to pH 9 with a suitable base, the MA groups are ionized and the backbone behaves like an anionic polyelectrolyte. The resulting expanded conformation of the polymer chains, together with the hydrophobic molecular associations, contributed to the formation of network structure that significantly enhances the thickening efficiency of HASE. The model HASE polymer used in this work was synthesized according to the method previously described by Jenkins et al. (6) and Tirtaatmadja et al. (7). The chemical structure of HASE is shown schematically in Figure 1. This copolymer was prepared by conventional semi-continuous emulsion polymerization. The composition of the MAA/EA/AM was kept at 49.05/50.04/0.91 mole percentage. The structure of the AM used in this study consists of a poly(oxyethylene) chain with an average 35 moles of ethoxylation (p=35) en-capped to an alkyl hydrophobic group (R=C20H41) at one end and a vinyl polymerisable group at the other end. This model polymer was denoted as HASE-(EO)35-C20H41 through out the text.

The model polymer latex at low pH (ca 3-4) was dialyzed against distilled de-ionized water for about 4 weeks to remove all the impurities and unreacted chemicals. A 3wt% stock solution in 10^{-4} M KCl was prepared and stored at 4°C prior to use. From the 3wt% stock HASE, samples of 1.0 wt% HASE in varying concentrations of surfactant solutions were prepared. The alkaline used to neutralize the polymer to the required pH value of approximately 9 was 2-amino-2-methyl-1-propanol (AMP). The non-ionic surfactant used in this study is

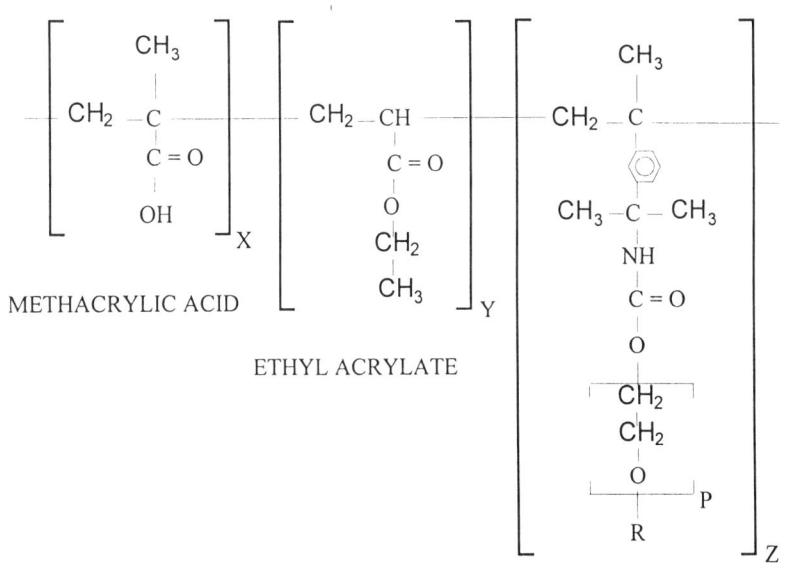

Figure 1. Molecular structure of HASE. R: hydrophobic group. p: number of moles of ethoxylation (spacer length) between polymer backbone and hydrophobic group. x : y : z = 49.05 : 50.04 : 0.9.

polyoxyethylene 100 Stearyl Ether (Brij700) $C_{18}H_{37}O(CH_2CH_2O)_{100}H$ from Sigma (denoted here as $C_{18}EO_{100}$). They were used as received without further purification.

Rheological Measurements

Rheological measurements were performed on a controlled stress Carri-Med CSL500 rheometer. The cone-and-plate measuring system, which has a diameter of 40 mm and a cone angle of 2 degree, was used on the CSL500. The rheological measurement was carried out over a temperature range of 10-40°C. The shear stress range was between 0.2 Pa to 500 Pa. Strain sweep measurements, at a strain range between 0.001 and 0.2 radian, were carried out to ensure that the selected strain amplitude yielded data that was within the linear viscoelastic region. The frequency dependent moduli were measured over frequency range of 0.05 to 200 rad/s.

Results and discussion

Steady shear

Figure 2 shows the dependence of steady shear viscosity on shear stress for 1wt% HASE-(EO)35-C20H41 in varying concentrations of $C_{18}EO_{100}$. From these steady shear results, low shear viscosity values were extracted and plotted in Figure 3a. The presence of surfactant modified the low shear viscosity of the polymer above 0.1mM $C_{18}EO_{100}$, which clearly suggests that polymer-surfactant interactions had occurred. The low shear viscosity increased to a maximum at a critical $C_{18}EO_{100}$ concentration, c* at around 1mM. Further addition of surfactant beyond this critical value caused the low shear viscosity to decrease. It is generally accepted that the later phenomenon is a result of the disruption of associative network caused by the isolation of polymer hydrophobes, which are solubilized by surfactant micelles (8). The *cmc* of $C_{18}EO_{100}$ was found to be 0.02mM using dynamic light scattering where the size of the micelle is 14-16nm. Hence the maximum occurred at approximately 50 times the *cmc* which was much higher than the results for SDS (9). However, the viscosity enhancement behavior below c* for non-ionic surfactant is unclear and subject to many debates.

From Figure 2, one would observe that at the surfactant concentration of around 10^{-3} M, the polymer exhibited shear-thickening behavior at moderate shear stresses, followed by a shear-thinning region at higher shear stresses. The shear-thickening phenomenon has been observed in some aqueous systems of associative polymers (10-24). A number of hypotheses have been advanced which remained controversial (25, 26).

Dynamic shear

The dependence of storage (G') and loss modulus (G") on angular frequency for 1wt.% HASE-(EO)35-C20H41 in varying concentrations of $C_{18}EO_{100}$ are

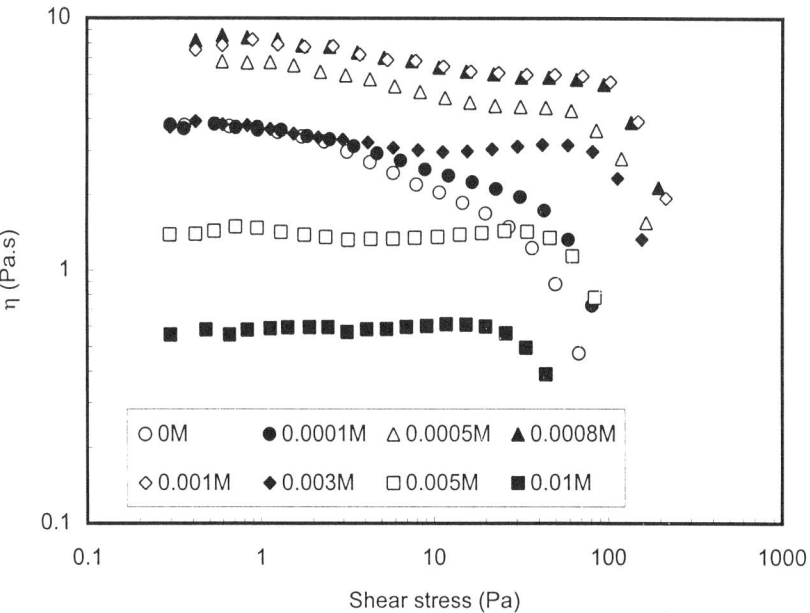

Figure 2. Dependence of shear viscosity on shear stress for 1wt% HASE-(EO)35-C20H41 in varying concentrations of $C_{18}EO_{100}$ at pH 9.

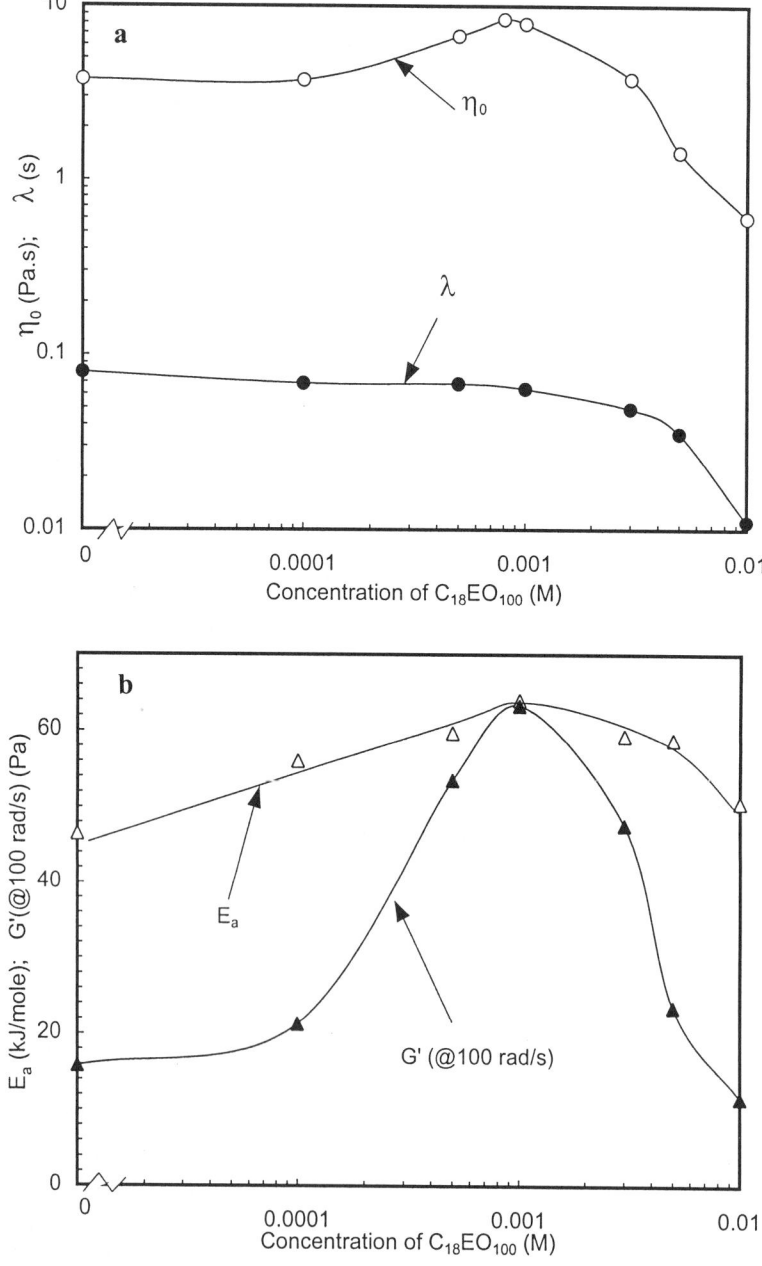

Figure 3. Dependence of rheological parameters of 1wt% HASE-(EO)35-C20H41 in varying $C_{18}EO_{100}$ concentrations (a) Low shear viscosity and relaxation time, (b) activation energy and storage modulus (at 100rad/s).

shown in Figures 4a and 4b respectively. Both the dynamic moduli increased as the surfactant concentration was increased, until a maximum value that corresponded to the critical concentration c*. The addition of surfactant beyond this critical concentration caused both the moduli to decrease.

The dynamic shear results also allowed the determination of an average characteristic relaxation time, λ, which is an important property characterizing the reptation movements of the polymer chains (27). The model system could not be characterized by a single relaxation time, as it did not correspond to the single-mode Maxwell model. The relaxation time for hydrophobically modified systems is generally orders of magnitude larger than the unmodified ones because of the existence of associating junctions which hinder the reptation process of the polymer chains (28). The inverse of the angular frequency that corresponded approximately to the intersection point of G' and G" gave an approximate value of λ. The relaxation time of 1wt% HASE-(EO)35-C20H41 was extracted from Figure 4a using the above approach and plotted in Figure 3a against concentration of $C_{18}EO_{100}$. The plot shows that λ remained reasonably constant up to c*. Beyond this critical value, λ decreased.

The G' values at moderate to high frequency region can be directly related to the plateau modulus G_N^0, where G_N^0 is directly proportional to the number of mechanically active hydrophobic junctions according to the Green-Tobolsky relationship (29):

$$G_N^0 = g\nu RT \qquad (1)$$

where ν is the number of elastically effective chains per unit volume (mol/m³), R is the gas constant and T is the temperature in Kelvin. Although this relationship is based on the classical theory of rubber elasticity for an ideal network, the parameter ν in this case is used to represent the number of mechanically active inter-molecular hydrophobic junctions. The constant g is assumed to be approximately equal to unity according to the transient network theory (23). In many systems, the plateau region lies outside the frequency range of many rheometers where the highest measurable frequency is usually in the region of less than 200rad/s. Beyond this limit, the accuracy of the measurement would be affected by fluid inertia. It is not possible to compute η_0 from the G' data at high angular frequency of 100 rad/s (denoted as $G'_{@100rad/s}$) and the crossover relaxation time numerically according to the expressions below(30),

$$\eta_0 = \int_0^\infty G(t)dt \qquad (2)$$

$$\eta_0 = G_N^0 \lambda \qquad (3)$$

since the $G'_{@100rad/s}$ obtained was not identical to the plateau modulus and λ determined was an average value. Nevertheless, the changes in these parameters followed a trend that correlated well with eq 3 and this would be discussed later. Besides, they provided us with information on the relative number of inter-molecular hydrophobic junctions and the overall relaxation dynamics of the network structure. Figure 3b indicated that the number of junctions increased until a maximum which corresponded to the critical surfactant concentration c* and beyond this critical point the number of junctions decreased.

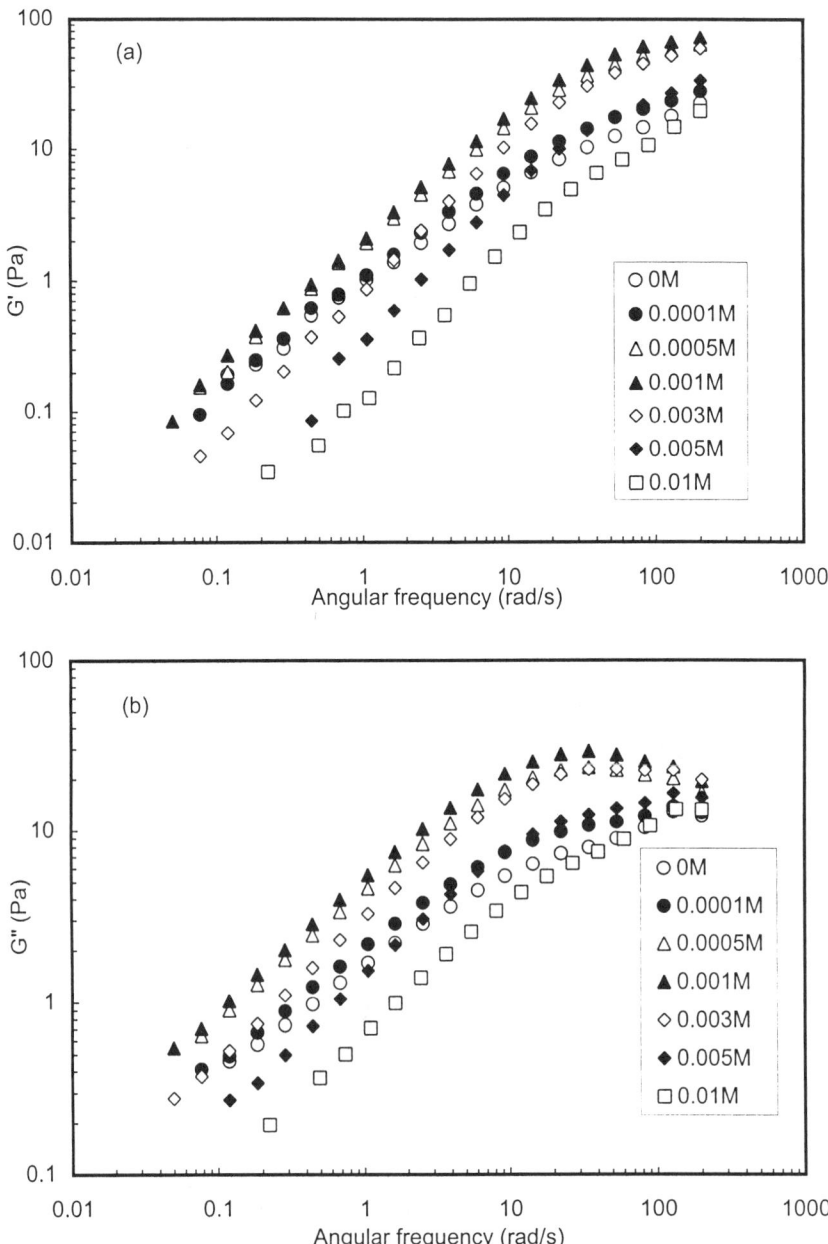

Figure 4. Dependence of (a) storage and (b) loss modulus on angular frequency for 1wt% HASE-(EO)35-C20H41 in varying concentrations of $C_{18}EO_{100}$ at pH 9.

Temperature dependence

The viscosity of polymer system generally obeys the Arrhenius relationship as given below (31):

$$\eta = A \exp^{E_a/RT} \qquad (4)$$

where η is the solution viscosity, E_a is the activation energy, R is the gas constant and T the absolute temperature. In order to use the Arrhenius relationship to obtain the activation energy E_a, viscosity measurements were carried out over several temperatures i.e. 10, 15, 25, 35 and 40°C. Figure 5 shows the dependence of steady shear viscosity on shear stress for 1wt% HASE-(EO)35-C20H41 in 0.003M $C_{18}EO_{100}$ at different temperatures. The viscosity decreased with increasing temperatures.

Based on the Arrhenius expression given in eq 4, a plot of $Ln(\eta_0)$ versus (1000/T) is plotted and shown in Figure 6. From the gradient of the Arrhenius plot, the activation energy was determined and plotted in Figure 3b against the concentrations of $C_{18}EO_{100}$. The activation energy of each polymer solution was approximately related to the overall network strength of the mechanically active junctions. The E_a relationship indicated that the strength of the network junctions increased initially due to the formation of mixed micelles consisting of polymer hydrophobes and surfactant monomers. The strength of the network reached a maximum at the critical surfactant concentration c*. Beyond this critical concentration, the strength of the network decreased.

Polymer-surfactant interactions

Figures 3a and 3b show the combined plots of low shear viscosity η_0, relaxation time λ, and activation energy E_a, storage modulus $G'_{@100rad/s}$ versus the concentration of surfactant $C_{18}EO_{100}$ respectively. In summary, the plots of E_a and $G'_{@100rad/s}$ followed a similar trend to that of η_0 and the interacting mechanisms in different surfactant concentration (Csurf) ranges were described as below.

At Csurf < *cmc*, individual $C_{18}EO_{100}$ molecules started to bind onto the hydrophobic junctions (both intra- and inter-). However, the binding in this case is very weak. The effect on the number of junctions and lifetime is negligible and thus the changes in the viscosity, activation energy, average relaxation time and storage modulus could not be observed. Preliminary results from isothermal titration calorimetry performed in our laboratory showed evidence of weak interactions between a non-ionic $C_{12}EO_{23}$ and HASE. Using the fluorescence technique, Chang et. al. (32) observed interactions between non-ionic Triton X-100 and a randomly-modified copolymer of acrylamide and dimethyldodecyl(2-acrylamidoethyl)ammonium bromide (DAMAB) below *cmc*, even when the reduced viscosity remained unchanged.

At *cmc* ≤ Csurf < 0.1mM, (where 0.1mM is the onset concentration when the low shear viscosity begins to increase), the HASE polymer formed a network structure with polydispersed junctions of varying aggregation number. Based on

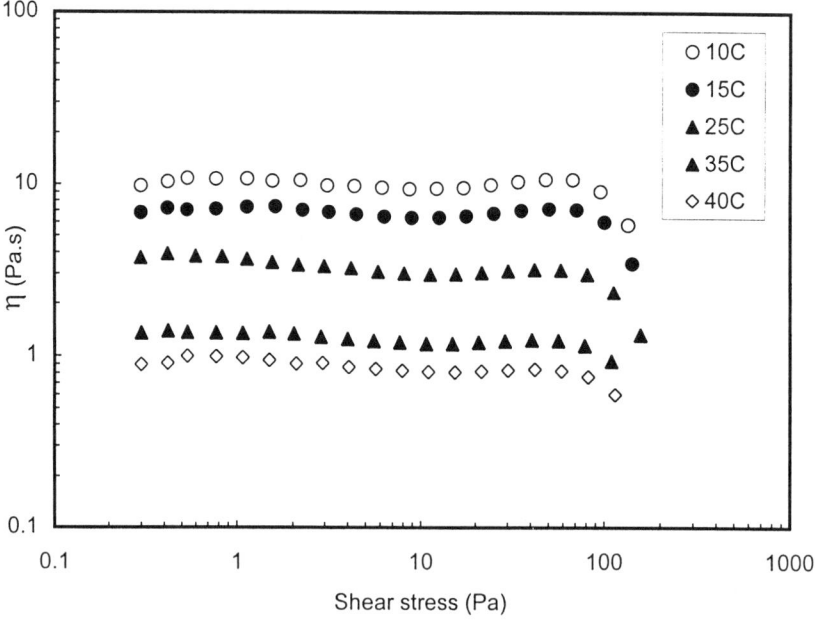

Figure 5. Dependence of viscosity on shear stress for 1wt% HASE-(EO)35-C20H41 in 0.003M $C_{18}EO_{100}$ at different temperatures.

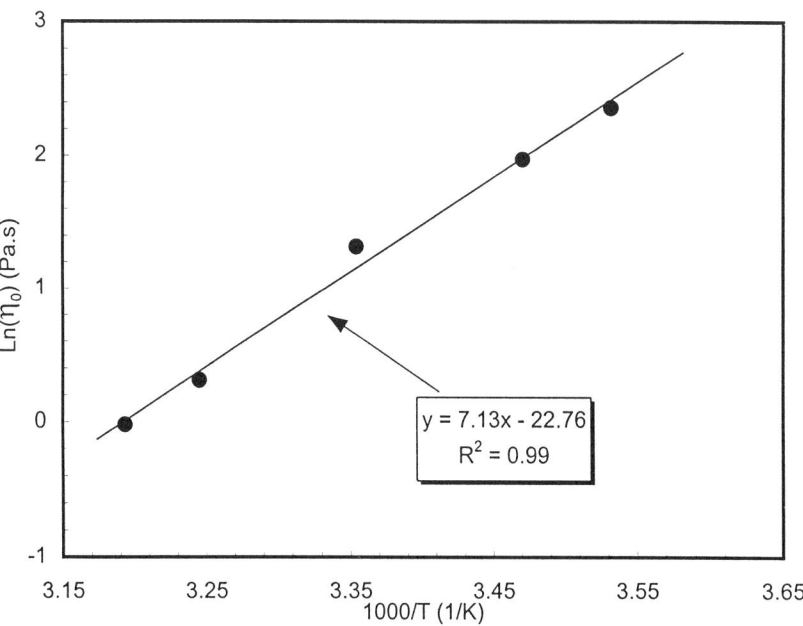

Figure 6. An example of Arrhenius plot of 1wt% HASE-(EO)35-C20H41 in 0.003M $C_{18}EO_{100}$.

entropic consideration, free spherical surfactant micelles would first bind onto junctions with lower aggregation number. The addition of surfactant to HASE at this stage shifted the polydisperse aggregation number junctions to ones, which were more monodisperse. The growth in the micellar size of these junctions would hence lead to stronger hydrophobic junctions as shown by the increase in the activation energy E_a. At 0.1mM < Csurf < c* (50 times *cmc*), lower aggregation junctions were saturated by the micelles and free micelles started to interact with the higher aggregation junctions. This caused a reduction in the functionality of the polymer, which created additional inter-molecular junctions to the network. At the same time, the electrostatic repulsive forces along the polymer backbone provided a more expanded conformation of the chains in the network. The resulting effect was a low shear viscosity enhancement brought about by the increase in the number of mechanically active junctions, network strength and the hydrodynamic volume. However, λ remained constant up to c*, which signified that $C_{18}EO_{100}$ had negligible effects on the structural relaxation dynamics. At Csurf > c*, there was a clear observation of network disruption due to saturation of mixed micellar junctions by the free micelles (8), as reflected by the decrease in low shear viscosity. Such decrease was attributed to a reduction in the relaxation time, number of mechanically active junctions and network strength.

In order to extend the above discussion further, three samples of 1wt% HASE-(EO)35-C20H41 containing different concentrations of surfactant (i.e. 0, 0.0001, 0.003M) but with **identical low shear viscosity** were chosen and compared in Figure 7. Note that the concentrations of 0.0001M and 0.003M lied in the regions before and after the critical concentration c* respectively. For the polymer in pure solvent and 10^{-4}M $C_{18}EO_{100}$ surfactant solution, the viscosity decreased with increasing shear stress. However, at surfactant concentration of 0.003M, shear-thickening behavior was observed between the stress level ranging from 20 to 50Pa. The values of low shear viscosity, relaxation time, activation energy and storage modulus (at 100rad/s) of these three samples were summarized in Table I.

It shows that, for samples with the same low shear viscosity, λ decreased slightly and $G'_{@100rad/s}$ increased as the surfactant concentration increased. This correlated well with eq 3. The activation energy and $G'_{@100rad/s}$ increased while the relaxation time decreased as the surfactant concentration was increased from 0 to 0.003M. This indicated that the incorporation of surfactant played an important role in providing additional inter-molecular hydrophobic junctions, which increased the strength of the overall network although the low shear viscosity was not affected. In the next section, we examined the technique of superposition of oscillation on steady shear performed on these three samples. This technique will provide further insight on the topology of network structure and various modes of chain relaxation under the effect of different shear conditions.

Superposition of oscillation on steady shear

The rheological technique of superposition of oscillation on steady shear flows has been used to study the effects of shear stress on the linear viscoelastic properties of associative polymer solutions (33,34). This technique has also been used by Vlastos and co-workers to investigate the properties of human blood and

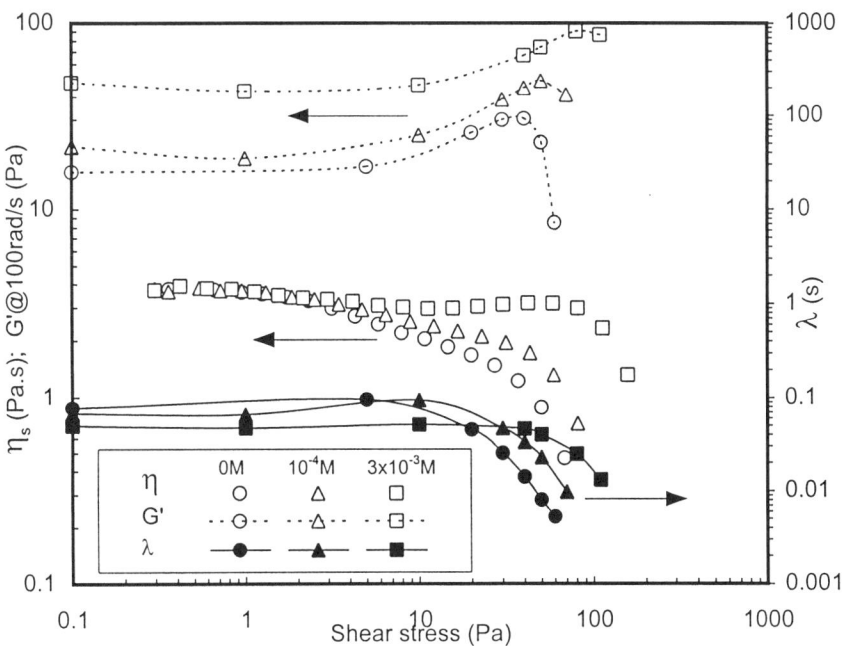

Figure 7. Dependence of shear viscosity, storage modulus (at 100rad/s) and relaxation time on applied shear stress for 1wt% HASE-(EO)35-C20H41 in 0, 0.0001 and 0.003M concentration of $C_{18}EO_{100}$ at pH 9.

Table I: Low shear viscosity, relaxation time, activation energy and storage modulus (at 100rad/s) of 1wt% HASE-(EO)35-C20H41 in a non-ionic surfactant $C_{18}EO_{100}$.

Concentration of $C_{18}EO_{100}$ (M)	η_0 (Pa.s)	λ (s)	Ea (kJ/mole)	G'$_{@100rad/s}$ (Pa)
0	3.77	0.077	46.5	15.8
0.0001	3.74	0.067	56.0	21.3
0.003	3.80	0.049	59.3	47.6

two other aqueous polymer solutions, namely polyacrylamide (AP 273E) and polysaccharide xanthan gum (35). It is a method whereby small amplitude oscillations at different frequencies are applied onto a sample that is continuously being subjected to a steady shear. Figure 8 shows the plots of storage and loss modulus versus angular frequency for 1wt% HASE-(EO)35-C20H41 in 0.0001M $C_{18}EO_{100}$ at various imposed stresses. Plots in 0.0, 0.001 and 0.003M surfactant were not shown here as they showed similar trends to Figure 8. The imposed stresses were varied between 1 to 150Pa.

At increasing applied stresses, the terminal region shifted to higher frequencies as the network structure of the polymer solution was perturbed by the increasing applied stress. Tirtaatmadja referred to this phenomenon as a "truncation" of the relaxation spectrum at the long time-scale end (33). This is to say that some of the longer modes of relaxation time were removed due to the perturbation on the associating network. At higher frequency region, G' curves became parallel to that at no applied stress. The values of G' increased slightly as the applied stress increased which signified an increase in the number of mechanically active hydrophobic junctions. The imposed deformation in this case caused a restructuring of the associative network.

At concentrations of $C_{18}EO_{100}$ greater than c*, it was observed that the G" modulus seemed to go through a maximum and then decreased as the frequency was increased. Under increasing applied stresses, the maximum value of G" also showed an initial decrease followed by a significant increase (the increase was about 3.3 times for solution in 0.001M and 3.5 times in 0.003M $C_{18}EO_{100}$). Comparing the solutions with no surfactant and in surfactant concentration below c*, the G" maxima could only be observed above an applied stress of about 10 Pa. Applying stress beyond this value would then result in an increase in the G" maxima. This increase in the G" maxima might be a consequence of the applied stress resulting in the perturbation or even disruption of the associative network. As a result, the polymer chains were stretched which caused the remaining network structure to occupy a larger hydrodynamic volume in the solution. And it is this volume expansion that leads to the increase in the G" maxima.

The crossover of G' and G" separating the terminal region and the "plateau" region gives a rough value of λ at each applied stress. This enables one to estimate the relaxation time of the polymer solution under various conditions of network perturbation. A plot of these relaxation times and steady shear viscosity as a function of stress is shown in Figure 7. It was observed that the phenomenon of shear-thinning together with reduction in relaxation time signified a disruption of associating network beyond a certain critical stress. In the absence of surfactant, viscosity profiles shear-thinned at a lower stress level and much faster rates than that in the presence of surfactant. The dependence of relaxation time on applied stress (filled symbols) followed the same trends as the shear viscosity. The critical stress for the onset of shear-thinning corresponded with the decrease in λ, i.e. the critical stress increased with surfactant concentrations. The storage modulus at 100rad/s is also plotted in Figure 7 against applied stresses and it shows that the number of inter-molecular hydrophobic junctions increased to a maximum corresponding to the critical stress for each of the 3 samples. Thereafter, the effective number of junctions decreased. The values of shear viscosity η, relaxation time, activation energy and storage modulus (at 100rad/s)

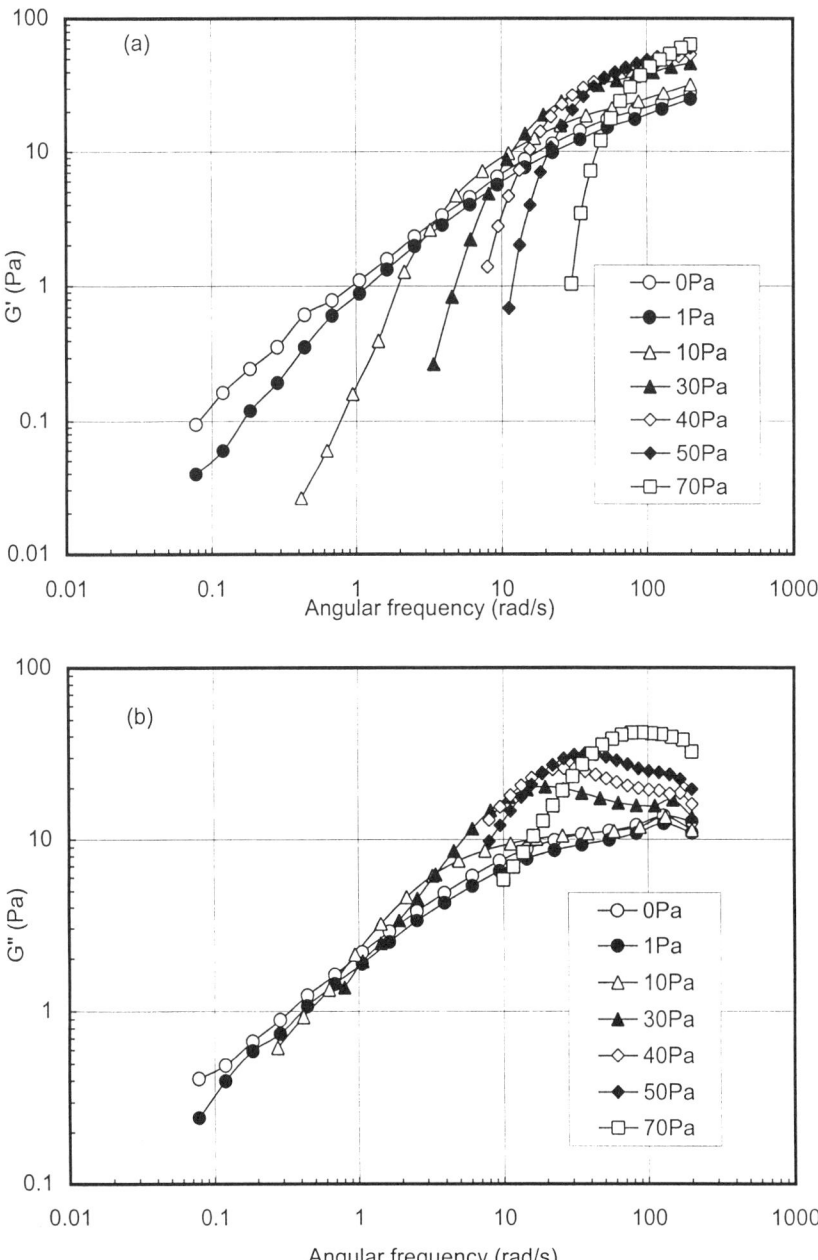

Figure 8. Dependence of (a) storage and (b) loss modulus on angular frequency for 1wt% HASE-(EO)35-C20H41 in 0.0001M $C_{18}EO_{100}$ at various applied stresses at pH 9.

of HASE-(EO)35-C20H41 in 0, 0.0001 and 0.003M $C_{18}EO_{100}$ at the same applied stress of 50 Pa are given in Table II.

At the applied stress of 50Pa, η and λ increased with surfactant concentrations. This is in contrast to the condition with no applied stress where the viscosity of the three solutions was identical (i.e. ~3.7-3.8 Pa.s). Similarly, E_a and $G'_{@100rad/s}$ increased with increasing surfactant concentrations. Thus, in the presence of increasing surfactant concentrations, the application of stress up to the critical value promoted the formation of associative network with topology that was driven predominantly by an increase in number and strength of the inter-molecular hydrophobic junctions. The decreasing trend in average relaxation time, strength and number of mechanically active junctions beyond the critical stress signified a disruption of associative network.

As discussed earlier, the application of stress plays an important role in promoting an associative network with different topology. From Table I and II, the polymer solution in 0.003M $C_{18}EO_{100}$, which exhibited shear-thickening were compared under different conditions. One would observe that in the shear-thickening region (at 50 Pa):

1. The relaxation time λ of the system was less affected by stress compared to those systems with no shear-thickening, indicating that the structural relaxation dynamics of the network remained consistent,
2. The activation energy E_a showed an increase which indicated an increase in the strength of the network,
3. The value of $G'_{@100rad/s}$ significantly increased which clearly signified an increase in the number of mechanically active junctions attributed to a shear-induced transformation of intra- to inter-molecular junctions.

It is clear from the above results that the observed shear-thickening behavior is due to the intra- to inter-molecular transformation yielding greater proportion of mechanically active junctions (10, 17).

Table II: Shear viscosity, relaxation time, activation energy and storage modulus (at 100rad/s) of 1wt% HASE-(EO)35-C20H41 in a non-ionic surfactant $C_{18}EO_{100}$ at an applied stress of 50 Pa.

Concentration of $C_{18}EO_{100}$ (M)	η_s (Pa.s)	λ (s)	Ea (kJ/mole)	$G'_{@100rad/s}$ (Pa)
0	0.88	0.008	45.0	22.7
0.0001	1.55	0.023	58.2	48.8
0.003	3.17	0.040	63.2	74.3

Conclusions

The interactions and rheological properties of hydrophobically modified alkali-soluble emulsion (HASE) model polymer in the presence of non-ionic surfactant could be correlated to the network strength and junction density. The interacting mechanisms for enhanced viscosifying properties involved the binding of free non-ionic surfactant micelles onto HASE hydrophobic junctions above *cmc* of free surfactant. The maximum thickening efficiency was attained at a concentration 50 times of *cmc*.

Acknowledgement

This work is supported by the Ontario-Singapore collaborative research program and the Ministry of Education. One of the authors (SWP) would like to acknowledge the financial support provided by the Ministry of Education, Singapore. We thank Professor Winnik for the many helpful discussions. We also thank the reviewers for their useful and constructive comments.

References

1. Lundberg, D. J.; Ma, Z.; Alahapperuna, K.; Glass, J. E. In *Polymers as Rheology Modifiers*; Schulz, D. N., Glass, J. E., Eds.; ACS Symp. Ser. 462; American Chemical Society: Washington, DC, 1991; pp 234-253.
2. Loyen, K.; Iliopoulos, I.; Olsson, U.; Audebert, R. *Progr. Colloid Polym. Sci.* **1995**, 98, 42-46.
3. Tirtaatmadja, V.; Tam, K. C.; Jenkins, R. D. *AIChE J.* **1998**, 44, No. 12, 2756-2765.
4. Jenkins, R. D.; Bassett, D. R. Synergistic Interactions among Associative Polymers and Surfactants, In *Polymeric Dispersions: Principles and Applications*; Asua, J. M., Ed.; Kluwer: Dordrecht, The Netherlands, 1997; pp 477.
5. Olesen, K. R.; Bassett, D. R.; Wilkerson, C. L. Surfactant Co-thickening in Model Associative Polymers. (Presented in the organic coatings conference, Athens, summer, 1998).
6. Jenkins, R. D.; Delong, L. M.; Bassett, D. R. In *Hydrophilic Polymers: Performance with Environmental Acceptability*; Glass, J. E., Ed.; Advances in Chemistry Series No. 248; American Chemical Society: Washington, DC, 1996; pp 425-447.
7. Tirtaatmadja, V.; Tam, K. C.; Jenkins, R. D. *Macromolecules* **1997**, 30, 3271-3282.
8. Sau, A. C.; Landoll, L. M. In *Polymers in Aqueous Media: Performance Through Association*; Glass, J. E., Ed.; Advances in Chemistry Series No. 223; American Chemical Society: Washington, DC, 1989; pp 343-364.
9. Seng, W. P.; Tam, K. C.; Jenkins, R. D. *Colloids and Surfaces* **1999**, *in press*.

10. Bock, J.; Siano, D. B.; Valint, Jr., P. L.; Pace, S. J. In *Polymers in Aqueous Media: Performance Through Association*; Glass, J. E., Ed.; Advances in Chemistry Series No. 223; American Chemical Society: Washington, DC, 1989; pp 411-424.
11. Tarng, M. R.; Kaczmarski, J. P.; Lundberg, D. J.; Glass, J. E. In *Hydrophilic Polymers: Performance with Environmental Acceptability*; Glass, J. E., Ed.; Advances in Chemistry Series No. 248; American Chemical Society: Washington, DC, 1996; pp 305-341.
12. Hu, Y.; Rajaram, C. V.; Wang, S. Q.; Jamieson, A. M. *Langmuir* **1994**, 10, 80-95.
13. Volpert, E.; Selb, J.; Candau, F. *Macromolecules* **1996**, 29, 1452-1463.
14. Hu, Y.; Wang, S. Q.; Jamieson, A. M. *Macromolecules* **1995**, 28, 1847-1853.
15. Francois, J.; Maite, S.; Rawiso, M.; Sarazin, D.; Beinert, G.; Isel, F. *Colloids and Surfaces A: Physicochemical and Engineering Aspects* **1996**, 112, 251-265.
16. Tam, K. C.; Jenkins, R. D.; Winnik, M. A.; Bassett, D. R. *Macromolecules* **1998**, 31, 4149-4159.
17. Tam, K. C.; Guo, L.; Jenkins, R. D. *Polymer* **1999**, in press.
18. Ballard, M. J.; Buscall, R.; Waite, F. A. *Polymer* **1988**, 29, 1287-1293.
19. Jenkins, R. D.; Silebi, C. A.; El-Aasser, M. S. In *Polymers as Rheology Modifiers*; Schulz, D. N.; Glass, J. E., Eds.; ACS Symp. Ser. 462; American Chemical Society: Washington, DC, 1991; pp 222-233.
20. Annable, T.; Buscall, R.; Ettelaie, R. *Colloids and Surfaces A: Physicochemical and Engineering Aspects* **1996**, 112, 97-116.
21. van den Brule, B. H. A. A.; Hoogerbrugge, P. J. *J. Non-Newtonian Fluid Mech.* **1995**, 60, 303-334.
22. Lundberg, D. J.; Glass, J. E.; Eley, R. R. *J. Rheol.* **1991**, 35(6), 1255-1274.
23. Jenkins, R. D. Ph.D. Thesis, Lehigh University, 1990.
24. Li, Z. Ph.D. Thesis, Lehigh University, 1995.
25. Glass, J. E.; Schulz, D. N.; Zukoski, C. F. In In *Polymers as Rheology Modifiers*; Schulz, D. N.; Glass, J. E., Eds.; ACS Symp. Ser. 462; American Chemical Society: Washington, DC, 1991; pp 2-17.
26. Volpert, E.; Selb, J.; Candau, F. *Polymer* **1998**, 39, No. 5, 1025-1033.
27. Leibler, L.; Rubinstein, M.; Colby, R. H. *Macromolecules* **1991**, 24, 4701-4707.
28. Aubry, T.; Moan, M. *J. Rheol.* **1996**, 40, 441-448.
29. Green, M. S.; Tobolsky, A. V. *J. Chem. Phys.* **1946**, 15, 80-89.
30. Macosko, C. W. *Rheology Principles, Measurements, and Applications*; VCH Publ., 1994; Chap 3.
31. Barnes, H. A.; Hutton, J. F.; Walters, K. *An Introduction to Rheology*; Elsevier: Armsterdam, The Netherlands, 1989.
32. Chang, Y.; Lochhead, R. Y.; McCormick, C. L. *Macromolecules* **1994**, 27, 2145-2150.
33. Tirtaatmadja, V.; Tam, K. C.; Jenkins, R. D. *Macromolecules* **1997**, 30, No.5, 1426-1433.
34. English, R. J.; Gulati, H. S.; Jenkins, R. D.; Khan, S. A. *J. Rheol.* **1996**, 41(2), 427-444.
35. G. Vlastos, D. Lerche, B. Koch, O. Samba, M. Pohl, Rheol Acta 36, (1997), 160-172.

Chapter 22

Rheology of a HASE Associative Polymer and Its Interaction with Non-Ionic Surfactants

R. J. English[1], R. D. Jenkins[2], D. R. Bassett[3], and Saad A. Khan[4,5]

[1]Department of Color Chemistry,
The University of Leeds, Leeds, United Kingdom
[2]Union Carbide Asia Pacific Inc., 16 Science Park Drive, Singapore 118227
[3]UCAR Emulsions Systems, Union Carbide Corporation, Cary, NC 27511
[4]Department of Chemical Engineering,
North Carolina State University, Raleigh, NC 27695-7905

Solutions of a hydrophobically modified alkali soluble emulsion (HASE) polymer in alkaline media are seen to behave as reversible networks in small amplitude oscillatory shear, but their response is seen to deviate from the simple Maxwellian response exhibited by telechelic HEUR associative polymers. In this polymer, which contains a relatively small number of complex alkylaryl hydrophobes bound to each polymer chain, stress relaxation is considered to be influenced by the both the disengagement rate of the hydrophobes from their junction domains and topological constraints arising from physical entanglements of the chains. Several unusual phenomena are observed in steady shear, including shear induced structuring and stress saturation. Possible microstructural interpretations of these phenomena are discussed. The hydrophile - lipophile balance of non-ionic surfactants is seen to exert a profound influence on the nature of the polymer - surfactant interaction. Rheological differences between solutions of the polymer containing a nonylphenol ethoxylate of higher HLB (NP + 10 EO) and a more hydrophobic surfactant (NP + 6 EO) are described and interpreted in terms of possible differences in surfactant phase behaviour.

Introduction

In spite of their commercial utilisation [1], understanding of the dynamics and topology of HASE (Hydrophobic Alkali Soluble Emulsion) polymer networks is less well developed than that of simple, telechelic HEUR (Hydrophobic Ethoxylated Urethane) associative polymers [2-7]. This arises largely as a consequence of the

[5]Corresponding author.

more complex architecture of HASE systems, lack of knowledge regarding the state of supramolecular assembly of the polymer in the dilute regime and the number of chain states possible at higher concentrations. In particular, the assignment of key parameters, such as the mean number of hydrophobes residing in a junction domain (~aggregation number) and their rate of disengagement, presents a difficult task. Recent experimental work on HASE polymers has considered the effects of hydrophobe size and side chain length on solution rheology [8-10], behaviour in parallel superposed steady/dynamic shear [11,12], physicochemical changes during solubilisation [13] interactions with surfactants [14,15] and the effects of added electrolytes on the behaviour in the dilute regime [16]. In the present study, we concentrate on some of the rheological phenomena observed with solutions of HASE polymers and how these are affected on addition of non-ionic surfactants.

As majority of the water-soluble, polymers are employed in formulations containing surfactants or other amphiphilic species, polymer-amphiphile interactions are of considerable technological and theoretical interest [17,18]. The nature of the polymer-surfactant interaction may involve conformational transitions, owing to adsorption of surfactant, microphase separation due to complexation, or modification of network formation via formation of mixed micelles at network junctions [19]. The latter form of interaction is particularly relevant to HASE polymers, as network junctions involve dynamic association of hydrophobic groups. In the case of non-ionic amphiphiles of low water solubility, the phase structure of the system may be more complex and modified by inclusion of polymeric species [20]. Recent studies have concentrated on the role of surfactant phase behavior on the rheology of polymer-amphiphile systems and also the development of systems that exhibit gelation. Panmai *et al.* [21] examined the rheology of hydrophobically modified hydroethyl cellulose and hydrophobically modified poly(acrylamide) in the presence of a range of surfactants. Under conditions where the surfactants formed spherical micelles, interchain hydrophobic interactions were screened at the highest surfactant concentrations. Kaczmarski *et al.* [22] found the presence of both SDS and an octylphenol ethoxylate to promote structuring of HEUR polymers bearing large terminal hydrophobes. Deguchi *et al.* [23] demonstrated the gelation of solutions of cholesterol modified pullulan in the presence of SDS. The thermoreversible gelation of hydrophobically modified poly(acrylic acid)s in the presence of linear alcohol ethoxylates was studied by Loyen *et al.* [20, 24]. Interactions of a HASE polymer with non-ionic surfactants is considered in the present study, using alkylphenol ethoxylates of different hydrophile-lipophile balance and different degrees of aqueous solubility. In this respect, we attempt to demonstrate more clearly the role of surfactant phase behavior in dictating the mode of interaction with a hydrophobically modified polymer of relatively complex architecture.

Experimental

The polymer examined here is identical to the material employed in our previous publications [11,14] and has the idealised constitution depicted in Figure 1. The hydrophobes comprise oligomeric condensates of nonylphenol, resulting in a relatively high hydrophobe molar volume. Full details of the emulsion polymerization process employed in the synthesis of model HASE polymers of this type are described in detail elsewhere [25]. The mean degree of ethoxylation in the macromer, n, is ~80 and the weight fractions of ethyl acrylate, methacrylic acid and macromer employed in the polymer synthesis were 0.4, 0.4 and 0.2, respectively. This corresponds to a mole fraction of macromer of ~ 0.0055. Dilute solution viscometry showed the intrinsic viscosity, [η] of the polymer to be ~ 4.8 dlg^{-1}, at 25°C in 0.05 M NaCl at pH 9. The HASE polymer latex was purified by dialysis [Spectropore 7 cellulosic membrane – 50000 Da M_w cut-off], in order to remove serum electrolyte and excess anionic stabilizer. The non-ionic surfactants employed were commercial materials supplied by Union Carbide – Tergitol NP6, a nonyl phenol ethoxylate with a mean degree of ethoxylation of ~6, and Tergitol NP10 a nonylphenol ethoxylate with a mean degree of ethoxylation of 10. Both surfactants were used without further purification. The polymer latex was solubilized in the presence of 2-amino-2-methyl-1-propanol (AMP-95), at a level of 6.0 x 10^{-3} mol. of the amine per gram of polymer (pH~9). Samples where prepared at a constant ionic strength (0.05 M NaCl), by combining the required amounts of purified latex, distilled-deionised water, 0.5 M NaCl and 1.0 M AMP and, where appropriate, non-ionic surfactant. Although HASE polymers are polyelectrolytes in their native state, any polyelectrolyte behavior (or electrostatic effects) were eliminated/screened under these experimental conditions [16]. The HASE polymer concentration of samples containing NP6 was fixed at 0.6 g dl^{-1}, whilst the concentration of samples containing NP10 was fixed at 10 gl^{-1}. The concentration of NP6 in the system, c_{NP6}, was varied between 0.5 and 15 gl^{-1} and the concentration of NP10, c_{NP10}, between 1 and 11 gl^{-1}. All samples were centrifuged (2500 rpm, 5 min.) in order to remove entrained air and allowed to stand for several days prior to rheometrical characterization.

Rheometrical experiments were carried out in steady and dynamic shear using a Rheometric Scientific DSR controlled stress rheometer, fitted with appropriate cone and plate and concentric cylinder geometries. Steady shear data at low rates of deformation ($\dot{\gamma} < 10^{-2}$ s^{-1}) where derived from sequential creep experiments, thus ensuring that the duration of the experiment was sufficient to for attainment of a steady state strain rate. High frequency dynamic data where obtained on a Rheometrics Scientific RMS800 controlled deformation rheometer, fitted with a coni-cylinder geometry. All experiments were conducted at 25 ± 0.1 °C. Pre-shearing was carried out for 180s, followed by a rest period of 120s, prior to commencing the steady shear experiment. Evaporation of the samples was prevented by coating the exposed edges with a thin film of low viscosity PDMS fluid (Dow Corning DC200, 10 cS).

Results and Discussion

We consider initially the behavior of the HASE polymer before embarking on the effects of adding non-ionic surfactants to the system. Figure 2 depicts the steady shear viscosity of the HASE polymer solutions as a function of the applied shear stress. For each of the polymer concentrations studied, a linear response (i.e. $\eta \rightarrow \eta_o$) is only

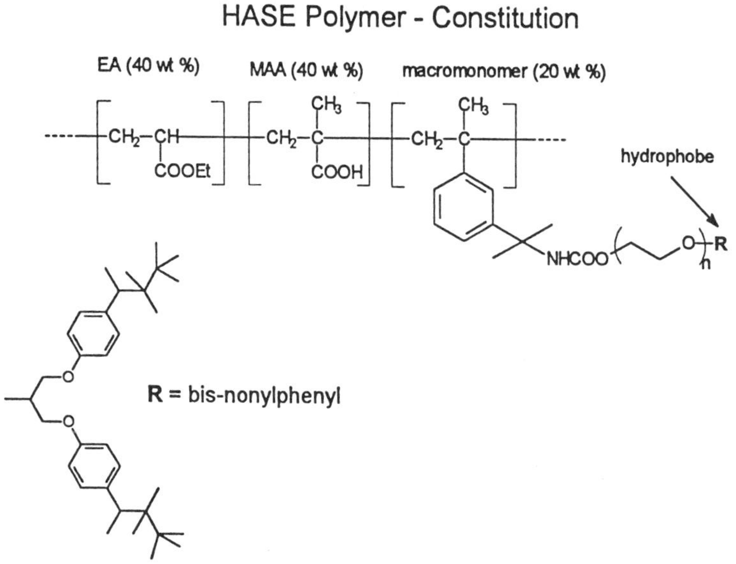

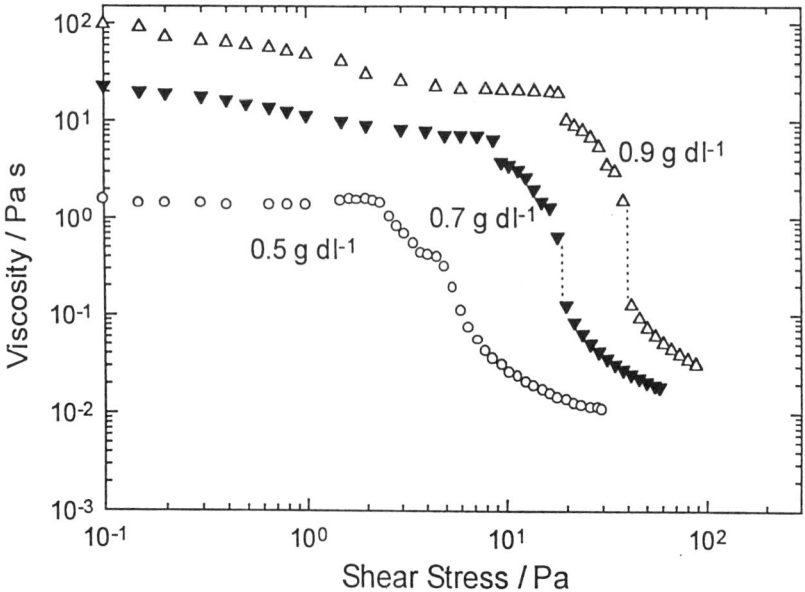

Figure 1. Idealized constitution of the HASE polymer considered in the present study.

Figure 2. Steady shear data for solutions of the HASE polymer: viscosity as a function of applied shear stress.

observed at the lowest applied shear stresses and the solutions exhibit rather complex non-linear behaviour, characterized by a number of features. At the lowest polymer concentrations studied, the zero shear plateau is seen to extend to intermediate levels of applied stress ($\sim 10^0$ Pa). At higher shear stresses and the lowest polymer concentration ($c = 0.5$ g dl^{-1}) the onset of non-linearity is marked by a transition to a region of weak shear thickening - attributable to shear induced structuring within the polymer solution [26-28]. At higher polymer concentrations, however, the onset of non-linearity is seen as a weak shear thinning. Shear induced structuring is also apparent at higher polymer concentrations, but is much less pronounced and is seen as a region where the viscosity becomes almost independent of the shear stress. The region of shear induced structuring in the HASE system precedes an abrupt decrease in the viscosity, consistent with pronounced shear thinning of the polymer solutions. At higher polymer concentrations stress saturation is noted, manifested as a marked discontinuity in the viscosity-stress curve.

The behaviour of the HASE polymer contrasts the response of telechelic HEUR systems, where there is typically a transition from linearity to a zone of pronounced shear thinning, separated by a well-defined maximum in the viscosity [3,4]. Thus, at higher concentrations of HASE polymer, it is likely that both topological entanglements and hydrophobic associations are present. Taking the value of 1/[η] as an estimate of the coil overlap concentration, c^* (~ 0.2 g dl^{-1}), each of the polymer solutions considered in Figure 2 would fall within the non-dilute (entangled) regime.

Shear induced structuring in polymeric systems is usually attributed to an increase in specific intermolecular associations, accompanying the conformational changes experienced in shear flow [27], whilst depletions of chain entanglements results in the shear thinning more typically associated with solutions of polymeric materials [29]. Thus, it is likely that the interplay between these effects is manifested in the overall form of the viscosity-stress curve. Such considerations, however, are not relevant to a discussion of telechelic HEUR systems, where physical entanglements of the chains are considered to be absent [3,4].

Data showing the response of the HASE polymer solutions in oscillatory shear are presented in Figure 3, showing the frequency dependence of the dynamic moduli. These results are in qualitative agreement with data previously reported for polymer networks with reversible interchain associations, in that the system does not respond as a physical gel [28, 30, 31] In contrast to systems in which interchain associations persist over long tome-scales, the dynamic response of the HASE polymer solutions is characteristic of a viscoelastic fluid, consistent with the transient nature of the hydrophobic associations. Thus, the mechanical spectra show features characteristic of polymers interacting by purely topological interactions, in that stresses are able to relax over long time-scales - i.e. the binding energy of the hydrophobes is sufficiently low to enable them to disengage at a finite rate [3,4]. Note, however, that the linear viscoelastic response differs significantly from that previously observed for HEUR systems, in that the network dynamics are no longer represented by a single Maxwell element. This is apparent in the weaker frequency dependence of the dynamic moduli in the terminal zone, with $G' \sim \omega^{0.8}$ and $G'' \sim \omega^{0.7}$, rather than the predicted exponents of 2 and 1, respectively. Such, a response presumably reflects the co-existence of both hydrophobic associations and topological entanglements. The linear viscoelastic

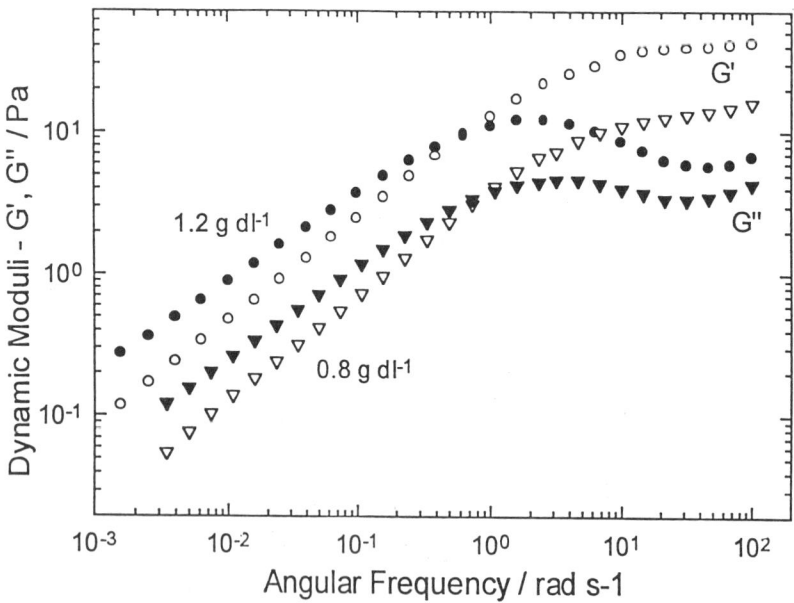

Figure 3. Frequency sweep data for solutions of HASE polymer - storage and less modulus as a function of angular frequency.

response may be described in terms of two characteristic parameters, a pseudoequilibrium modulus, G_N, and a characteristic time, T_d. Quantitatively, the magnitude of G_N may be interpreted in terms of the number density of elasticity active chains, ν [32]. A more convenient procedure involves the estimation of high frequency plateau modulus, G'_p, given by the value of G' at the frequency corresponding to the maximum in G'' [32]. The reciprocal of this characteristic frequency, ω_m, gives a measure of T_d, interpreted here as a mean longest relaxation time.

We present rheometrical data for solutions of the HASE polymer, prepared in the presence of the water-soluble surfactant NP10 in Figure 4. The effect of the more hydrophilic surfactant, NP10, is seen as a pronounced reduction in the magnitude of the high frequency value of the storage modulus, with G'_p decreasing by over two decades at the highest levels of surfactant addition. Correspondingly, the zero shear viscosity is seen to decrease by ~3 orders of magnitude over the same range of surfactant addition. Previous studies of hydrophobically associating polymers, in the presence of both anionic and non-ionic surfactants at concentrations greatly in excess of their critical micelle concentration (c_{CMC}), have also noted such effects [4, 33, 34]. Above c_{CMC} NP10 is likely to form small, spherical micelles. Microstructurally, one may envisage that polymer bound hydrophobes are incorporated into these micelles, forming mixed junction domains [4, 33, 34]. Consequently, as the concentration of surfactant micelles in the system is increased, the overall functionality of the junction domains (i.e. the mean number of polymer-bound hydrophobes present in a given junction domain) is progressively reduced. This loss in connectivity is reflected macroscopically as a decrease in the magnitude of G'_p and η_o.

Another apparent effect on increasing c_{NP10} is a profound reduction in the tendency for shear-induced structuring. At low/no surfactant concentrations, the steady shear viscosity is larger than the complex viscosity and shows a plateau region at high strain rates, corresponding to shear induced structuring. These effects are significantly reduced with increased surfactant addition consistent with the formation of mixed junction domains. The effect of low concentrations of added surfactant ($c > c_{CMC}$) to solutions of comb-like associative polymers is to initially cause a entropically driven change in network topology, leading to an increase in intermolecular hydrophobic associations at the expense of intramolecular [33]. This represents an overall increase in the connectivity of the dynamic network and there is an attendant increase in η_o. Our previous photophysical studies have indicated the CMC of NP10 to be well below the range of concentrations employed in the current study. However, this effect has been demonstrated by Moan and coworkers [34], when considering the interactions of a hydrophobically-modified galactomannan ether with non-ionic surfactant. Further increase in c_{NP10} leads to the depletion in network connectivity described above. Thus, this effective reduction in intramolecular polymer-polymer associations in the presence of surfactant is consistent with the data in Figure 4A and the apparent elimination of shear induced structuring. The attendant loss in connectivity is also reflected in the steady shear response becoming more linear at higher values of c_{NP10}, with the disappearance of the pronounced shear thinning seen in the absence of surfactant. Thus, interactions of the HASE polymer with NP10 demonstrate the profound effect of hydrophobic interactions on the

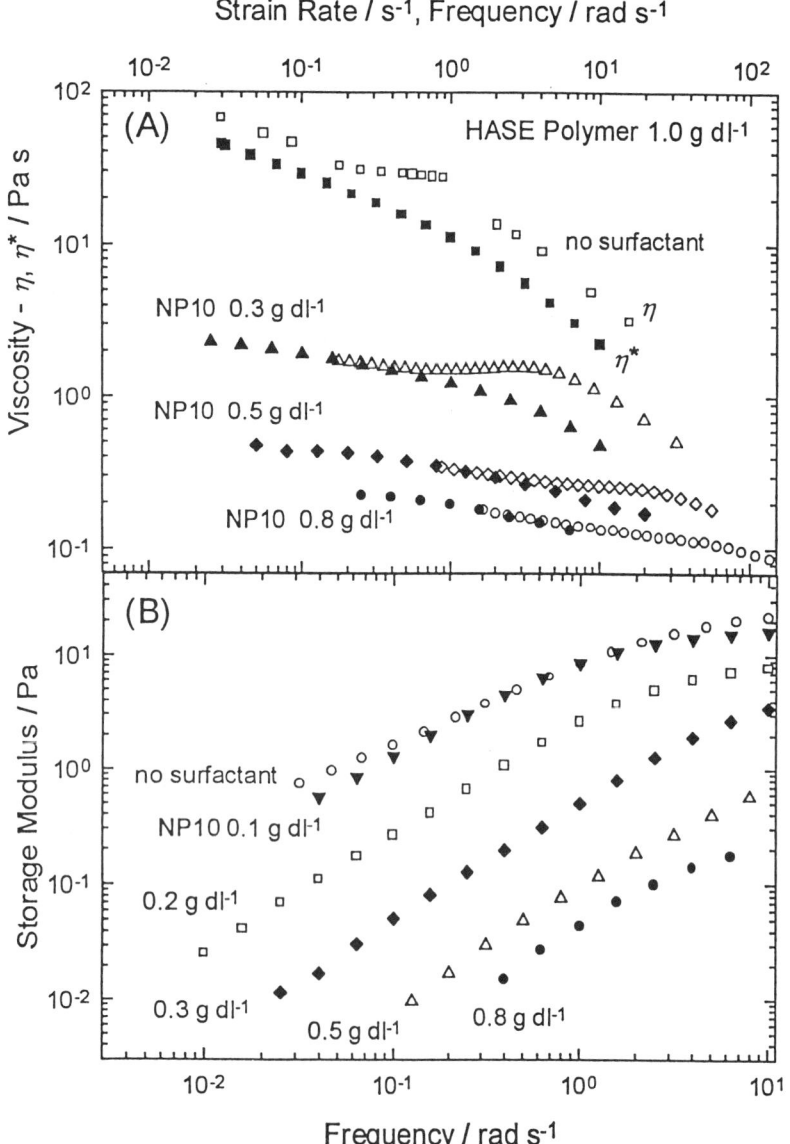

Figure 4. Influence of NP10 surfactant on the rheology of HASE polymer solutions - (A) comparison of steady and complex dynamic viscosities. (B) frequency dependence of the storage modulus.

network connectivity and dynamics in associative polymer systems. In the presence of high concentrations of NP10 the steady shear response is analogous to the behavior of alkali-swellable EA-*co*-MAA latex prepared without macromer [8, 10].

Selected steady shear data of the HASE polymer in the presence of NP6 [c_{NP6} = 1 gl^{-1}], a surfactant of limited water solubility, are shown in Figure 5A and contrasted with equivalent data obtained in the absence of surfactant. Here the presence of NP6 exerts a profoundly different effect, in that an attendant structuring (increase in η_0) is apparent and shear-induced structuring is preserved. It is interesting to note that higher amounts of NP6 [$c_{NP6} \rightarrow$ 15 gl^{-1}] resulted in systems with progressively more structured and gel-like behavior. The tendency towards shear fracture and the generation of flow instabilities precluded reliable characterization in steady shear, at higher levels of NP6, and we resorted to dynamic experiments to provide additional microstructural information. The contrasting behavior, associated with addition of NP6 to the system is also shown in Figure 5B. Increased network connectivity is again reflected in an increase in the value of G'_p with increasing c_{NP6}. There is also an attendant increase in the characteristic time, T_d reflected in the shifting of the terminal region to progressively lower frequencies, as the concentration of the more hydrophobic surfactant is increased. There is approximately a twofold increase in G'_p as c_{NP6} is increased from 1 to 15 gl^{-1}. More significantly, the characteristic time, T_d, increases by two orders of magnitude over the same concentration regime.

It is apparent that the above trends may be underpinned by more complex phase changes induced by incorporation of relatively large amounts of an amphiphilic species of low aqueous solubility [20, 35]. Surfactant phase structure has also been considered with respect to structuring of in HASE polymer solutions containing a primary alcohol exthoxylate of low HLB ($C_{12}EO_4$) [15]. Non-ionic surfactants of low degree of ethoxylation have been shown to exhibit a rich variety of phase structures in the presence of water soluble polymers. Specifically, large vesicular structures may be produced (Lamellar dispersion, L^+_α), or "sponge" type structures involving formation of a continuous bilayer (L_3) [20, 35]. The thermodynamically favored phase structure will be influenced by the concentration of polymer and surfactant, temperature and the surfactant hydrophile-lipophile balance. It is plausible, therefore, that the pronounced structuring seen in the present of higher concentrations of NP6 arises as a consequence of polymer bridging, with incorporation of polymer hydrophobes into the surfactant bilayers. The HASE-NP6 system is thus worthy of further study, employing LALS or cryo-TEM techniques. Parallels may be drawn with previous studies involving hydrophobically-modified poly(acrylic acid) and $C_{12}EO_n$ surfactants. In this case, thermally induced gelation was interpreted in terms of the generation of vesicular phases at elevated temperatures [24,36].

Conclusions

The solution rheology of the HASE polymer exhibits a non-linear response characterized by shear inducing structuring and stress saturation, and a dynamic response not adequately described by a single Maxwell element. The presence of non-ionic surfactants in the polymer indicates a pronounced difference in the rheological

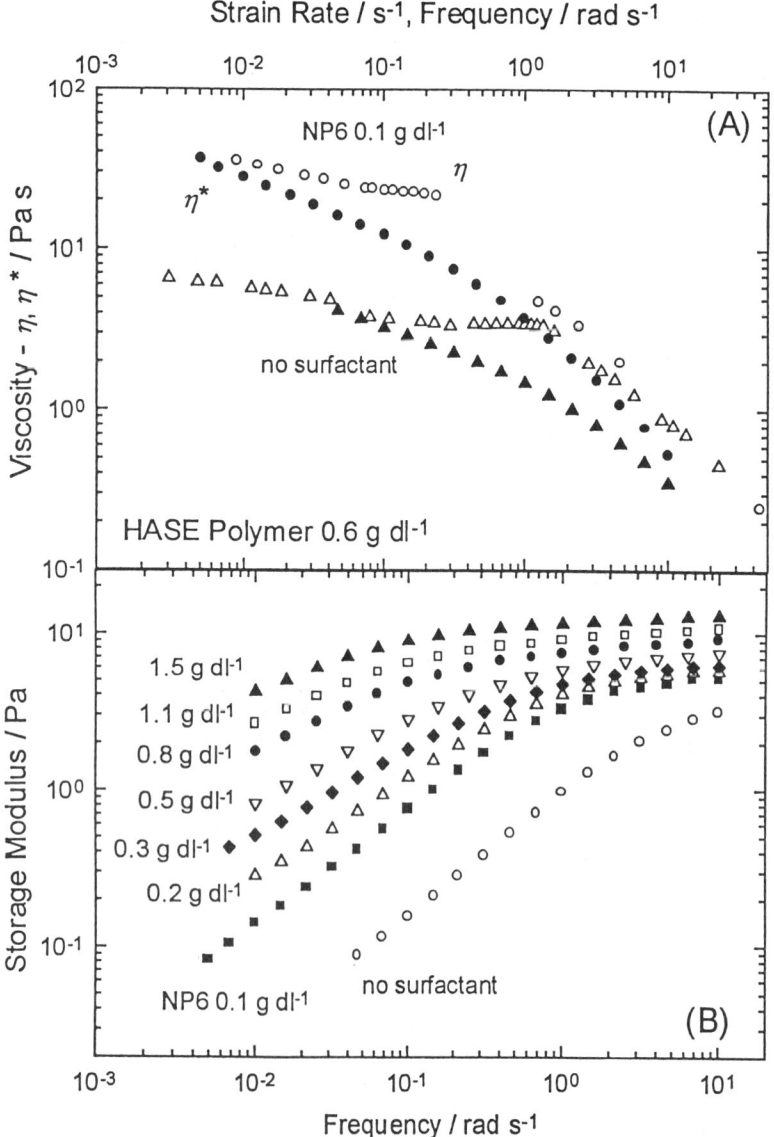

Figure 5. Influence of NP6 surfactant on the rheology of HASE polymer solutions - (A) comparison of steady (η) and complex dynamic (η^*) viscosities. (B) frequency dependence of the storage modulus.

behaviour of the systems depending on the type of surfactants used. In the case of surfactants that are completely water soluble, the system exhibits a reduction in structuring and a decrease in viscosity and modulus. In the case of surfactants with limited water solubility, we observe an enhancement in structuring and a progressive shift to gel-like behavior. The observed differences in the rheology of the HASE polymer may only be adequately interpreted in terms of possible differences in surfactant phase behaviour, rather than a simple incorporation of the surfactant into the hydrophobic junction domains of the polymer.

References

1. Shay, G. D. and Rich, A. F., *J. Coatings Tech.*, **58**(1986), 43.
2. Jenkins, R. D. *"The Fundamental Thickening Mechanism of Associative Polymers in Latex Systems: A Rheological Study"*, Ph.D. Dissertation, Lehigh Univ. (1990)
3. Annable, T., Buscall, R., Ettelaie, R. and Whittlestone, D., *J. Rheol.*, **37** (1993), 695.
4. Annable, T.; Buscall, R. and Ettelaie, R., *Colloids Surfaces A: Physicochem. Eng. Aspects"*, **122** (1996), 97.
5. Jenkins, R. D.; Silebi, C. A. and El-Aasser, M. S. *"Steady Shear and Linear Viscoelastic Properties of Model Associative Polymer Solutions"*, in *"Polymers as Rheology Modifiers"*, Glass, J. E., Ed., ACS Symp. Series #462., pp 222, American Chemical Society (1991).
6. Semenov, A. N.; Joanny, J.-F. and Khokhlov, A. R., Macromolecules, **28** (1995), 1066.
7. Groot, R. D. and Agterof, W. G. M, Macromolecules, **28** (1995), 6824.
8. Tirtaatmadja, V; Tam, K. C. and Jenkins, R. D, Macromolecules, **30** (1997), 3271.
9. Tam, K.C.; Farmer, M. L.; Jenkins, R. D. and Bassett, D. R, J. Polym. Sci. B - Polym. Phys., **36** (1998), 2275.
10. English, R. J.; Raghavan, S. R., Khan, S. A. and Jenkins, R. D., J. Rheol (in press)
11. English, R. J., Gulati, H. S.; Jenkins, R. D. and Khan, S. A., J. Rheol., **41** (1997), 427.
12. Tirtaatmadja, V., Tam, K. C. and Jenkins, R. D, Macromolecules, **30** (1997), 1426.
13. Kumacheva, E.; Rharbi, Y.; Winnik, M.A., Guo, L.; Tam, K. C. and Jenkins, R. D., Langmuir, **13** (1997), 182.
14. English, R. J., Gulati, H. S.; Smith, S.; Khan, S. A. and Jenkins, R. D, Proc. Int. Congr. Rheol., Quebec City, Canada (1996).
15. Tirtaarmadja, V., Tam, K. C. and Jenkins, R. D., AIChE J., **44** (1998), 2756.
16. Guo, L.; Tam, K. C. and Jenkins, R. D., Macromolecular Chem. Phys., **199** (1998), 1175.
17. Goddard, E. D. and Ananthapadmanabhan, K. P. *"Interactions of Surfactants with Polymers and Proteins"*, CRC Press, Boca Raton FL (1993).

18. Winnik, F. M. and Regismond, S. T. A., Colloids Surfaces A: Physicochem. Eng. Aspects, **118** (1996), 1.
19. Tanaka, F. and Ishida, M., Macromolecules, **30** (1997), 1836.
20. Loyen, K.; Iliopoulos, I.; Olsson, U. and Audebert, R., Progr. Colloid Polym. Sci., **98** (1995), 42.
21. Panmai, S; Prudhomme, R. K. and Peiffer, D. G., *Colloids Surfaces A - Phsicochem. Eng. Aspects*, **147**(1999), 3.
22. Kaczmarski, J. P.; Tarng, M. R. Ma, Z. Y. and Glass, J. E., *Colloids Surfaces A - Phsicochem. Eng. Aspects,* **147**(1999), 39
23. Deguchi, S.; Kuroda, K.; Akiyoshi, K.; Lindman, B. and Sunamoto, S., *Colloids Surfaces A - Phsicochem. Eng. Aspects*, **147**(1999), 203.
24. Loyen, K.; Iliopoulos, I.; Audebert, R. and Olsson, U., Langmuir, **11** (1995), 1053.
25. Jenkins, R. D. and Bassett, D. R.,, Sterlen, R. A. and Daniels, W. B., US Patent 5561189, 1996
26. Ahn, K. H. and Osaki, K., J. Non-Newtonian Fluid Mech, **55** (1994), 215.
27. Ballard, M. J.; Buscall, R. and Waite F. A., Polymer, **29** (1988), 1287.
28. Aubry, T. and Moan, M., *J. Rheol.*, **38** (1994), 1681.
29. Graessley, W. W. *"The Entanglement Concept in Polymer Rheology"*, *Adv. Polym Chem.*, Vol 17. (1974)
30. Maerker, J. M. and Sinton, S. W., J. Rheol., **30** (1986), 77.
31. Pezron, E.; Ricard, A. and Leibler, L., J. Polym. Sci., **28** (1990), 2445.
32. Ferry, J. D. *"Viscoelastic Properties of Polymers"*, (Third Ed.), J. Wiley & Sons Inc., New York (1980).
33. Annable, T.; Buscall, R.; Ettalaie, R; Shepard, P. and Whittlestone, D., Langmuir, **10** (1994), 1070.
34. Aubry, T. and Moan, M., J. Rheol., **40** (1996), 441.
35. Jonstromer, M. and Strey, R., J. Phys. Chem., **96** (1992), 5993.
36. Sarrazin-Cartalas, A.; Iliopoulos, I., Audebert, R. and Olsson, U., Langmuir, **10** (1994), 1421.

Indexes

Author Index

Allgaier, J., 21
Anghel, Dan F., 286
Argillier, J. -F., 239
Audibert, A., 239
Bassett, D. R., 351, 369
Beers, Kathryn, 52
Borisov, O.V., 109
Candau, Françoise, 95
Chu, Benjamin, 2
Coca, Simion, 52
Cooper, Stuart L., 127
Davis, H. T., 239
Davis, Kelly, 52
Elliott, Peter T., 163
English, R. J., 369
Fu, Zengli, 206
Gaynor, Scott G., 52
Gerst, Matthias, 37
Gitsov, Ivan, 72
Glass, J. Edward, 163, 254
Halperin, A., 109
Hansen, Finn Knut, 303
Hogen-Esch, Thieo E., 179
Huff, Jürgen, 271
Jenkins, R. D., 351, 369
Khan, Saad, A., 369
Lau, W., 221
Lecourtier, J., 239
Liang, Dehai, 2
Lindman, Björn, 303
Liu, Tianbo, 2
Ma, Sharon, 127
Ma, Zeying, 254
Matyjaszewski, Krzysztof, 52
Miller, C. M., 338
Mubarekyan, Ervin, 206
Muhlebach, Andreas, 52
Myrvold, Rolf, 303
Nace, Vaughn M., 2

Nilsson, Susanne, 317
Nomula, Srinivas, 127
Olesen, K. R., 338
Pham, Q. T., 221
Piculell, Lennart, 317
Poppe, A., 21
Qiu, Jian, 52
Regismond, Sudarshi T. A., 286
Richter, D., 21
Russel, W. B., 221
Santore, Maria M., 206
Schuch, Horst, 37
Selb, Joseph, 95
Seng, W. P., 351
Shay, G. D., 338
Sjöström, Jesper, 317
Starkey, P. T., 239
Stellbrink, J., 21
Tam, K. C., 351
Thibeault, J. C., 221
Thuresson, Krister, 317
Tirrell, M. V., 239
Urban, Dieter, 37
Vorobyova, Olga, 143
Weickenmeier, Meik, 271
Wenz, Gerhard, 271
Wetzel, Wylie H., 163
Willner, L., 21
Winnik, Françoise M., 286
Winnik, Mitchell A., 143
Wu, Chunhung, 2
Xia, Jianhui, 52
Xie, David, 179
Xie, Yi, 2
Xing, Linlin, 163
Zhang, Huashi, 179
Zhang, Xuan, 52
Zhou, Shuiqin, 2

Subject Index

A

Acrylamides. *See* Associating polymers in water; Force study of adsorbed layers on hydrophobically modified polyacrylamide

Adsorption
hydrophobic ethoxylated urethanes (HEURs) on latices, 263–267
See also β-Cyclodextrin complexations with surfactants and hydrophobically modified ethoxylated urethanes (HEUR); Dispersions containing poly(ethylene oxide) (PEO) with C_{16} hydrophobes; Dynamics of adsorbed polymer layers; Ethylated hydroxyethyl cellulose (EHEC); Force study of adsorbed layers on hydrophobically modified polyacrylamide

Aggregation
intramolecular in polysoaps, 111–114
surface tension of aqueous solutions as function of hydrophobic block length, 40, 41*f*
See also Amphiphilic block copolymers as surfactants; Associating polymers in water

Aggregation numbers
calculating from micelle concentration, 146–147
dependence of mean aggregation number of hydrophobic modified ethoxylated urethane (HEUR) solutions on polymer concentration, 169–171
determination methods, 144
determination of N_{agg} and N_R by fluorescence quenching, 145–148, 159–160
diblock and triblock copolymers, 3
estimation from steady-state fluorescence quenching measurements, 147–148
function of concentration of HEURs with various hydrophobes, 170*f*
hexadecyl isocyanate end-capped urethane (HDU), 152–153
hydrophobically modified alkali soluble emulsion (HASE) polymer, 155–159
hydrophobically modified hydroxyethylcellulose, 153–155
mean number of hydrophobic substituents N_R, 144
N_{agg} for surfactants and N_R for associative polymers, 146–147
octadecyl isocyanate end-capped urethane (ODU), 152
polyoxyalkylene triblock copolymers, 4, 6
surfactant micelles, N_{agg}, 144
urethane end-capped poly(ethylene oxide), 151–153
See also Fluorescence quenching

Aging
effects on self-exchange for longer poly(ethylene oxide) (PEO) chains, 215
establishing surface mobility of adsorbed chains, 216
self exchange kinetics after various aging periods for short and long PEO layers, 216*f*
See also Dynamics of adsorbed polymer layers

Alkali soluble emulsion (ASE)
general viscosity profiles as

function of neutralization degree for methacrylic acid containing polymer emulsions, 340
swelling and dissolution behavior during neutralization, 340
variety of practical applications, 339
viscosity increase upon neutralization of methacrylic acid-*co*-alkyl acrylate emulsion copolymers (ASE), 339–340
See also Thickening mechanism of hydrophobically modified alkali soluble emulsion (HASE)
Alkali-swellable and alkali-soluble thickeners (ASTs)
ionic form, 339
preparation, 339
Amphiphilic block copolymers as surfactants
aggregation number and micellar diameter by static and dynamic light scattering, 43–44
aggregation of block copolymers, 40–45
aqueous block copolymer solutions preparation, 40
average hydrodynamic diameter of colloidal particles by dynamic light scattering experiments, 44
block copolymer data, 39t
confocal fluorescence correlation spectroscopy (FCS) for determining critical micelle concentration (cmc), 41, 42f
core formation by hydrophobic chains, 45
corona of spherical micelle, 45
dynamic mechanical analysis (DMA) of butyl acrylate/methyl methacrylate (BA/MMA) film, 47, 49f
emulsion polymerization, 46–47
film properties, 47–50
measuring rate of exchange between aggregates by time resolved fluorescence spectroscopy, 43
model of spherical micelle by block copolymer $P(A_m\text{-}b\text{-}B_n)$, 45$f$
number of polymer particles per kg dispersion as function of block copolymer concentration, 47f
reduced aggregation number of micelles in aqueous solution as function of hydrophobic block length m, 46f
static light measurements for average absolute molar mass of micelles and radius of gyration, 45
surface tension of aqueous $P(MMA_m\text{-}b\text{-}AA_n)$ (acrylic acid, AA) solution as function of hydrophobic block length m, 40, 41f
switchover from cationic to anionic chain end in synthesis, 40
synthesis of block copolymers, 38–40
synthesis of PMMA–PAA and PMMA–PMAA (methacrylic acid, MAA) block copolymers, 39f
transmission electron micrograph (TEM) analysis of BA/MMA film before and after annealing, 47, 48f
values for cmc for $P(IB_m\text{-}b\text{-}MAA_n)$ (isobutylene, IB), $P(MMA_m\text{-}b\text{-}AA_n)$, and $P(MMA_m\text{-}b\text{-}MAA_n)$, 43$t$
water uptake of BA/MMA/AA (49%/49%/2%) copolymer film before and after annealing, 50f
Amphiphilic copolymers with linear dendritic architecture
aqueous SEC of ABA copolymers (poly-3,5-dihydroxybenzilic dendrimers as A blocks and poly(ethylene glycol) (PEG) as B block) [G-2]-PEG5,000-[G-2] at different concentrations, 81f

aqueous SEC of [G-2]-PEG5,000-[G-2] solutions in water after heating, 82f
binding of polyaromatic hydrocarbons (PAHs) and C_{60} by [G-2]-PEG5,000-[G-2], 89f
dissolution of PAHs in water, 75
encapsulation of PAHs by [G-2]-PEG5,000-[G-2] in water, 83, 88
encapsulation studies, 75
experimental materials, 75
fluorescence spectroscopy of mixtures of [G-2]-PEG5,000-[G-2] and pyrene in water at different copolymer concentrations, 86f
hybrid copolymer solutions, 75
hydrophilic block PEG chosen, 73
instrumentation, 75
limitations of micelle forming materials, 73
optical microscopy of [G-2]-PEG5,000-[G-2] crystallized from aqueous solutions, 85f
polarized optical micrograph of [G-2]-PEG5,000-[G-2] crystallized from melt, 84f
polarized optical micrographs of [G-2]-PEG5,000-[G-2]/pyrene and [G-2]-PEG5,000-[G-2]/C_{60} mixtures crystallized from aqueous solution, 84–85
poly-3,5-dihydroxybenzilic dendrimers as A blocks, 73, 75
schematic of pyrene encapsulation by [G-2]-PEG5,000-[G-2] in water, 87f
solid state properties of [G-2]-PEG5,000-[G-2], 80, 83
solution behavior of [G-2]-PEG5,000-[G-2], 80
spectroscopic procedures, 77
synthesis of ABA copolymers using monodendrons as initiators and end-capping agents, 78
synthesis of ABA copolymers using PEG as template for divergent dendrimer growth, 79
synthesis of linear-dendritic block copolymers containing PEG, 77, 80
synthesis of second-generation monodendrons [G-2]-Z and subsequent formation of ABA copolymers by reaction of preformed blocks, 76
unimolecular dendritic micelle versus conventional micelle, 74
Amphiphilic polymers, two broad classes, 286
Amphiphilic polymers in water
aggregation, 286
chemical structure and composition of polymers studied, 287, 288f
chemical structure of hydrophobically modified pyrene-labeled cationic (hydroxyethyl)cellulose (LM200-Py) indicating two sites for pyrene incorporation, 291f
emission spectra of PAA-Py in water and in 0.1M aqueous NaCl at pH 3 and 6, 294f
experimental materials, 287, 289
experimental methods, 289
fluorescence spectra of PAA-Py-Np (pyrene and naphthalene labeled poly(acrylic acid)) in 0.1M aqueous NaCl at pH 3 and 6, 296f
fluorescence spectra of PAA-Py-Np in dimethylformamide, 297f
fluorescence spectra of pyrene-labeled cationic (hydroxyethyl)cellulose (JR400-Py) and LM200-Py in water, 292f
fluorescence spectroscopy for investigating solution properties, 287
linear Stern–Volmer plots for quenching pyrene monomer and excimer emissions in aqueous solutions of JR400-Py, 299
materials and spectroscopy, 289–293

non-radiative energy transfer
  measurements for PAA-Py-Np,
  295
non-radiative energy transfer, 290,
  293
photophysical tools as measure of
  interactions between cationic
  surfactants and cationic cellulose
  ethers, 295, 299–301
photophysical tools as measure of
  pH-induced coil expansion of
  labeled PAAs, 293–295
physical properties of polymers
  studied, 291t
plot of ratio of pyrene monomer to
  excimer emission intensities for
  aqueous solutions of LM200-Py
  in water as function of added
  surfactant, 300f
plot of ratio of pyrene to
  naphthalene emission intensities
  and ratio of pyrene excimer to
  monomer emission for 0.1M
  aqueous NaCl solutions of PAA-
  Py-Np as function of pH, 298f
pyrene excimer emission of PAA-
  Py, 293, 295
pyrene monomer and excimer
  emission, 290
spectroscopy of polymers in
  solution, 290, 293
Stern–Volmer constants for
  quenching pyrene-labeled
  polymers with various
  pyridinium chloride derivatives,
  300t
synthesis and structure of
  polymers, 289–290
See also Associating polymers in
  water
Amphiphilic polyoxyalkylene
  triblock copolymers
apparent hydrodynamic radius
  $R_{h,app}$, 7
association behavior of $B_nE_mB_n$, 6,
  8f
cloud-point temperature (T_{cl}), 13,
  15

critical micelle concentration (cmc)
  and aggregation number (nw), 4,
  6
entropy penalty, 7
hydrodynamic radius (R_h) and
  intermicellar interactions, 6–7
micellization of block copolymers
  in presence of electrolytes, 3
micellization of two mixed triblock
  copolymers in aqueous solution,
  7, 10
phase behavior, 10, 13–16
phase diagram in mixture of
  miscible solvents, 15
phase diagram in presence of
  inorganic salts, 15–16
phase diagram of copolymer/water
  systems with added immiscible
  solvent, 15
plot of logarithmic cmc values of
  $B_nE_mB_n$ (oxybutylene–
  oxyethylene–oxybutylene)
  triblock copolymers in aqueous
  solution versus number of B
  units in polymer chain, 8f
plot of Rh of $B_nE_mB_n$ triblock
  copolymers versus average
  monomer unit, 9f
plots of cmc values of pure
  $E_{45}B_{15}E_{45}$ and $E_{99}P_{69}E_{99}$
  (oxyethylene–oxypropylene–
  oxyethylene) copolymers versus
  1/T(K) to determine co-
  association point, 11f
schematic of phase behaviors in
  aqueous solution, 5f
self-association, 4–10
separation medium for DNA
  capillary electrophoresis, 16, 17f
temperature-concentration phase
  diagrams in aqueous solution, 4
temperature dependence of 1:1
  weight ratio mixed $E_{45}B_{15}E_{45}$
  and $E_{99}P_{69}E_{99}$ solutions, 12f
thermodynamic process of
  micellization, 6
three-dimensional diagram of T_{cl} of
  1 wt% $B_nE_mB_n$ triblock

copolymers in aqueous solution, 14*f*
Architecture. *See* Comb architecture
Associating polymers in water
  compositional heterogeneity for polyacrylamides modified with alkylacrylamide, 96, 98–99
  compositional heterogeneity for polyacrylamides modified with polymerizable surfactant, 102, 104
  copolymer synthesis and microstructure, 96
  influence of method of synthesis on properties of acrylamide (AM)–polymerizable surfactant N16 copolymers, 104*t*
  polyacrylamides modified with alkylacrylamide, 96, 98–100
  polyacrylamides modified with polymerizable surfactant, 100–105
  polymerizable surfactant (N16) for copolymerization process II and non-polymerizable analog (B16), 101*f*
  pyrene fluorescence emission spectra for aqueous solutions of AM–N16 copolymers from pure N16 micelles and N16–B16 mixed micelles, 106*f*
  relationship between molecular structure and properties in solution for polyacrylamide modified with alkylacrylamide, 100
  relationship between molecular structure and properties in solution for polyacrylamide modified with polymerizable surfactant, 104–105
  schematic of copolymerization processes, 97*f*
  schematic of formation of random copolymer from mixed micelles containing polymerizable and non-polymerizable surfactants, 101*f*
  schematic of microstructure and conformation of AM–N16 copolymers by different experimental conditions, 106*f*
  shear thickening, 133–138
  structure and nomenclature of monomers for copolymerization process I, 98*f*
  variation of copolymer composition as function of conversion for copolymerization of AM with N16, 103*f*
  variation of copolymer composition as function of conversion for micellar copolymerization (process I) of AM with N-hexylacrylamide (HexAM) or N-methyl,N-hexylacrylamide (MeHexAM), 98*f*
  variation of zero-shear viscosity for various copolymers by micellar copolymerization (process I) at constant hydrophobe content but different surfactant concentrations, 99*f*
  See also Amphiphilic polymers in water; Hydrophobically associating polymers (HAP); Inter-molecular aggregations; Intra-molecular aggregations; Polysoaps
Association behavior
  polyoxyethylene triblock copolymers, 6, 8*f*
  studies of poly(ethylene oxide) derivatives by ^{19}F NMR, 184, 187, 190
Associative polymers
  adsorption onto polymer latices via hydrophobic interactions, 221
  rheological technique studying effects of shear stress on linear viscoelastic properties, 362, 364
  See also Dispersions containing poly(ethylene oxide) (PEO) with C_{16} hydrophobes; Hydrophobically modified alkali

soluble emulsion (HASE)
Associative thickeners (ATs)
    comparisons between narrow molecular weight, linear, and branched hydrophobic modified ethoxylated urethanes (HEUR), 177
    environmental and industrial acceptance, 352
    general synthetic reaction for uni-HEUR associative thickeners with variable terminal hydrophobe sizes, 165
    hydrophobe aggregation number, 144
    interaction between β-cyclodextrin and thickeners, 262, 263f
    model ATs in complexation studies of β-cyclodextrin with HEURs, 261
    polyelectrolytes with hydrophobic pendant groups, 155–159
    viscosity profile, 144
    See also Aggregation numbers; Associative thickener by host-guest interaction; β-Cyclodextrin complexations with surfactants and hydrophobically modified ethoxylated urethanes (HEUR); Hydrophobically modified alkali soluble emulsion (HASE); Hydrophobically modified alkali soluble emulsion (HASE) polymer in non-ionic surfactant solutions; Hydrophobic ethoxylated urethanes (HEUR); Telechelic polymers
Associative thickeners by host-guest interaction
    cyclodextrins, 271–272
    dependence of viscosity on mole fraction of guest polymer with pronounced maximum at 0.5 mol fraction, 278, 279f
    evaluating chain length effect on gelation by intrinsic viscosity of guest polymers, 279–280
    experimental materials and methods, 273
    increase in viscosity upon mixing host and guest solutions, 278, 279f
    interaction of 4-t-butyl-benzoic acid with host polymers as measured by titration microcalorimetry, 277t
    interaction of cyclodextrins polymers with guest polymers, 278–280
    interaction of guest polymers with monomeric cyclodextrins, 277–278
    interaction of host polymers with monomeric guests, 277
    side chain guest polymers, 272
    switching the thickening effect, 281
    synthesis and properties of cyclodextrin polymers, 275t
    synthesis and properties of guest polymers, 276t
    synthesis of cyclodextrin polymers, 273–274
    synthesis of cyclodextrin polymers allowing design of two component associative thickeners, 272
    synthesis of cyclodextrin polymers by polymer analogous reaction, 275
    synthesis of guest polymers, 274, 276–277
    synthesis of host polymers, 274–275
    synthesis of poly[(maleic anhydride)-co-(N-vinyl-pyrrolidone)], 273
    thickening effect strongly increasing with intrinsic viscosity of guest polymer, 280
    viscosity of mixture of host and guest polymers versus molar

ratio of added hydroxypropyl-b-cyclodextrin at various shear rates, 281f
Atom transfer radical polymerization (ATRP)
  mechanism, 53
  persistent radical effect, 53
  See also Water soluble polymers by atom transfer radical polymerization (ATRP)
Axisymmetric drop shape analysis. *See* Ethylated hydroxyethyl cellulose (EHEC)

B

Biopolymers
  applications, 303–304
  See also Ethylated hydroxyethyl cellulose (EHEC)
Block copolymers
  micellization in presence of electrolytes, 3
  phase diagram of copolymer in presence of inorganic salts, 15–16
  See also Poly(ethylene-propylene)-b-polyethylene oxide (PEP–PEO) block copolymers
Branched hydrophobes
  terminal multi-branched HEURs, 172, 174
  See also Hydrophobic ethoxylated urethanes (HEUR)
Branched surfactants, studies, 164–165
Buckminsterfullerene, C_{60}
  binding by amphiphilic hybrid block copolymer in water, 89f
  crystalline structures of micellar solutions containing, 84–85
  encapsulation by amphiphilic hybrid block copolymer, 83, 88
  See also Amphiphilic copolymers with linear dendritic architecture
Butyl acrylate (BA), block copolymers with 2-hydroxyethyl acrylate (HEA) using trimethylsilyl–HEA macroinitiator, 65–66
t-Butyl acrylate
  kinetics of atom transfer radical polymerization in dimethoxybenzene, 62f
  molecular weight byproduct of atom transfer radical polymerization in dimethoxybenzene, 62f
  polymerization, 60–61

C

C_{60}. *See* Buckminsterfullerene, C_{60}
Capillary electrophoresis, DNA, separation medium, 16, 17f
Cationic surfactants, photophysical tools as measure of interactions with cationic cellulose ethers, 295, 299–301
Cellulose derivatives. *See* Ethylated hydroxyethyl cellulose (EHEC); Hydroxyethyl cellulose (HEC)
Cellulose ethers, cationic, photophysical tools as measure of interactions with cationic surfactants, 295, 299–301
Cetyltrimethylammonium chloride (CTAC), association of pyrene-labeled hydrophobically modified (hydroxyethyl)cellulose LM200-Py, 295, 299
Chain extension
  extensional flow of globular polysoaps, 118–119
  uniform flow of globular polysoaps, 119–120, 122
  See also Polysoaps
Closed association process, surfactant micelle formation in water, 144
Cloud-point temperature (T_{cl})
  polyoxyalkylene triblock copolymers in aqueous solution, 13, 15
  three-dimensional diagram of 1

wt% polyoxyalkylene triblock copolymers in aqueous solution, 14f
Coil expansion, pH-induced, labeled poly(acrylic acid)s, 293–295
Collapsed polysoaps. *See* Polysoaps
Comb architecture
  low shear viscosity dependence on concentration of step-growth internal comb HEUR thickeners with and without $C_{18}H_{37}$ endcap, 177f
  multiple branching in hydrophobic ethoxylated urethanes (HEUR), 175–176
  synthesis of model internal comb hydrophobe thickener, 175
  synthesis of step-growth internal comb hydrophobe HEUR thickener, 176
  See also β-Cyclodextrin complexations with surfactants and hydrophobically modified ethoxylated urethanes (HEUR)
Confocal fluorescence correlation spectroscopy (FCS), critical micelle concentration (cmc), 41, 42f
Contrast variation
  absolute small angle neutron scattering (SANS) data under core and shell contrasts for PEP5PEO5, 28f
  absolute small angle neutron scattering (SANS) data under core and shell contrasts for PEP5'PEO50, 35f
  small angle neutron scattering (SANS) technique, 22
  See also Poly(ethylene-propylene)-*b*-polyethylene oxide (PEP–PEO) block copolymers
Controlled/living polymerization
  well-defined polymers, 52–53
  See also Water soluble polymers
Critical aggregation concentration (cac)
  determination by fluorescence, 166
  effect of chain length and hydrophobe structures, 168–169
  polymers binding surfactants, 304
  See also Hydrophobic ethoxylated urethanes (HEUR)
Critical micelle concentration (cmc)
  branched versus linear sodium dodecyl sulfate, 164
  confocal fluorescence correlation spectroscopy (FCS), 41, 42f
  diblock and triblock copolymers, 3
  effect of cyclodextrin associations with surfactants, 255
  polyoxyalkylene triblock copolymers, 4, 6
  values for P(IB-*b*-MAA), P(MMA-*b*-AA), and P(MMA-*b*-MAA), 43t
Crosslinking. *See* Hydrophobically modified hydroxyethyl cellulose (HMHEC) with sodium dodecyl sulfate (SDS)
β-Cyclodextrin complexations with surfactants and hydrophobically modified ethoxylated urethanes (HEUR)
  absorbance of 554 nm as function of total concentration of nonionic surfactants, non-branched $C_{16}H_{33}$- and $C_{18}H_{37}$-, octyl phenyl (OP), and nonyl phenyl (NP), 259f
  absorbance of 554 nm phenolphthalein band as function of total concentration of anionic surfactants, sodium C_8H_{17}- to $C_{14}H_{29}$- alkyl sulfonates, 258f
  adsorption isotherms exhibiting rapid increase in amount adsorbed at low concentrations, 264
  adsorption isotherms of $C_{12}H_{25}$-H_{12}MDI on P(MMA–MAA) latices at pH 9.5, 266f
  adsorption isotherms of C_8H_{17}-H_{12}MDI and C_6H_{13}-H_{12}MDI on P(MMA–MAA) latex, 266f

adsorption isotherms of nonionic surfactant NP-O(EtO)$_{12}$H on PMMA latex, 265f
adsorption isotherms of NP comb HEUR on P(MMA–MAA) and vinyl acetate (VA) latices at pH 6, 264f
adsorption isotherms of NP-O(EtO)$_{12}$H on MMA/MAA (96/4) latex in presence of (C$_{18}$H$_{37}$HN-COO-)$_{1.6}$(EtO)$_{182}$ at pH 9.5, 267f
adsorption of step-growth HEUR with H$_{12}$MDI urethane linkages on PMMA latices, 265, 267
associations of CDs with surfactants, 255
β-CD/surfactants complexations, 257
C$_{18}$H$_{37}$- hydroprobe inhibiting displacement of phenolphthalein from β-CD cavity, 263f
chemical structures of model associative thickeners with H$_{12}$MDI internal hydrophobes and different aliphatic terminal hydrophobes, 261f
colorimetric method for thickener adsorption studies, 263–264
complexation of β-CD with HEURs, 261–263
cyclodextrins (CDs), 255
dependence on absorbance at 554 nm on molar concentration of surfactants with NP hydrophobe but different poly(ethylene oxide) (PEO) chain length, 260f
dependence on absorbance at 554 nm on molar concentration of surfactants with same dodecyl hydrophobic tail but with anionic, cationic, and nonionic hydrophilic groups, 260f
effect of polar head groups of surfactant, 259, 261
effect of surfactant hydrophobe size, 258–259
experimental, 256–257
HEUR adsorption by size exclusion chromatography (SEC), 267
HEUR adsorption on latices, 263–267
latex substrates for adsorption study, 256–257
model compound HEURs, 255–256
preparation of individual samples, 256
preparation of stock solutions, 256
spacer length effect on competitive adsorption behavior, 267, 268f
stronger interaction between β-CD and thickeners as terminal hydrophobe size increases from C$_6$H$_{13}$- to C$_{12}$H$_{25}$-, 262f
surfactant-modified water-soluble polymers containing multiple hydrophobes, 255–256
UV measurements, 256
validity of displacement technique, 264–265
β-Cyclodextrin polymers
synthesis, 273–274
synthesis and properties, 275t
See also Associative thickeners by host-guest interaction

D

Daoud–Cotton model, intramolecular micelles, 112–113
Deformation behavior
globular polysoap, 115–118
weak strains deforming sphere into ellipsoid, 115
See also Polysoaps
Dendritic architecture, linear. See Amphiphilic copolymers with linear dendritic architecture
Density, junction. See Hydrophobically modified alkali

soluble emulsion (HASE) polymer in non-ionic surfactant solutions
Diffusion coefficient, equation for observed for surfactant, 325
Diffusion exchange model exchange kinetics, 213–214
See also Dynamics of adsorbed polymer layers
Dimerization model
poly(ethylene glycol) (PEG) derivatives, 197
simulation of one-ended and telechelic poly(*N,N*-dimethylacrylamides), 200–201
2-(Dimethylamino)ethyl methacrylate (DMAEMA)
copolymers, 66–67
effect of ligand on polymerization, 59*t*
effect of solvent on polymerization, 58*t*
effect of temperature on polymerization, 60*t*
molecular weight evolution with conversion versus conversion for polymerization of DMAEMA initiated by 2-bromopropionitrile (BPN), 60*f*
polymerization by atom transfer radical polymerization (ATRP), 58–60
SEC (size exclusion chromatography) traces, 61*f*
SEC traces of difunctional poly(methyl methacrylate) (PMMA) macroinitiator and resulting triblock copolymer poly(DMAEMA-*b*-MMA-*b*-DMAEMA), 68*f*
SEC traces of PMMA macroinitiator and poly(MMA-*b*-DMAEMA), 68*f*
SEC traces of poly(methyl acrylate) (PMA) macroinitiator and poly(MA-*b*-DMAEMA), 67*f*
triblock copolymers by ATRP, 66–67

N,N-Dimethyl-*N*-dodecylamine oxide (DDAO)
addition to β-cyclodextrin/phenolphthalein complex, 257
See also Surfactants
Dispersions containing poly(ethylene oxide) (PEO) with C_{16} hydrophobes
adsorbed amounts and layer thicknesses, 222*t*
adsorption, 222–224
adsorption isotherms for hexadecyl unimers (HDU), octadecyl unimers (ODU), and PEO on poly(methyl methacrylate) (PMMA) latices, 225*f*
apparent intrinsic viscosity and Huggins coefficient of dispersions, 224*t*
associative polymers adsorbing by hydrophobe endcap, 223
behavior resembling viscoelasticity of crosslinking polymers near gel point, 231
characteristic time denoting transition or crossover between discrete and power law relaxation regimes, 232, 235
characterizing fraction of thickening due to solution viscosity, 228
comparing HDU and ODU effects on viscosity, 224
correlating low shear viscosity across range of particle and polymer concentrations, 226
decomposition of viscoelastic spectrum, full data and residuals after subtraction of high frequency relaxation, 233*f*
determining moduli as integrals over sum of single rapid relaxation and power law distribution of slower ones, 231
earlier work with C_{18} terminal hydrophobes adsorbing on

PMMA latices from water, 221–222
enhancing dispersion rheology by increased particle interactions with HDU, 237
estimating layer thicknesses by simple theories, 223
fitting parameters, 232t
fractional enhancement of dispersion viscosities due to solution, 230f
gel strength corresponding to degree of crosslinking and functionality, 235
HDU adsorbing with both hydrophobes on particle surface, 235, 237
high frequency modulus of dispersion normalized by polymer solution, 234f
hydrodynamic diameters of particles with adsorbed layers, 223–224
key differences for HDU, 228, 231
larger hydrophobes imparting greater viscosity and viscoelasticity to associated solution, 235
limiting lower relaxation time for power law spectrum, 236f
linear regression of low shear viscosities with line indicating least squares fit, 227f
linear viscoelasticity, 228, 231
measured low shear viscosities as function of total polymer concentration and effective particle volume fraction compared with calculations, 226, 229f
model, 231–235
number of chains per unit area, 223
ratio of viscosity of dispersion with associative polymer to that without, 229f
relaxation spectrum of form, 231
shortest relaxation time normalized by solution, 234f
steady shear viscosity of dispersion containing 0.5 wt% HDU, 227f
steady shear viscosity of dispersions with 3 wt% HDU, 225f
storage and loss moduli of dispersions containing 3.5 wt% HDU, 230f
strength of hydrophobic interaction affecting aspects of rheology of dispersions, 224, 226
strength of power law relaxation, 235, 236f
thicker layers for associative polymers, 223
thickness calculation on spherical particle, 223
unmodified PEO adsorption on latex particles, 222–223
viscoelasticity of PMMA+HDU exhibiting multiple relaxation modes, 228, 231
viscosity, 224, 226, 228
DNA capillary electrophoresis, separation medium, 16, 17f
Dodecyltrimethylammonium chloride (DTAC), association of pyrene-labeled hydrophobically modified (hydroxyethyl)cellulose LM200-Py, 295, 299
Dynamic light scattering (DLS)
aggregation number and micellar diameter, 43–44
investigation of micellar solutions, 22
measurements of poly(ethylene-propylene)-b-polyethylene oxide (PEP–PEO) micelles in water, 32–33
See also Poly(ethylene-propylene)-b-polyethylene oxide (PEP–PEO) block copolymers
Dynamic shear. *See* Viscosity
Dynamics of adsorbed polymer layers

addressing relative time scales of adsorption and relaxation for poly(ethylene oxide) (PEO) on silica in aqueous environment, 208–209
adsorption equation for surface with large adsorption capacity, 211
adsorption isotherms, 210–211
adsorption isotherms for PEO and hydroxyethyl cellulose (HEC), 210f
adsorption kinetics for mixture of long and short PEO chains, 211, 212f
cases not exhibiting linear absorption kinetics and possible reasons, 208
coadsorption run at diffusion limited rate, 212–213
critical aspects of interfacial dynamics, 206–207
driving force for diffusion away from surface, 207
effects of aging on exchange of 120K PEO chains, 215
effects of molecular weight on adsorption and coadsorption kinetics, 211–213
exchange kinetics in context of diffusion exchange model, 213–214
exchange rate, 213
experimental, 209–210
explaining initial stages of self-exchange and exchange of young layers of higher molecular weight PEO chains, 215–216
explanation of relative time scales, 207
HEC aging, 218
implications for interpretation of HEC adsorption isotherms, 218
influence of layer age on displacement of coumarin-tagged PEO by unlabeled chains, 214–215

kinetics of surface relaxations and influence on exchange via fluorescence self-exchange studies, 213
layer age and molecular weight establishing surface mobility of adsorbed chains, 216
minimal self exchange for narrow molecular weight PEO standards, 219
quantitative reports of adsorption kinetics, 207–208
relaxation behavior of HEC, 218
rounded absorption kinetics, 208
self-exchange experiments for 33K PEO layers in forward and reverse protocols, 214, 215f
self-exchange kinetics after various aging periods for 33K and 120K PEO layers, 216f
self-exchange kinetics for HEC layers of various ages, 217f
self-exchange reaction leading to fluorescence decay, 214
self exchange step of HEC chains, 218
single component adsorption kinetics for narrow molecular weight PEO standards, 212f
step-wise evolution of total surface coverage for bimodal mixture, 213
studies shifting to dynamic behavior, 207
summary of surface exchange kinetics, 217t
surface exchange step between labeled and unlabeled chains, 214
surface relaxations, 213–219
total internal reflectance fluorescence (TIRF) procedures, 209–210
Dynamic viscosity measurements
experimental method for calorimeter, 342
See also Thickening mechanism of

hydrophobically modified alkali soluble emulsion (HASE)

E

Emulsion. *See* Thickening mechanism of hydrophobically modified alkali soluble emulsion (HASE)
Emulsion polymerization, amphiphilic block copolymers as surfactants, 46–47
End-functionalized hydrocarbon and perfluorocarbon derivatives. *See* Poly(ethylene oxide) (PEO); Poly(N,N-dimethylacrylamide) (PDMA); Poly(ethylene oxide) (PEO)
Ethylated hydroxyethyl cellulose (EHEC)
adsorption of EHEC and hydrophobically modified EHEC (HM-EHEC) to air-water interface, 308f
automatic Langmuir surface balance for recording isotherms, 305
axisymmetric drop shape analysis instrument, 304–305
calculation of surface rheological parameters, 306–307
complex surface dilatational modulus, 307
dependence of rate of surface tension reduction on sodium dodecyl sulfate (SDS) concentration, 309
dynamic surface tension and surface rheological measurements, 304–305
EHEC and HM-EHEC, 305–306
excess surface pressure/area isotherms of EHEC and HM-EHEC monolayers spread on SDS solution with different concentration, 311f
experimental chemicals, 305–306
frequency dependence of elastic and viscous moduli of adsorbed EHEC/SDS and HM-EHEC/SDS layers, 312–313, 314f
governing surface tension reduction by molecular rearrangements/internal relaxation processes, 310
lagtime period for increasing SDS concentration, 308–309
measuring growth in surface pressure as function of time with Langmuir surface balance, 312, 313f
power law, 309
power law exponent equation, 309
power law exponent from data as function of SDS concentration, 310, 312f
procedure and data analysis methods, 306–307
surface activity of polymer/SDS system reaching maximum value in certain SDS concentration range, 310, 311f
surface elastic moduli for EHEC and HM-EHEC adsorbed layers as function of SDS concentration, 314, 315f
surface pressure of EHEC and HM-EHEC as function of SDS concentration in subphase, 311f
three distinct regimes of adsorption of macromolecules to air-water interface, 307
Exchange process. *See* Dynamics of adsorbed polymer layers
Extended polysoaps. *See* Polysoaps

F

Films
dynamic mechanical analysis (DMA) of butyl acrylate/methyl methacrylate (BA/MMA) film, 47, 49f

properties, 47–50
transmission electron microscopy (TEM) analysis of BA/MAA films before and after annealing, 47, 48f
water uptake before and after annealing, 50f
See also Amphiphilic block copolymers as surfactants

Flow fields
coexistence of different configurational states possible in polysoap unfolded by uniform flow, 121f
extensional flow, 118–119
extension of globular polysoaps, 118–120, 122
uniform flow, 119–120, 122
See also Polysoaps

Flower micelles, formation, 148–149

Fluorescence correlation spectroscopy (FCS), critical micelle concentration (cmc), 41, 42f

Fluorescence quenching
critical aggregation concentration (CAC), 146
data analysis, 146
determining aggregation numbers, 145–148, 159–160
effect of chain length and hydrophobe structures on CAC, 168–169, 170f
equation for mean number of quenchers per micelle, 146
estimation of aggregation number from steady-state measurements, 147–148
expected behavior for parameters in Poisson quenching model, 147f
hydrophobically modified alkali soluble emulsions (HASE) polymer, 155–159
hydrophobically modified hydroxyethylcellulose, 153–155
influence of hydrophobe on intensity ratio (I_1/I_3) of aqueous pyrene solutions as function of hydrophobic ethoxylated urethanes (HEUR) concentration, 170f
measurements for determining N_R for $C_{12}H_{25}$ end-capped PEO, 151
methodology, 145–146
polyelectrolyte with hydrophobic pendant groups, 155–159
principle of method, 146
procedure for HEUR, 166
simulations of experiments in micelles, 158–159
See also Aggregation numbers; Hydrophobic ethoxylated urethanes (HEUR)

Fluorescence spectroscopy
investigating solution properties of amphiphilic polymers, 287
See also Amphiphilic polymers in water

Fluorescence spectroscopy, time-resolved, measuring rate of exchange of block copolymers between aggregates, 43

Fluorocarbon groups. *See* Poly(N,N-dimethylacrylamide) (PDMA)

Force study of adsorbed layers of hydrophobically modified polyacrylamide
adsorbed amount of hydrophobically associative polyacrylamide (HAPAM) with salt and polymer concentration, 243f
adsorption isotherm for non-modified polyacrylamide (PAM), 243f
close-up of separation cycle for 760 ppm HAPAM adsorbed in 0.9M NaCl, 245, 246f
concentration of salt affecting adsorbed amount, 243
conducting experiments with multi-step compression/separation cycles, 244–245
copolymer of acrylamide and ~0.75% nonyl methacrylate, 241f

enhancement of adsorption of hydrophobically modified polymers versus non-modified version, 248
examining how changes in polymer concentration in solution affects adsorption behavior, 246
examining separation profiles for one-sided experiments, 251
force-distance profile for bare mica surface and 2000 ppm HAPAM adsorbed in 0.9M NaCl, 248f
force-distance profile for 2000 ppm HAPAM adsorbed in 0.3M NaCl, 246f
force-distance profile for 760 ppm HAPAM adsorbed in 0.9M NaCl, 244f
force-distance profile for bare mica surface and 760 ppm HAPAM adsorbed in 0.9M NaCl, 247f
force measurements by surface force apparatus (SFA), 241, 252
forces between adsorbed layers at very small separations by surface force apparatus (SFA), 240
fringes of equal chromatic order (FECO), 241–242
HAPAM polymer of choice, 240–241
hydrophobic modification leading to multilayer build-up, 248–249
increasing polymer concentration increasing adsorbed amount, 243
insight on interactions at mica surface as well as layer/layer hydrophobic interactions, 249
longest relaxation time of hydrophobe/hydrophobe interactions by rheology, 251–252
materials and methods, 241–242
multi-step compression force-distance profile for 760 ppm HAPAM adsorbed in 0.9M LiCl, 245f

network of clusters by interactions of hydrophobes, 251
performing experiments with HAPAM absorbed on only one surface, 247
relating addition of hydrophobic moieties to changes in layer-layer interactions by SFA measurements, 244
repulsive force initially due to polymer-polymer overlap increasing with decreasing distance, 249–250
research focusing on dependence of adsorbed amount of modified polymer on polymer concentration, 240
rheological experiments, 242
rheological experiments to understand effectiveness of hydrophobic interactions within adsorbed layer, 247–248
schematic of adsorbed layers formed by HAPAM at low and high polymer concentration, 250f
SFA experiments, 242
surface force measurements of PAM for comparison with modified version, 245
unoccupied binding sites on substrate to explain irreversibility of first compression of HAPAM layers, 250
weak, long-range attraction for some systems, 250–251

G

Gelation pH, viscosity maxima for alkali soluble emulsion (ASE), 340
Gel swelling experiments. *See* Hydrophobically modified hydroxyethyl cellulose (HM-HEC) with sodium dodecyl sulfate (SDS)

Globular polysoaps. *See* Polysoaps
Green–Tobolsky relationship, 357
Guest-host interactions. *See* Associative thickeners by host-guest interaction

H

HASE. *See* Hydrophobically modified alkali soluble emulsion (HASE)
Hexadecyl-terminated poly(ethylene oxide)
  effect of polymer chain length and concentration on shear thickening, 139f
  shear thickening, 137–138
  viscoelastic model fitting experimental data, 136, 137f
Hexadecyl unimers (HDU). *See* Dispersions containing poly(ethylene oxide) (PEO) with C_{16} hydrophobes
Hexamethylene diisocyanate (HDI). *See* Hydrophobic ethoxylated urethanes (HEUR)
Host-guest interaction. *See* Associative thickeners by host-guest interaction
Huggins constant
  dispersions containing poly(ethylene oxide) with C_{16} hydrophobes, 224t
  end-functionalized poly(ethylene glycol)s (PEGs), 191
  poly(N,N-dimethylacrylamide) (PDMA) polymers, 200
Hydrocarbon groups. *See* Poly(N,N-dimethylacrylamide) (PDMA)
Hydrodynamic radius (R_h)
  calculating from slow mode diffusion coefficients, 131–132
  entropy penalty, 7
  intermicellar interactions, 6–7
  polyoxyalkylene triblock copolymer micelles, 6–7
  relation of apparent R_h to apparent diffusion coefficient, 7
  See also Ionomers
Hydrophobes. *See* Dispersions containing poly(ethylene oxide) (PEO) with C_{16} hydrophobes
Hydrophobic ethoxylated urethanes (HEUR)
  addition of multiple branch hydrophobes to polyoxyethylene (POE) with hexamethylene diisocyanate (HDI), 174
  adsorption on latices, 263–267
  aggregation number as function of concentration of H_{12}MDI-based HEURs with various hydrophobes, 170f
  branched sodium dodecyl sulfate (SBDS) versus linear analog (SDS), 164
  comparisons between narrow molecular weight, linear, and branched H_{12}MDI-based uniHEUR associative thickeners, 177
  complexation with β-cyclodextrin, 261–263
  dependence of mean aggregation number of HEUR solutions on polymer concentration, 169–171
  effect of chain length and hydrophobe structures on critical aggregation concentration (CAC), 168–169
  effect of hydrophobe structure on low shear viscosity, 167–168
  experimental, 165–167
  fluorescence studies procedures, 166
  general synthetic route for narrow molecular weight HEUR (uni-HEUR) associative thickeners with variable terminal hydrophobe sizes, 165–166
  influence of hydrophobe on intensity ratio (I_1/I_3) of aqueous pyrene solutions as function of

H_{12}MDI-based HEUR concentration, 170f
influence of hydrophobe structure on low shear rate viscosity of aqueous solutions of HDI-based uniHEURs as function of thickener concentration, 175f
larger linear and branched hydrophobes with HDI as diisocyanate, 177–178
linear and branched Guerbet surfactants, 164–165
low shear rate viscosity of H_{12}MDI-based uniHEUR thickeners with different hydrophobe structures as function of concentration, 168f
low shear viscosity dependence on concentration of step-growth internal comb HEUR thickeners with and without $C_{18}H_{37}$ endcap, 177f
multiple-branching in comb architecture HEURs, 175–176
oscillatory measurements, 172
polymer preparation, 148
prior studies, 165
representation of branched b-$C_{16}H_{34}$-H_{12}MDI EO_{670}-uniHEUR (oxyethylene EO=670) micelle, 171f
rheological measurements, 167
storage and loss moduli dependence on frequency for H_{12}MDI-based uniHEURs with different size hydrophobes, 173f
surfactant structures from previous branching studies, 164f
synthesis of model internal comb hydrophobe thickener, 175
synthesis of multiple branch hydrophobe, 174
synthesis of step-growth internal comb hydrophobe HEUR thickener, 176
synthesis using H_{12}MDI (4,4'-methylenebis(cyclohexyl isocyanate)), 167
terminal multi-branched HEURs, 172, 174
See also β-Cyclodextrin complexations with surfactants and hydrophobically modified ethoxylated urethanes (HEUR); Telechelic polymers

Hydrophobically associating polymers (HAP)
distinguishing between HAP-dominated, transitional, and surfactant-dominated composition regimes, 330
effect of deformation on structure and chain stretching of hexadecyl-terminated PEO, 139–140
effect of polymer length and concentration on shear thickening of hexadecyl-terminated PEO with M_w of 8000, 139f
effects of added surfactant, 317–318
estimating critical shear rate by non-Gaussian chain stretching model, 138, 139f
non-Gaussian chain stretching, 137 138
polymer solution at rest, 135–136
polymer solution under shear, 137–138
predicting zero shear viscosity, 136
pseudo-phase separation model, 320
Raoult's law of partial pressure, 318
relation between concentration of free surfactant monomers and mixed micellar stoichiometry, 318
relation between microscopic structure and viscosity, 327, 330
schematic of different species

present in HAP/surfactant mixtures, 318, 319f
shear thickening, 133–138
shear thinning at high shear rate, 137
simple rubber elasticity theory, 136
simple viscoelastic model fitting experimental data, 136, 137f
substitution pattern of HAP preventing formation of mixed micelles, 320
transient network theory, 136
See also Associating polymers in water; Hydrophobically modified hydroxyethyl cellulose (HMHEC) with sodium dodecyl sulfate (SDS); Ionomers

Hydrophobically modified alkali soluble emulsion (HASE)
apparent aggregation number for micelles by hydrophobes of HASE polymer, 157–158
behavior of HASE polymer before adding non-ionic surfactants, 371, 373
behavior of HASE polymer versus telechelic hydrophobically modified ethoxylated urethane (HEUR) systems, 373
class of alkali soluble emulsions (ASE) in ionic form, 339
description, 156
effect of hydrophilic surfactant (NP10), 376f
effect of hydrophobic interactions on network connectivity and dynamics of system, 375, 377
effect of increasing concentration of NP10 on shear-induced structuring, 375, 377
experimental polymer preparation, 370–371
experiments determining hydrophobe aggregation numbers N_R, 156
frequency sweep data for solutions of HASE polymer, 374f
idealized constitution of HASE polymer, 372f
influence of NP10 surfactant on rheology of HASE polymer solutions, 375, 376f
influence of NP6 surfactant on rheology of HASE polymer solutions, 378f
lacking understanding of dynamics and topology of HASE networks, 369–370
nature of polymer-surfactant interaction, 370
response of HASE polymer solutions in oscillatory shear, 373, 375
rheometrical data for solutions of HASE prepared in presence of NP10, 375
rheometrical experiments, 371
shear thickening, 138
shear thinning, 137
simulations of fluorescence quenching experiments in micelles, 158–159
steady shear data for solutions of HASE polymer, 372f
steady shear data of HASE in presence of NP6 surfactant of limited water solubility, 377
structure, 155
variations of fitting parameters with probe concentration for pyrene and ethylpyrene in aqueous solutions of HASE polymer, 157f
variety of practical applications, 339
See also Thickening mechanism of hydrophobically modified alkali soluble emulsion (HASE)

Hydrophobically modified alkali soluble emulsion (HASE) polymer in non-ionic surfactant solutions
$C_{18}EO_{100}$ non-ionic surfactant, 352, 354
chemical structure of HASE, 353f

comparing three 1 wt% HASE samples containing different surfactant concentrations but identical low shear viscosity, 362, 363f

crossover of storage and loss moduli separating terminal region and plateau region for rough value of relaxation time, 364, 366

dependence of rheological parameters of 1wt% HASE in varying $C_{18}EO_{100}$ concentrations, 356f

dependence of shear viscosity, storage modulus, and relaxation time on applied shear stress for 1 wt% HASE in $C_{18}EO_{100}$ at pH 9, 362, 363f

dependence of shear viscosity on shear stress for 1 wt% HASE in varying concentrations of $C_{18}EO_{100}$ at pH 9, 355f

dependence of storage and loss modulus on angular frequency, 354, 357

dependence of storage and loss modulus on angular frequency at various applied stresses at pH 9, 365f

dependence of storage and loss modulus on angular frequency in varying $C_{18}EO_{100}$ concentrations at pH 9, 358f

dependence of viscosity on shear stress at different temperatures, 360f

dynamic shear, 354, 357

dynamic shear results allowing determination of average characteristic relaxation time, 357

effect of increasing water-soluble surfactant NP10 concentration on shear-induced structuring, 375, 377

example of Arrhenius plot, 361f

experimental materials, 352, 354

low shear viscosity, relaxation time, activation energy and storage modulus for 1 wt% HASE in $C_{18}EO_{100}$, 363t

number of mechanically active hydrophobic junctions according to Green–Tobolsky relationship, 357

observations in shear-thickening region, 366

parameter representing number of mechanically active intermolecular hydrophobic junctions, 357

polymer-surfactant interactions, 359, 362

possible complex phase changes by amphiphilic species incorporation, 377

rheological measurements, 354

rheometrical data for HASE solutions containing NP10, 375, 376f

role of applied stress in promotion of associative network with different topology, 366

shear viscosity, relaxation time, activation energy, and storage modulus of 1 wt% HASE in $C_{18}EO_{100}$ at 50 Pa applied stress, 366t

steady shear, 354

steady shear data of HASE with NP6 surfactant of limited solubility in water, 377, 378f

structuring for higher concentrations of NP6, 377

studying effects of shear stress on linear viscoelastic properties of associative polymer solutions, 362, 364

superposition of oscillation on steady shear, 362, 364–366

temperature dependence, 359

viscosity of polymer system obeying Arrhenius relationship, 359
Hydrophobically modified hydroxyethyl cellulose (HMHEC) with sodium dodecyl sulfate (SDS)
  absence of swelling of hydrophobically modified gels at low concentration, 333
  concentrations of free and bound SDS in HEC and HMHEC solutions, 326f
  constructing binding isotherm for SDS to HMHEC, 327
  degree of crosslinking affecting extent of gel swelling, 331
  distinguishing between HAP-dominated, transitional, and surfactant-dominated composition regimes, 330
  equation for observed diffusion coefficients of surfactant, 325
  equilibrium swelling of gels of covalently crosslinked HEC or HMHEC in solutions of SDS at varying concentrations, 332f
  experimental materials, 321
  gel swelling experimental methods, 322–323
  gel swelling experiments, 331–333
  HMHEC and HEC slowing down diffusion of SDS, 323, 325
  interaction between HEC/HMHEC and SDS, 333–334
  microscopic parameters describing micelles, 329f
  NMR measurements, 322
  ratio of bound SDS to HMHEC hydrophobes in mixed micelles as function of free SDS concentration, 328f
  relation between microscopic structure and viscosity, 327, 330
  sample preparation, 321
  self-diffusion coefficients for SDS in solutions of HEC and HMHEC at varying SDS concentrations, 326f
  surfactant binding, 325, 327
  surfactant response for HEC and HMHEC, 331
  swelling of HAP gels in ionic surfactant solutions reflecting interplay between hydrophobic crosslinking, ionic swelling, and electrostatic screening, 333
  variation of viscosity of semi-dilute solutions of HMHEC at different total concentrations of added SDS, 324f
  viscometry, 321–322
  viscosity and self-diffusion, 323, 325
Hydrophobically modified polymers
  hydroxyethylcellulose, 153–155
  preparation by chemical modification of water-soluble polymers, 180
  See also Force study of adsorbed layers on hydrophobically modified polyacrylamide; Poly(*N,N*-dimethylacrylamide) (PDMA); Poly(ethylene oxide) (PEO); Telechelic polymers
2-Hydroxyethyl acrylate (HEA)
  block copolymers with butyl acrylate (BA) using trimethylsilyl (TMS)-HEA macroinitiator, 65–66
  molecular weight evolution with conversion for bulk polymerization, 55f
  polymerization, 54
  TMS protected HEA, 56, 58
  zero and first order kinetic plots of bulk polymerization, 55f
2-Hydroxyethyl methacrylate (HEMA)
  molecular weight evolution with conversion for polymerization, 57f
  molecular weight evolution with conversion for polymerization of trimethylsilyl (TMS)-HEMA, 59f
  polymerization, 56

TMS protected HEMA, 56, 58
zero and first order kinetic plots for polymerization, 57f
Hydroxyethyl cellulose (HEC)
adsorption isotherms, 210–211
aging, 218
experimental, 209–210
implications for interpretation of HEC adsorption isotherms, 218
relaxation behavior, 218
self exchange kinetics for layers of various ages, 216, 217f
self exchange step, 218
See also Dynamics of adsorbed polymer layers; Ethylated hydroxyethyl cellulose (EHEC)
Hydroxyethyl cellulose, hydrophobically modified
adsorption onto mica by surface force measurements, 240
experiments for aggregation numbers, 154–155
fluorescence spectra, 292f
physical properties of pyrene-labeled, 291t
preparation, 153–154
pyrene monomer and excimer emission, 290
spectroscopy of polymers in solution, 290, 293
structure, 154
synthesis and structure of pyrene-labeled, 289–290, 291f
See also Amphiphilic polymers in water; Hydrophobically modified hydroxyethyl cellulose (HMHEC) with sodium dodecyl sulfate (SDS)

I

Inorganic salts, phase diagrams of block copolymers in presence of, 15–16
Intermicellar interactions, triblock copolymer micelles in aqueous solution, 6–7

Inter-molecular aggregations
compositional heterogeneity, 96, 98–99
copolymer synthesis, 96, 97f
polyacrylamides modified with alkylacrylamide, 96, 98–100
relationship between molecular structure and properties in solution, 100
structure and nomenclature of monomers, 98f
variation of copolymer composition as function of conversion for micellar copolymerization of acrylamide (AM) with N-hexylacrylamide (HexAM) or N-methyl,N-hexylacrylamide (MeHexAM), 98f
variation of zero-shear viscosity for various copolymers at constant hydrophobe content but different surfactant concentrations, 99f
See also Associating polymers in water
Intrachain self assembly, polysoaps, 111
Intra-molecular aggregations
copolymer synthesis and microstructure, 96, 97f
heterogeneity in composition, 102, 104
influence of method of synthesis on properties of acrylamide (AM)-polymerizable surfactant (N16) copolymers, 104t
polyacrylamides modified with polymerizable surfactant, 100–105
polymerizable surfactant (N16) and non-polymerizable analog (B16), 101f
polysoaps, 111–114
pyrene fluorescence emission spectra for aqueous solutions of AM–N16 copolymers, 106f
relationship between molecular structure and properties in

405

solution, 104–105
schematic of formation of random copolymer from mixed micelles containing polymerizable and non-polymerizable surfactants, 101*f*
schematic of microstructure and conformation of AM–N16 copolymers, 106*f*
variation of copolymer composition as function of conversion for copolymerizations of AM with N16, 103*f*
See also Associating polymers in water

Intrinsic viscosity. *See* Associative thickeners by host-guest interaction

Ionomers
association in low-polarity solvents, 128
behavior of solutions, 127–128
critical factors for determining solution behavior, 139
diffusion coefficients corresponding to fast model for polyurethane PU4 in different polar solvents, 131*t*
effect of polymer length and concentration on shear thickening of hexadecyl-terminated PEO with M_w of 8000, 139*f*
effect of water content on shear thickening of magnesium-neutralized dicarboxy-polybutadiene in toluene, 135*f*
effect of water content on zero-shear viscosity of magnesium-neutralized dicarboxy-polybutadiene in toluene, 135*f*
fluorescence measurements, 129
hydrodynamic radii corresponding to slow mode for PU4 in different polar solvents, 132*t*
lacking quantitative comparison between experimental and theoretical, 129
light scattering measurements, 129
model polyurethane ionomer, 130*f*
monitoring change in loose aggregate as function of solvent polarity in terms of hydrodynamic radius, 131–132
normalized fluorescence intensities as function of concentration PU1 and PU4 in different solvents, 133*f*
polyelectrolyte behavior, 133
polymer solution at rest, 135–136
polymer solution under shear, 137–138
previous studies using random copolymer and model telechelic ionomers, 128
rheology testing, 129
shear thickening of associating polymers, 133–138
structure of polyurethane ionomers in solvents of varying polarity, 130–133
synthesis of hydrocarbon-terminated poly(ethylene oxide) (PEO), 135*f*
unique solution behavior by forming three-dimensional network, 128–129
See also Hydrophobically associating polymers (HAP)

J

Junction density. *See* Hydrophobically modified alkali soluble emulsion (HASE) polymer in non-ionic surfactant solutions

L

Latex dispersions
rheology, 221
See also Dispersions containing poly(ethylene oxide) (PEO) with C_{16} hydrophobes

Linear dendritic architecture. *See* Amphiphilic copolymers with linear dendritic architecture
Linear hydrophobes. *See* Hydrophobic ethoxylated urethanes (HEUR)
Linear viscoelasticity, hydrophobes in associative polymers, 228, 231
Loss modulus
  crossover of storage and loss moduli separating terminal region and plateau region for rough value of relaxation time, 364, 366
  dependence of storage and loss modulus on angular frequency, 354, 357
  dependence of storage and loss modulus on angular frequency at various applied stresses at pH 9, 365*f*
  dependence of storage and loss modulus on angular frequency in varying $C_{18}EO_{100}$ concentrations at pH 9, 358*f*
  dependence on frequency for hydrophobe ethoxylated urethanes (HEUR), 172, 173*f*
  storage and loss moduli dependence on frequency for $H_{12}MDI$-based uniHEURs with different size hydrophobes, 173*f*
  storage and loss moduli of dispersions containing 3.5 wt% hexadecyl unimers, 230*f*

M

Macroinitiators, synthesis of poly(ethylene-propylene)-*b*-polyethylene oxide (PEP–PEO) block copolymers, 23–24
Macromonomers, graft copolymers of *N*-vinylpyrolidinone and polystyrene, 69
Macrophase separation, clouding, 15

Marangoni effect, surface properties under dynamic conditions, 304
Mechanism, thickening. *See* Thickening mechanism of hydrophobically modified alkali soluble emulsion (HASE)
4,4'-Methylenebis(cyclohexyl isocyanate) ($H_{12}MDI$). *See* Hydrophobic ethoxylated urethanes (HEUR)
Micellar free energy
  equation, 33
  interfacial contribution, 33–34
Micellar polymerization processes. *See* Associating polymers in water
Micelle aggregation number, equation, 29
Micelle concentration. *See* Critical micelle concentration (cmc)
Micelles
  analyzing structure using model', 27, 29
  closed association process for surfactant micelle formation, 144
  Daoud–Cotton model, 112–113
  factors determining size and shape of traditional surfactant, 145
  flower formation, 148–149
  hydrodynamic radius and intermicellar interactions, 6–7
  model of spherical by block copolymer, 45*f*
  polymerization in, and applications, 95
  simulations of fluorescence quenching experiments, 158–159
  spherical forming gel-like system for separation medium, 16, 17f
  structure, 145
Micellization
  block copolymers in presence of electrolytes, 3
  model systems, 21
  polyoxyalkylene triblock copolymers, 3

thermodynamic parameters, 6
two mixed triblock copolymers in aqueous solution, 7, 10
See also Poly(ethylene-propylene)-*b*-polyethylene oxide (PEP–PEO) block copolymers

Microphase separation, micellization, 15

Miscible solvents, phase diagrams of triblock copolymers, 15

Models
  describing steady shear or dynamic behavior of associating polymer solutions, 133–134
  diffusion exchange, 213–214
  dimerization for poly(ethylene glycol) derivatives, 197
  dimerization for poly(N,N-dimethylacrylamide) (PDMA) derivatives, 200–201
  Poisson quenching, 146, 147f
  polymer solutions at rest, 135–136
  polymer solutions under shear, 137–138
  polysoap consisting of amphiphilic monomers joined by flexible spacer chains, 111
  pseudo-phase separation model for hydrophobically associating polymer (HAP)/surfactant micelle, 320
  simulations of fluorescence quenching experiments in micelles, 158–159
  spherical micelle by block copolymer, 45f
  viscoelasticity of dispersions, 231–235
  See also Dynamics of adsorbed polymer layers; Hydrophobically modified alkali soluble emulsion (HASE) polymer in non-ionic surfactant solutions; Ionomers; Poly(ethylene-propylene)-*b*-polyethylene oxide (PEP–PEO) block copolymers; Polysoaps; Thickening mechanism of hydrophobically modified alkali soluble emulsion (HASE)

Molar mass of micelles, static light scattering, 45

Molecular architectures, waterborne associating polymers, 109

Molecular structure. *See* Associating polymers in water

Multi-branched hydrophobes
  comb architecture hydrophobic ethoxylated urethanes (HEUR), 175–176
  terminal HEUR, 172, 174

N

Naphthalene. *See* Amphiphilic polymers in water

Network strength. *See* Hydrophobically modified alkali soluble emulsion (HASE) polymer in non-ionic surfactant solutions

Neutralization. *See* Thickening mechanism of hydrophobically modified alkali soluble emulsion (HASE)

Non-Gaussian chain stretching
  associating polymer solution under shear, 137–138
  estimating critical shear rate, 138, 139f

Non-ionic micelles, structure, 145

Non-ionic surfactants. *See* Hydrophobically modified alkali soluble emulsion (HASE) polymer in non-ionic surfactant solutions

Nonyl phenol (NP). *See* β-Cyclodextrin complexations with surfactants and hydrophobically modified ethoxylated urethanes (HEUR)

O

Octadecyldeca(ethyleneoxy)-2-phenylacrylate (ODPA) synthesis, 180

See also Poly(*N,N*-dimethylacrylamide) (PDMA)
Octadecyl unimers (ODU). *See* Dispersions containing poly(ethylene oxide) (PEO) with C_{16} hydrophobes
Oscillating bubble technique, measuring dynamic properties of surface films, 304
Oscillatory measurements, linear and branched hydrophobic modified ethoxylated urethanes (HEURs), 172, 173f
Oxybutylene. *See* Amphiphilic polyoxyalkylene triblock copolymers
Oxyethylene. *See* Amphiphilic polyoxyalkylene triblock copolymers
Oxypropylene. *See* Amphiphilic polyoxyalkylene triblock copolymers

P

Perfluorocarbon groups. *See* Poly(*N,N*-dimethylacrylamide) (PDMA)
Persistent radical effect, atom transfer radical polymerization (ATRP), 53
Phase behavior
  cloud-point temperature (T_{cl}), 13, 15
  phase diagrams of polyoxyalkylene triblock copolymers, 15–16
  polyoxyalkylene triblock copolymers in aqueous solutions, 4, 5f, 10, 13
  three-dimensional diagram of T_{cl} of 1 wt% polyoxyalkylene triblock copolymers in aqueous solution, 14f
  See also Amphiphilic polyoxyalkylene triblock copolymers

Phase diagrams
  block copolymers in presence of inorganic salts, 15–16
  polyoxyalkylene triblock copolymers, 15
Phase separation, solutions of polysoaps, 122–123
Phenolphthalein
  complexation with cyclodextrin, 257
  stock solution preparation, 256
  See also β-Cyclodextrin complexations with surfactants and hydrophobically modified ethoxylated urethanes (HEUR)
Pincus law, deformation of globular polysoaps, 117–118
Poisson quenching model, expected behavior, 146, 147f
Polyacrylamide, hydrophobically modified. *See* Force study of adsorbed layers on hydrophobically modified polyacrylamide
Polyacrylamides. *See* Associating polymers in water
Poly(acrylic acid) (PAA)
  emission spectra of pyrene labeled PAA (PAA-Py) in water and in 0.1M aqueous NaCl at pH 3 and 6, 294f
  fluorescence spectra of doubly-labeled PAA, 295, 298f
  non-radiative energy transfer measurements, 295
  photophysical tools as measure of pH-induced coil expansion of labeled PAAs, 293–295
  pyrene excimer emission, 293, 295
  synthesis and structure of PAA-Py, 289–290
  See also Amphiphilic polymers in water
Polyaromatic hydrocarbons (PAHs)
  binding of PAHs and C_{60} by

amphiphilic hybrid block
copolymer in water, 89f
dissolution of PAHs in water, 75
encapsulation of pyrene and C_{60} by
amphiphilic hybrid ABA block
copolymer, 83, 88
encapsulation studies, 75
fluorescence spectroscopy of
amphiphilic hybrid block
copolymer and pyrene in water
at different copolymer
concentrations, 86f
schematic of pyrene encapsulation
by amphiphilic hybrid
copolymer in water, 87f
See also Amphiphilic copolymers
with linear dendritic architecture
Poly(benzyl ether) monodendrons
choice as A blocks of ABA
copolymers, 73, 75
See also Amphiphilic copolymers
with linear dendritic architecture
Poly(N,N-dimethylacrylamide)
(PDMA)
characterization methods, 181
effect of molecular weight on
association of telechelic
PDMAs, 201
end functionalization of living
PDMA with fluorocarbon or
hydrocarbon groups, 183
end-functionalized PDMA
derivatives, 191, 194
expression for reduced viscosity,
200
Huggins constant of one-ended
C_8F_{17} PDMA derivative, 200
PDMA polymers, 200–201
reduced viscosity as function of
concentration of
octadecyldeca(ethyleneoxy)-2-
phenylacrylate (ODPA) bis-
functionalized PDMAs and
homopolymer precursors, 194,
196f
reduced viscosity as function of
concentration of PDMA
homopolymer precursor,
matching end-capped
perfluorooctanoyl polymer, and
telechelic R_F end-capped
PDMA, 193f
reduced viscosity as function of
concentrations of PDMA
precursor and its two-ended
octadecyl-functionalized PDMA,
ODPA bisfunctionalized PDMA,
194, 195f
reduced viscosity of 2 wt% two-
ended perfluorooctanoyl
functionalized PDMA as
function of ammonium
perfluorooctanoate (APFO)
concentration, 198f
simulation via dimerization model,
200–201
spacer effects, 201
surfactant effects, 197
surfactants effects, 201–202
synthesis and characterization of
octadecanoyl or ODPA end-
functionalized PDMAs, 186t
synthesis and characterization of
perfluorooctanoyl end-
functionalized PDMAs, 185t
synthesis of mono- and
difunctionalized PDMA
derivatives, 181, 183–184
synthesis procedures for PDMA
derivatives, 180–181
Polyelectrolyte solutions
ionomers as model system, 128
model polyurethane ionomers in
solvents of varying polarity,
130–133
See also Ionomers
Poly(ethylene glycol) (PEG)
association modes of
hydrophobically modified
telechelic polymers, 199
hydrophilic block (B) in ABA
copolymers, 73, 75
influence of perfluorocarbon size
on hydrophobic association of

one-ended fluorocarbon modified PEG derivatives, 197, 200
modes of association of R_F end-functionalized PEGs, 199
PEG derivatives, 197, 200
synthesis and characterization of perfluorocarbon end-functionalized PEGs, 181, 182t
See also Amphiphilic copolymers with linear dendritic architecture
Poly(ethylene oxide) (PEO)
adsorption isotherms, 210–211
aggregation numbers NR of hexadecyl isocyanate end-capped PEO, 152–153
aggregation numbers NR of octadecyl isocyanate end-capped PEO, 152
association studies of PEO derivatives by ^{19}F NMR, 184, 187, 190
characterization methods, 181
coadsorption kinetics for mixture of long and short chains, 211–212
concentration dependence of ^{19}F NMR resonance of CF_3 group in aqueous solutions of FP605M (fluorocarbon functionalized PEG derivative), 188f
effect of deformation on structure and chain stretching of hexadecyl-terminated, 139–140
effect of polymer chain length and concentration on shear thickening of hexadecyl-terminated PEO, 139f
forward and reverse self-exchange experiments, 214, 215f
Huggins constants for end-functionalized PEGs, 191
log-log plot of intensities of up- and downfield CF_3 resonance's of FP605M as function of concentration, 189f
poly(ethylene glycol) (PEG) derivatives, 197, 200

quantitative reports of adsorption kinetics, 207–208
reduced viscosity versus concentration profile of telechelic C_8F_{17} end-functionalized PEO, 191, 193f
shear thickening of hexadecyl end-capped PEO, 137–138
synthesis of hydrophobically end-capped PEO, 134, 135f
synthesis of PEO derivatives, 181, 182t
synthesis procedures for PEO derivatives, 180
viscoelastic model fitting experimental data for hexadecyl-terminated PEO, 136, 137f
viscosity studies of end-functionalized PEGs, 190–191
See also Dispersions containing poly(ethylene oxide) (PEO) with C_{16} hydrophobes; Dynamics of adsorbed polymer layers
Poly(ethylene-propylene)-*b*-polyethylene oxide (PEP–PEO) block copolymers
absolute SANS data under core contrast and shell contrast for PEP5PEO5, 28f
absolute SANS data under core contrast and shell contrast for PEP5'PEO50, 35f
calculating micelle aggregation number, 29
characterization of PEP–PEO micelles in water, 30t
comparing neutron scattering with light scattering, 27
dependence of aggregation number on PEP polymerization degree, 35f
deprotonation of hydroxyl polymers and process to macroinitiators PEP–OK, 23–24
dynamic light scattering (DLS) measurements of PEP–PEO micelles in water, 32–33
experimental, 22

fitting contrasts of PEP5'PEO30 and PEP5'PEO50 micelles simultaneously, 32
fitting scattering curve of core contrast independently, 29
interfacial contribution to free energy, 33–34
investigating micelles by PEP5PEO15, PEP5'PEO30, and PEP5'PEO50, 29–30
labeling PEP in block copolymer with deuterium, 26–27
macroinitiators PEP–OK for polymerizing ethylene oxide (EO), 24
micellar free energy, 33
micellar properties of PEP–PEO in water, 26–34
model fitting structure of micelle, 27, 29
molecular weight characterization, 26t
molecular weight characterization of PEP–OH precursors, 23t
radial density profile, 30, 32
relative PEO density profiles of micelle shells for different PEP–PEO block copolymers, 31f
SANS measurements of PEP–PEO micelles in water, 27–32
SANS (small angle neutron scattering), 26–27
scattering amplitude equation, 27
scattering length densities for PEP, PEO, and water, 28
size exclusion chromatography (SEC) traces of block copolymers and corresponding precursors, 25f
synthesis, 23–24
synthesis of hydroxyl end functionalized PEP–OH, 23, 25
techniques for investigation of micellar solutions, 22
thermodynamic reflection of micellar parameters, 33–34
Poly(isobutylene-b-methacrylic acid) P(IB-b-MAA). See Amphiphilic block copolymers as surfactants
Polymerizable surfactants. See Surfactants, polymerizable
Polymer-surfactant interactions
 nature of, 370
 non-ionic surfactant with hydrophobically modified alkali soluble emulsion (HASE), 359, 362
 See also Hydrophobically modified alkali soluble emulsion (HASE) polymer in non-ionic surfactant solutions
Poly(methyl methacrylate) (PMMA)
 adsorption of C_{18} terminal hydrophobes on PMMA latices from water, 221–222
 monodisperse latices for adsorption study, 256–257
 steady state profiles of PMMA dispersions thickened with hexadecyl unimers, 224, 225f
 See also β-Cyclodextrin complexations with surfactants and hydrophobically modified ethoxylated urethanes (HEUR); Dispersions containing poly(ethylene oxide) (PEO) with C_{16} hydrophobes
Poly(methyl methacrylate-b-acrylic acid) P(MMA-b-AA). See Amphiphilic block copolymers as surfactants
Poly(methyl methacrylate-b-methacrylic acid) P(MMA-b-MAA). See Amphiphilic block copolymers as surfactants
Polyoxyalkylene triblock copolymers
 micellization, 3
 phase behavior, 4, 5f
 See also Amphiphilic polyoxyalkylene triblock copolymers
Polyoxyethylene (POE)
 terminal multi-branch hydrophobes, 172, 174
 See also Hydrophobic ethoxylated urethanes (HEUR)

Polysoaps
  accommodating stronger extensions by modifying secondary, micellar structure, 117–118
  assumptions underlying theoretical model, 123
  attributes of aggregated amphiphile, 112
  chain span by Pincus law, 117
  coexistence of different configurational states possible for polysoap unfolded by uniform flow, 121f
  consisting of amphiphilic monomers joined by flexible spacer chains, 110f
  Daoud–Cotton model, 112
  deformation behavior of globular polysoap, 115–118
  elastic energy leading to criterion for onset of micellar dissociation, 117
  extensional flow, 118–119
  extension of globular polysoaps by flow fields, 118–120, 122
  force law characterizing extension of globular polysoap, 116f
  interaction free energy, 114
  interchain and intrachain self assembly, 111
  intramolecular aggregation, 111–114
  intramolecular micelles, 112–113
  limitations of assumed molecular architecture, 123–124
  model, 111
  model not explicitly allowing for distinctive aspects of aqueous solutions, 124
  phase separation in solutions, 122–123
  Pincus law, 117–118
  radius of spherical globule of closed packed intrachain micelles, 114
  role of exchange attraction in large-scale configurations, 113–114
  uniform flow, 119–120, 122
  weak strains deforming globule from sphere to ellipsoid, 115
Polystyrene, graft copolymers with N-vinylpyrolidinone, 69
Polyurethane ionomers
  critical factors for determining solution behavior, 139
  exhibiting polyelectrolyte behavior in poor solvent, 133
  ionomer PU4 in different polar solvents, 131
  model correlating structure and properties, 128
  monitoring change in loose aggregate as function of solvent polarity in terms of hydrodynamic radius, 131–132
  normalized fluorescence intensities as function of concentration, 133f
  structure in solvents of varying polarity, 130–133
  structure of model, 130f
  viscometric measurements, 132–133
  See also Ionomers
Poly(vinyl acetate) (PVA), adsorption isotherms of nonyl phenol comb hydrophobic ethoxylated urethane on latex, 264
Pyrene
  crystalline structures of micellar solutions containing, 84–85
  encapsulation by amphiphilic hybrid block copolymer, 83, 88
  fluorescent probe, 150, 153f, 166
  See also Amphiphilic copolymers with linear dendritic architecture; Amphiphilic polymers in water; Fluorescence quenching

R

Radial density profile, equation, 30, 32

Radius of gyration
  mean squared for water-soluble polymer, 145
  static light scattering, 45
Rheology
  effect of hydrophobe structure on low shear viscosity, 167–168
  low shear viscosity dependence on concentration of step-growth internal comb hydrophobic ethoxylated urethanes (HEUR) thickeners with and without $C_{18}H_{37}$ endcap, 177f
  measurements, 167
  oscillatory measurements, 172
  See also Dispersions containing poly(ethylene oxide) (PEO) with C_{16} hydrophobes; Ethylated hydroxyethyl cellulose (EHEC); Hydrophobically modified alkali soluble emulsion (HASE); Hydrophobically modified alkali soluble emulsion (HASE) polymer in non-ionic surfactant solutions; Hydrophobic ethoxylated urethanes (HEUR); Viscosity
Rubber elasticity theory, approximating plateau modulus, 136

S

Salting-in effect, increasing solubility of copolymers in aqueous solution, 3
Salting-out effect, decreasing solubility of copolymers in aqueous solution, 3
Scattering amplitude, equation, 27
Scattering length density
  expression, 26
  poly(ethylene-propylene) (PEP), poly(ethylene oxide), and water, 28
Self-assembly
  amphiphilic polymers in water, 286
  applications for polymers capable of, 72–73
  association behavior of polyoxyethylene triblock copolymers, 6, 8f
  behavior of block copolymers, 3–4
  critical micelle concentration and aggregation number, 4, 6
  hydrodynamic radius (R_h) and intermicellar interactions, 6–7
  limitations of micelle forming materials, 73
  micellization of two mixed triblock copolymers in aqueous solution, 7, 10
  thermodynamic process of micellization, 6
  See also Amphiphilic polymers in water; Amphiphilic polyoxyalkylene triblock copolymers
Self-diffusion. *See* Hydrophobically modified hydroxyethyl cellulose (HMHEC) with sodium dodecyl sulfate (SDS)
Self exchange. *See* Dynamics of adsorbed polymer layers
Separation medium, DNA capillary electrophoresis, 16, 17f
Shear thickening
  hexadecyl end-capped poly(ethylene oxide), 137, 139f
  hydrophobically modified alkali soluble emulsion (HASE), 138
  role of applied stress in region, 366
  See also Hydrophobically associating polymers (HAP); Hydrophobically modified alkali soluble emulsion (HASE) polymer in non-ionic surfactant solutions; Ionomers
Shear thinning, hydrophobically modified alkali soluble emulsion (HASE), 137
Size exclusion chromatography (SEC), hydrophobic ethoxylated urethane (HEUR) adsorption, 267, 268f

Small angle neutron scattering (SANS)
   contrast variation technique, 22
   measurements of PEP–PEO micelles in water, 27–32
   See also Poly(ethylene-propylene)-*b*-polyethylene oxide (PEP–PEO) block copolymers
Sodium dodecyl sulfate (SDS)
   addition to β-cyclodextrin/phenolphthalein complex, 257
   See also Ethylated hydroxyethyl cellulose (EHEC); Hydrophobically modified hydroxyethyl cellulose (HMHEC) with sodium dodecyl sulfate (SDS); Surfactants
Solid state properties
   amphiphilic linear-dendritic copolymers, 80, 83
   optical microscopy of amphiphilic hybrid block copolymer crystallized from aqueous solutions, 85f
   polarized optical micrograph of amphiphilic hybrid block copolymer crystallized from melt, 84f
Solubility, solution behavior of amphiphilic linear-dendritic copolymers in aqueous media, 80
Solution behavior, formation of three-dimensional network, 128–129
Solvent polarity. *See* Ionomers
Spacer effects, poly(N,N-dimethylacrylamide) (PDMA) derivatives, 201
Static light scattering
   aggregation number and micellar diameter, 43–44
   molar mass of micelles and radius of gyration, 45
Steady shear
   superposition of oscillation on, 362, 364–366
   See also Viscosity
Stern–Volmer quenching model constants for quenching pyrene-labeled polymers with various pyridinium chloride derivatives, 300t
   linear Stern–Volmer plots for quenching of pyrene monomer and excimer emissions in aqueous solutions of pyrene-labeled (hydroxyethyl)cellulose, 299
   quenching results, 299
   See also Amphiphilic polymers in water
Storage modulus
   crossover of storage and loss moduli separating terminal region and plateau region for rough value of relaxation time, 364, 366
   dependence of storage and loss modulus on angular frequency, 354, 357
   dependence of storage and loss modulus on angular frequency at various applied stresses at pH 9, 365f
   dependence of storage and loss modulus on angular frequency in varying $C_{18}EO_{100}$ concentrations at pH 9, 358f
   dependence on frequency for hydrophobe ethoxylated urethanes (HEUR), 172, 173f
   storage and loss moduli dependence on frequency for H_{12}MDI-based uniHEURs with different size hydrophobes, 173f
   storage and loss moduli of dispersions containing 3.5 wt% hexadecyl unimers, 230f
Surface dilatational rheology surface properties under dynamic conditions, 304
   See also Ethylated hydroxyethyl cellulose (EHEC)
Surface exchange
   summary of kinetics, 217t
   See also Dynamics of adsorbed

polymer layers
Surface force apparatus (SFA)
  means to measure forces between adsorbed layers at very small separations, 240
  See also Force study of adsorbed layers on hydrophobically modified polyacrylamide
Surface relaxations. See Dynamics of adsorbed polymer layers
Surfactants
  addition to β-cyclodextrin/phenolphthalein complex, 257
  amphiphilic block copolymers in emulsion polymerization, 46–47
  associations with cyclodextrins, 255
  binding of sodium dodecyl sulfate (SDS) to hydroxyethyl cellulose (HEC) and hydrophobically modified HEC (HMHEC), 325, 327
  branching studies, 164–165
  effect of hydrophobe size on cyclodextrin complexation, 258–259
  effect of polar head groups of surfactant on cyclodextrin complexation, 259–261
  effects of addition for poly(N,N-dimethylacrylamide) (PDMA) derivatives, 201–202
  See also Amphiphilic block copolymers as surfactants; β-Cyclodextrin complexations with surfactants and hydrophobically modified ethoxylated urethanes (HEUR); Hydrophobically modified alkali soluble emulsion (HASE) polymer in non-ionic surfactant solutions; Hydrophobically modified hydroxyethyl cellulose (HMHEC) with sodium dodecyl sulfate (SDS); Micelles
Surfactants, cationic, photophysical tools as measure of interactions with cationic cellulose ethers, 295, 299–301
Surfactants, polymerizable. See Associating polymers in water; Intra-molecular aggregations
Swellable polymers. See Water soluble polymers by atom transfer radical polymerization (ATRP)

T

Telechelic ionomers
  solution properties studies, 128
  See also Ionomers
Telechelic polymers
  aggregation numbers in flower micelles from diffusion coefficient measurements, 150–151
  association mechanism, 148–149
  effect of chain length on aggregation number N_R values, 150
  electron paramagnetic resonance (EPR) spectroscopy method determining N_R value for poly(ethylene oxide) (PEO) polymer with $C_{12}H_{25}$ end groups, 150
  end-group aggregation numbers N_R, 149–151
  fitting parameter n plotted versus probe-to-polymer ratio in hexadecyl isocyanate end-capped urethane (HDU) solutions for pyrene and ethylpyrene, 153f
  flower micelle formation, 148–149
  hexadecyl isocyanate end-capped urethane (HDU), 152–153
  hydrophobic ethoxylated urethane (HEUR) polymer preparation, 148
  N_R values for HEUR polymers with $C_{16}H_{33}O$- end groups, 149–150
  octadecyl isocyanate end-capped

urethane (ODU), 152
structure of urethane end-capped PEOs, 151
urethane end-capped PEO, 151–153
See also Poly(ethylene oxide) (PEO)
Terminal hydrophobes. *See* Hydrophobic ethoxylated urethanes (HEUR)
m-Tetramethylxylene diisocyanate (TMXDI)
coupling reaction of two nonyl phenol (NP) surfactant, 261, 263
See also β-Cyclodextrin complexations with surfactants and hydrophobically modified ethoxylated urethanes (HEUR)
Thermodynamic process, micellization, 6
Thermodynamics, micellar parameters, 33–34
Thickening mechanism of hydrophobically modified alkali soluble emulsion (HASE)
different equilibration behavior at different degrees of neutralization, 345
dynamic viscosity measurements, 342
effect of concentration of HASE thickener, 342–343
effect of degree of neutralization on thickening mechanism, 348
effect of rate of neutralization, 344–345
effect of rate of neutralization on viscosity as function of degree of neutralization, 344*f*
external measurements, 342
general profiles of viscosity as function of neutralization degree for methacrylic acid containing copolymer emulsions, 340
HASE thickener for study, HASE–EO40NP, 341
importance of dynamics of neutralization process, 344
measured instantaneous viscosity versus feed rate for 10% HASE–EO40NP solutions, 345*f*
monitoring viscosity of HASE thickener as function of degree and rate of neutralization, 349–350
percent change in viscosity and equilibration time for 10% HASE–EO40NP solutions neutralized to different extents, 346, 347*f*
potentiometric and conductometric titration techniques determining effect of acid distribution on dissolution behavior, 340
relaxation behavior for HASE–EO40NP neutralized to various extents at 25 g/hr feed rate, 347*f*
reviews of mechanism for ASE latex to polymer-solution transition, 339–340
structure of HASE–EO40NP, 341*f*
swelling and dissolution behavior of ethyl acrylate/methacrylic acid ASE copolymers during neutralization, 340
turbidity and particle size of fully equilibrated samples of 10% HASE–EO40NP solutions at different extents of neutralization, 346, 348*f*
validation of experimental technique, 348–349
viscosity as function of percent neutralization for various HASE–EO40NP solutions in water, 343*f*
viscosity as function of shear rate for fully equilibrated samples, 349*f*
viscosity maxima at gelation pH, 340
viscosity relaxation of HASE thickener at different degrees of neutralization, 346, 348
Time-resolved fluorescence spectroscopy, measuring rate of

exchange of block copolymers between aggregates, 43
Transient network theory
  behavior of associating polymer, 135–136
  plateau modulus equation, 357
Triblock copolymers
  phase diagram of copolymer in miscible solvents, 15
  See also Amphiphilic polyoxyalkylene triblock copolymers

U

Urethane end-capped poly(ethylene oxide)
  determination of aggregation number N_R, 151–153
  hexadecyl isocyanate end-capped, 152–153
  octadecyl isocyanate end-capped, 152
Urethanes. *See* β-Cyclodextrin complexations with surfactants and hydrophobically modified ethoxylated urethanes (HEUR); Hydrophobic ethoxylated urethanes (HEUR)

V

4-Vinylpyridine
  comparison of first order kinetic plots for polymerization using different initiators, 64f
  evolution of molecular weight with conversion for polymerization with different initiators, 64f
  polymerization, 61, 63
  size exclusion chromatography (SEC) traces for polymerization using 1-phenylethyl chloride, 65f
  tris(2-(dimethylamino)ethyl)amine (Me_6-TREN) ligand, 63f

N-Vinylpyrolidinone, graft copolymers with polystyrene, 69
Viscoelastic model, model fitting experimental data, 136, 137f
Viscoelasticity, linear, hydrophobes in associative polymers, 228, 231
Viscosity
  dynamic shear for hydrophobically modified alkali soluble emulsion (HASE) in non-ionic surfactant solution, 354, 357
  effect of concentration of HASE thickener, 342–343
  effect of hydrophobe structure on low shear, 167–168
  effect of rate of neutralization for HASE aqueous solutions, 344–345
  HASE thickener as function of degree and rate of neutralization, 349–350
  predicting zero shear with transient network theory, 136
  relaxation of HASE thickener at different degrees of neutralization, 346–348
  steady shear for HASE in non-ionic surfactant solution, 354
  studies of end-functionalized poly(ethylene glycol)s (PEGs), 190–191
  superposition of oscillation on steady shear, 362, 364–366
  See also Associative thickeners by host-guest interaction; Dispersions containing poly(ethylene oxide) (PEO) with C_{16} hydrophobes; Hydrophobically modified alkali soluble emulsion (HASE) polymer in non-ionic surfactant solutions; Hydrophobically modified hydroxyethyl cellulose (HMHEC) with sodium dodecyl sulfate (SDS); Poly(N,N-dimethylacrylamide) (PDMA); Poly(ethylene oxide) (PEO);

Rheology; Thickening mechanism of hydrophobically modified alkali soluble emulsion (HASE)

W

Water-soluble biopolymers, applications, 303–304
Water soluble polymers, surfactant-modified containing multiple hydrophobes, 255–256
Water soluble polymers by atom transfer radical polymerization (ATRP)
  t-butyl acrylate, 60–61, 62f
  copolymers of DMAEMA, 66–67, 68f
  2-(dimethylamino)ethyl methacrylate (DMAEMA), 58–60, 61f
  graft copolymers of N-vinylpyrolidinone and polystyrene, 69
  2-hydroxyethyl acrylate (HEA), 54, 55f
  2-hydroxyethyl acrylate (HEA)/butyl acrylate block copolymers, 65–66
  2-hydroxyethyl methacrylate (HEMA), 56, 57f
  trimethylsilyl protected HEA/HEMA, 56, 58, 59f
  4-vinylpyridine, 61, 63, 64f, 65f
Well-defined polymers. *See* Water soluble polymers by atom transfer radical polymerization (ATRP)
Williamson ether synthesis, synthesis of linear-dendritic block copolymers containing poly(ethylene glycol) (PEG), 77, 78

Z

Zero shear viscosity, predicting with transient network theory, 136

Highlights from ACS Books

Desk Reference of Functional Polymers: Syntheses and Applications
Reza Arshady, Editor
832 pages, clothbound, ISBN 0-8412-3469-8

Chemical Engineering for Chemists
Richard G. Griskey
352 pages, clothbound, ISBN 0-8412-2215-0

Controlled Drug Delivery: Challenges and Strategies
Kinam Park, Editor
720 pages, clothbound, ISBN 0-8412-3470-1

Chemistry Today and Tomorrow: The Central, Useful, and Creative Science
Ronald Breslow
144 pages, paperbound, ISBN 0-8412-3460-4

A Practical Guide to Combinatorial Chemistry
Anthony W. Czarnik and Sheila H. DeWitt
462 pages, clothbound, ISBN 0-8412-3485-X

Chiral Separations: Applications and Technology
Satinder Ahuja, Editor
368 pages, clothbound, ISBN 0-8412-3407-8

Molecular Diversity and Combinatorial Chemistry: Libraries and Drug Discovery
Irwin M. Chaiken and Kim D. Janda, Editors
336 pages, clothbound, ISBN 0-8412-3450-7

A Lifetime of Synergy with Theory and Experiment
Andrew Streitwieser, Jr.
320 pages, clothbound, ISBN 0-8412-1836-6

Chemical Research Faculties, An International Directory
1,300 pages, clothbound, ISBN 0-8412-3301-2

For further information contact:
Order Department
Oxford University Press
2001 Evans Road
Cary, NC 27513
Phone: 1-800-445-9714 or 919-677-0977
Fax: 919-677-1303